D1348865

PLASTICITY OF CRYSTALS
WITH SPECIAL REFERENCE TO METALS

PLASTICITY OF CRYSTALS

WITH SPECIAL REFERENCE TO METALS

BY

DR. E. SCHMID

AND

DR.-ING. W. BOAS

LONDON

CHAPMAN & HALL LTD

11 NEW FETTER LANE EC4

Kristallplastizitaet mit besonderer
Berücksichtigung der Metalle
Originally published in Germany
by Springer-Verlag
Copyright Springer Verlag

English translation copyright
Magnesium Elektron Limited
First published (F. A. Hughes & Co. Limited) 1950
Reissued by Chapman & Hall Limited 1968
Reproduced and Printed in Great Britian by
Latimer Trend & Co. Ltd., Whitstable, Kent

SBN   412   09130   5

Distributed in USA
by Barnes & Noble Inc.

TRANSLATORS' PREFACE

When Schmid and Boas decided to publish in book form the results of their widespread investigations into the mechanism of the deformation of metals, they rendered a great service to all those in science and industry interested in studying the same problem.

This book, with its lucid exposition and wide range, is cited as the first reference in innumerable metallurgical papers, and became a classic within a year or two of its publication.

The major problems resulting from the complex deformation behaviour of magnesium led the scientists concerned with that metal to welcome the concentrated knowledge on the subject contained in Schmid and Boas' publication.

In accordance with our resolve to make available to the magnesium and other industries new publications of value to their research and development, we decided to add the translation of this book to our previous publications of Beck and of Bulian and Fahrenhorst.

Since the appearance of Schmid and Boas' book there has been a great development in the field treated, and many aspects of strength and plasticity have changed considerably. However, no text-book has appeared that could replace Schmid–Boas in every respect, and it has remained as indispensable to the research worker as it was when first published. Information about the most important changes that have taken place in the meantime can be obtained from a less detailed treatment of the subject in the book *An Introduction to the Physics of Metals and Alloys*, by Dr. W. Boas (Melbourne University Press), while the structural aspects of plastic deformation are dealt with in C. S. Barrett's book, *Structure of Metals* (McGraw-Hill Book Company, Inc.).

The translation of *Kristallplastizitaet* has taken a long time, but we believe its contents to be of great value, and it is our hope that the provision of an English text will materially assist all those researchers who are interested in the deformation and plasticity of crystals.

In conclusion we wish to express our thanks to Dr. W. H. Taylor, Dr. E. Orowan and Mr. R. W. K. Honeycombe of the Cavendish Laboratory, Cambridge, for revising the translation and aiding us to prepare it for the press, and to Mr. L. H. Tripp (the translator), who has carried the main burden.

F. A. HUGHES & CO. LIMITED.

January 1950.

FOREWORD

Plasticity is that property of solids by virtue of which they change their shape permanently under the influence of external forces. Although this property has been exploited since the earliest days of human history—an exploitation which, thanks to modern technical methods, has now reached a very high level of perfection—and although unceasing efforts have been made to obtain a clear picture and a scientific explanation of the processes involved in the phenomenon of deformation, so far neither a full description nor an entirely adequate theoretical interpretation has been possible.

In the present work, which is based on lectures which I delivered in 1930–31 at the Technical High School in Berlin, we describe what is known about the plastic behaviour of a specially important class of solids—crystals. Since the deformation of crystal aggregates is mainly governed by the deformation of the individual grain, the latter provides the foundation for our knowledge of the plasticity of crystalline materials in general. In the last twenty years we have learnt a great deal about crystal plasticity, in the first place, owing to the development of methods for growing crystals, which have enormously increased the experimental material available, and secondly, as a result of the application of X-ray diffraction methods to the investigation of solids.

This book is addressed to a large circle of readers. The experimental data which it presents in classified form, and which it attempts to interpret, should assist the physicist to evolve a theory of plasticity. It brings to the notice of the crystallographer and mineralogist those researches into metal crystals which have for their particular object the dynamics of crystal deformation. The geologist will discover, in the development of textures in cast and wrought metals, analogies with similar phenomena in his own field of enquiry, and he should therefore find our tentative explanations instructive. Workers in the field of metals research and technology will find in this book the crystallographic and physical principles underlying the plastic behaviour of their material, and they will be shown by examples how our knowledge of the polycrystalline state can be both increased and applied. Technologist and designer will become familiar with that mass of data from which the technological characteristics of metals are derived. In this way the fundamental significance of the constants employed, and, in particular, the possibility of changing them during operations, will be made clear. Last but not least, it is hoped that all those who themselves are studying the plasticity of crystalline materials will be helped by this book in their choice of experimental technique and methods of research.

I wish to express my gratitude to Mr. M. Polanyi, who introduced me to this subject many years ago, who has since been of great assistance to me, and to whose inspiration this book is due. Sincere thanks are also extended to all my collaborators during the happy years of work at the Kaiser Wilhelm Institute for Fibre Chemistry and Metal Research in Berlin, and especially to Messrs. S. Wassermann, W. Boas (co-author of this book), W. Fahrenhorst and G. Siebel (Bitterfeld). I am also grateful to the Notgemeinschaft der Deutschen Wissenschaft for their continued assistance.

I should like to thank numerous colleagues for permission to reproduce illustrations and diagrams from their works; and I am also indebted to the publishers for their co-operation, and for the very helpful way in which they have met my wishes.

Fribourg, Switzerland.   ERICH SCHMID
January 1935.

CONTENTS

Contents

LIST OF TABLES

INTRODUCTION

Recent achievements in the field of crystal plasticity receive considerable prominence in the present work. This is largely due to the extension of researches on plasticity to include metal crystals. Methods of producing such crystals and the determination of their orientation are therefore described in detail in Chapters III and IV. On the other hand, in view of the excellent treatises which already exist on the subjects of crystallography and crystal elasticity, the two introductory chapters under these headings have been severely limited. A description of the mechanisms of deformation, the geometry of which is expounded in Chapter V, is followed, in Chapter VI, by a fully detailed account of the application of these principles to metal crystals. Metal crystals have been accorded this preferential treatment on account of their usefulness in the experimental investigation of plasticity, and because much of our recent knowledge on the subject has been obtained with them. With the aid of the general principles which have been developed in this chapter the behaviour of ionic crystals is briefly treated in Chapter VII. The amount of experimental data collected in each of these chapters is certainly large, but, in view of the present unsatisfactory nature of the theories of crystal plasticity, this seemed unavoidable if the available material was to be adequately surveyed. Chapter VIII discusses a number of modern hypotheses, which it is hoped will soon be replaced by a single comprehensive theory. In the final chapter the knowledge acquired in our study of the single crystal is applied to elucidate the behaviour of polycrystalline material. Unfortunately the practical significance of this undertaking is still imperfectly realized by the technician.

The bracketed figures interspersed throughout the text refer to the list of publications at the end of the book, where the material has been classified by chapters to enable students who desire further information on certain aspects of the subject to find the appropriate references. It is, of course, inevitable that by this method the same work should sometimes appear under different numbers. Equations in the text bear first the number of the section in which they appear, and are then numbered consecutively.

Readers are referred to the following general treatises on crystal plasticity :

xiii

G. Sachs, " Plastic Deformation ", *Handbuch der Experimental-physik*, Vol. 5/1, 1930.

A. Smekal, " Cohesion of Solids ", *Handbuch der physikalischen und technischen Mechanik*, Vol. 4/2, 1931, and " Structure Sensitive Properties of Crystals ", *Handbuch der Physik*, 2nd Edition, Vol. 24/2.

W. D. Kusnetzow, *The Physics of Solids*, Tomsk, 1932 (in Russian).

1. *Crystalline and Amorphous Solids*

Primarily, solids are commonly contrasted with liquids and gases, and are then divided into two fundamentally distinct groups : on the one hand, solids characterized by regular atomic arrangement (crystals), and on the other, amorphous materials of completely irregular structure. It has been found that a regular structure is by no means confined to those solids which, owing to their delimitation by plane surfaces, had already been recognized as crystals, but that it is, in fact, of very general occurrence throughout Nature.

The nature of the structure is of prime importance for the properties of a material, and the distinction between crystalline and amorphous states is revealed especially in plastic behaviour. With amorphous solids, deformation appears to occur by a mechanism of atomic migration under the influence of thermal movement, in the course of which the external forces merely bring about a preferential selection of those migrations which contribute to the relief of the imposed stresses. On the other hand, with crystals (which are the most important type of solids) the properties peculiar to a regular structure are also revealed in their plastic behaviour.

plane $A_2B_2C_2$ or of other planes parallel to it.[1] The symbol of these crystallographically equivalent planes is written (hkl). The unit plane, and the array of planes parallel to it, is designated (111). A plane with a zero index, therefore, is parallel to the corresponding axis (intercepts it at infinity). Planes of co-ordinates passing through two axes are designated by two zero indices. When a plane makes equal intercepts along each of the three axes, the Miller indices are proportional to the direction-cosines of the plane normal.

Four axes are used in the hexagonal system, of which three are equivalent and lie in the basal plane. Since a plane is completely represented by the ratios of three figures, the four indices of the

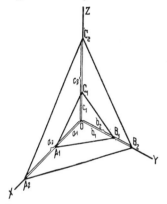

FIG. 4.—Diagram showing the Crystallographic Indexing of Faces.

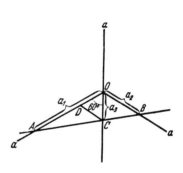

FIG. 5. — Intercepts on the Secondary Axes of the Hexagonal System.

hexagonal system are not independent. The relationship between indices, referred to the three two-fold secondary axes, can be easily deduced from Fig. 5, which shows the hexagonal basal plane. Let AB be the trace of the plane the indices of which are required, and a_1, a_2 and a_3 the intercepts on the digonal axes. Draw through C a parallel to OB and let it intercept the axis OA at D. We then obtain $OA : AD = OB : CD$, or $a_1 : (a_1 - a_3) = a_2 : a_3$, so that

$$a_3 = \frac{a_1 \times a_2}{a_1 + a_2}.$$

Transforming to the indices $h = \dfrac{1}{a_1}, k = \dfrac{1}{a_2}, i = \dfrac{1}{a_3}$, we have $\dfrac{1}{i} = \dfrac{1}{h+k}$.

[1] The reason for adopting these reciprocal intercepts is that they simplify the formulæ when calculating with crystallographic symbols.

Since the intercept on the a_3 axis is negative, we obtain finally $i = \overline{(h + k)}$,[1] *i.e.*, the index applicable to the third axis is always equal to the negative sum of the first two.[2]

The crystallographic notation for *directions* is also based on the ratios of three figures. The line representing the direction passes through the origin of co-ordinates, and the co-ordinates are then determined for a given point lying on the line. These values are reduced to prime whole numbers, u, v and w, which are distinguished from the indices of a plane by being placed in square ʹbrackets [uvw]. One zero index indicates that the direction is parallel to one of the co-ordinate planes. The co-ordinate axes are expressed by the indices [100], [010] and [001].[3]

The method of indexing planes and directions can now be described anew with the aid of the cubic crystal shown in Fig. 6 and the four-axial hexagonal crystal shown in Fig. 7. The cube faces in Fig. 6, which are parallel to the axial planes, have indices (100), (010) and (001). Of the four octahedral planes BDE has the indices (111), while BDG making an intercept -1 on the Z axis has indices (11$\overline{1}$).

---

[1] The minus sign is always written *above* the corresponding index.

[2] In addition, the following relations hold between the hexagonal indices ($hkil$) and the rhombohedral indices (pqr) with first-order pyramidal plane (10$\overline{1}$1) as rhombohedral plane :

$$p = 2h + k + l; \quad q = k - h + l; \quad r = -2k - h + l;$$

$$h = \frac{p - q}{3}; \quad k = \frac{q - r}{3}; \quad i = \overline{(h + k)}; \quad l = \frac{p + q + r}{3}.$$

[3] In order to specify directions in the *four-axial* hexagonal system of co-ordinates, assume the direction Z through the origin and a given point P to be divided into four vector components :

$$Z = ua_1 + va_2 + ta_3 + wc.$$

This expression must naturally be identical with one which uses only three axes, for instance a_1, a_2, c.

$$Z = ma_1 + na_2 + wc.$$

For the secondary axes selected it will be true to say that the sum of their unit vectors, which form an equilateral triangle, disappears :

$$a_1 + a_2 + a_3 = 0.$$

If therefore in the above expression for Z, a_3 is replaced by $-(a_1 + a_2)$, then by comparing the coefficients we obtain :

$$u - t = m; \quad v - t = n,$$

u, v and t are still not clearly defined by these two equations (a vector can be resolved into three co-planar components in an infinite number of ways). The equation $u + v + t = 0$ is added as an arbitrary condition in the same way as when specifying the planes. It is now obvious that

$$u = \frac{2m - n}{3}; \quad v = -\frac{m - 2n}{3}; \quad t = -\frac{m + n}{3}.$$

At this stage, however, the indices no longer have any obvious geometrical significance.

The two remaining octahedral planes are indicated by ($\bar{1}11$) and ($1\bar{1}1$). Of the six dodecahedral planes the plane $BDHF$ is specified by (110), and the plane $ACGE$, which may be assumed to pass in a parallel direction through the point B or D, is marked ($1\bar{1}0$) or ($\bar{1}10$). The two sets of indices become identical on applying the reduction factor (-1), which, as mentioned above, is always permissible.

Of the simple directions, reference should be made to $AB = [100]$ as one of the three edges of the cube, $AG = [111]$ as one of the four body diagonals, and $AF = [101]$ as one of the six face diagonals.

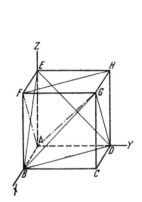

FIG. 6.—Method of Indexing the Faces and Directions of a Cubic Crystal.

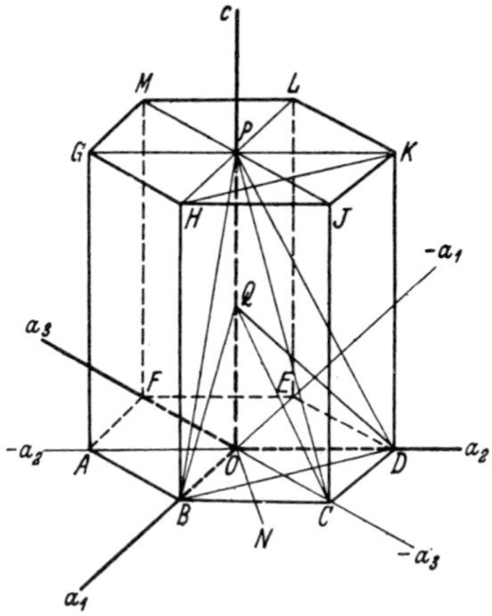

FIG. 7.—Method of Indexing the Faces and Directions of a Hexagonal Crystal.

The indices of the other crystallographically identical directions are obtained by transposing the indices (in cyclic order) and by using the negative sign. Other crystallographically important directions are those with indices [112]: these twelve identical directions connect a corner of the cube with a face centre on the opposite side.

The indices of some of the important planes of hexagonal crystals can now be easily stated (Fig. 7). The basal plane $ABCDEF$ is described by the symbol (0001); while the three prism planes type I ($BCJH$, $CDKJ$, $ABHG$) parallel to the digonal axes are indicated by ($10\bar{1}0$), ($01\bar{1}0$) and ($\bar{1}100$). The prism planes type II perpendicular to the digonal axes are indicated by ($11\bar{2}0$) for $BDKH$, etc. The pyramidal planes type I which pass through the edges of the basal hexagon are indicated by ($10\bar{1}l$), where l denotes the order of the pyramid (*BCP* pyramid type I, order 1; *BCQ* pyramid type I, order 2). Pyramidal planes of type II correspond to the indices (11$\bar{2}l$). (*BDP* pyramid type II, order 1; *BDQ* pyramid type II, order 2.) Among the directions to be noted are the hexagonal axis with the symbol [0001], the digonal axes type I— *OB*, *OD* and *OF*—with indices [$\bar{2}$110], [1$\bar{2}$10] and [11$\bar{2}$0], and finally one of the digonal axes type II, *ON*, with the indices [10$\bar{1}$0].

5. *Crystal Projection*

Visual representation of the relationships between the angles of crystals, and simple methods for the performance of crystallographic calculations, are provided by means of projections. There are two principal methods of projection, spherical and stereographic, both of which will now be briefly described.

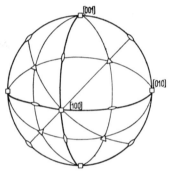

FIG. 8.—Spherical .Projection of a Cubic Crystal.

□ () △ : points at which the four-, three- and two-fold axes intersect the surface.

In *spherical* projection a point on the crystal is assumed to be at the centre of an imaginary sphere. The crystal is then set up in such a way that a principal crystal axis emerges at the North and South poles. The projection of a *direction* is the point at which the line which has been drawn parallel to that direction and through the centre of the sphere meets the surface. The angle between two directions is therefore given by the angular distance between the representative points on the reference sphere. *Planes*, too, are represented by a point on the reference sphere, known as the " pole " of the plane, which is the point at which a plane normal drawn from the centre of the sphere intersects the surface. The angle between two planes is given by the distance between the two poles. The sum of all planes passing through one direction (a zone) is shown on the polar sphere by a great circle perpendicular to the common direction or zone axis. By representing the principal planes and directions in this way the symmetry of the crystals is impressed on the projection sphere (Fig. 8).

Crystallographic problems are solved by connecting the projection points (of planes and directions) on the polar sphere by great circles, the required angles being then calculated from convenient triangles according to the formulæ of spherical trigonometry. The most commonly used formulæ are as follows, where a, b and c are the sides, and α, β and γ the angles of the triangle (Fig. 9) :

sine relation—

$$\frac{\sin a}{\sin \alpha} - \frac{\sin b}{\sin \beta} - \frac{\sin c}{\sin \gamma}$$

cosine relation—

$$\cos a = \cos b \cos c + \sin b \sin c \cos \alpha$$
$$\cos \alpha = - \cos \beta \cos \gamma + \sin \beta \sin \gamma \cos a.$$

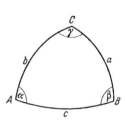

Fig. 9.—Spherical Triangle.

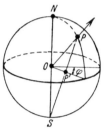

Fig. 10.—Spherical and Stereographic Projection of a Direction.

For the right-angled triangle ($\gamma = 90°$)

$$\sin b = \sin c \sin \beta$$
$$\sin a = \sin c \sin \alpha$$
$$\cos c = \cos a \cos b.$$

In *stereographic projection* the polar sphere used in spherical projection is projected on to the equatorial plane, the northern hemisphere being viewed from the South pole, and the southern from the North pole (cf. Fig. 10).

The picture thus obtained is accurate in regard to angles, but not in regard to planes. Great circles of the polar sphere (crystallographic zones) project into circular arcs, and, when they pass through the point of projection, into diameters of the reference circle. Fig. 11, which should be compared with Fig. 8, contains the stereographic projection of a cubic crystal.

In this case the required angles are determined graphically with the aid of a ruled net. This consists of a number of equidistant meridians and parallel circles, with axis lying in the *equatorial plane* of the reference sphere (Fig. 12). In order to determine the angle between two directions (*A* and *B*), the sphere (with ruled net) is rotated (see the arrow in Fig. 12*a*) until the points *A* and *B* are

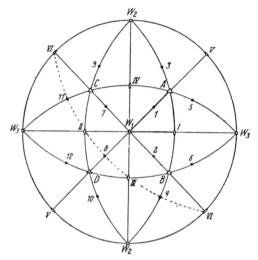

Fig. 11.—Stereographic Projection of a Cubic Crystal.

$W_1 \ldots W_3$: cubic axes; $A \ldots D$: body diagonals; I...VI: face diagonals; 1...12: [112] directions.

connected by a great circle; the number of parallel circles between them gives the required angle. In stereographic representation, the

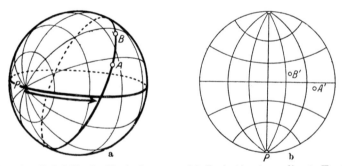

Fig. 12.—Ruled Net for Use in Stereographic Projection (according to Ewald).

Wulff net shown in Fig. 13 corresponds to the ruled sphere shown in Fig. 12*b*. If the Wulff net is placed beneath the transparent paper on which the projection has been traced and then turned about its centre, this will correspond to rotation of the ruled sphere. Angles can be measured in this way to within about $\frac{1}{4}°$, the accuracy attained depending on the distance between the circles on the net.

6. Simple Crystallographic Theorems

The use of crystallographic indices in calculations is illustrated below, by examples in common use.

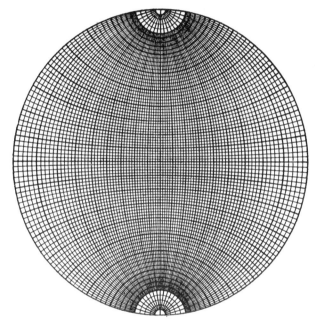

Fig. 13.—Wulff's Net.

(a) The direction $[uvw]$ lies in the plane (hkl); the plane (hkl) belongs to the zone $[uvw]$.

It follows from the analytical representation of planes and directions that the relation [1] which must be satisfied for coincidence is $hu + kv + lw = 0$.

Thus, for example, the plane $(11\bar{2})$ belongs to the zone with axis in the direction $[111]$ and not to that with direction $[100]$.

(b) The intersection $[uvw]$ of two planes $(h_1k_1l_1)$ and $(h_2k_2l_2)$.

---

[1] For the four-figure indices of the hexagonal axis the analogous relation is
$$hu + kv + it + lw = 0.$$

From $\qquad h_1 u + k_1 v + l_1 w = 0$

and $\qquad h_2 u + k_2 v + l_2 w = 0$

it follows that

$$u : v : w = \begin{vmatrix} k_1 l_1 \\ k_2 l_2 \end{vmatrix} : \begin{vmatrix} l_1 h_1 \\ l_2 h_2 \end{vmatrix} : \begin{vmatrix} h_1 k_1 \\ h_2 k_2 \end{vmatrix}$$

$$= (k_1 l_2 - l_1 k_2) : (l_1 h_2 - h_1 l_2) : (h_1 k_2 - k_1 h_2).$$

The planes (310) and (11$\bar{1}$) intersect in the direction [$\bar{1}$32].

(c) Plane (hkl) through the two directions [$u_1 v_1 w_1$] and [$u_2 v_2 w_2$]. From the conditions of coincidence

$$u_1 h + v_1 k + w_1 l = 0$$
$$\text{and } u_2 h + v_2 k + w_2 l = 0$$

it follows that $h : k : l = \begin{vmatrix} v_1 w_1 \\ v_2 w_2 \end{vmatrix} : \begin{vmatrix} w_1 u_1 \\ w_2 u_2 \end{vmatrix} : \begin{vmatrix} u_1 v_1 \\ u_2 v_2 \end{vmatrix}$

[123] and [311] determine the plane (1$\bar{5}$5).

(d) Transformation of indices to new axes.

It may sometimes be necessary to describe a crystal with reference to some axis other than the natural crystallographic axis. Examples of this are encountered in the indexing of layer-line diagrams. If, with reference to the original axes [100], [010] and [001], the new axes have indices [$u_1 v_1 w_1$], [$u_2 v_2 w_2$] and [$u_3 v_3 w_3$], then a plane with indices (hkl) in the original system has indices (h'k'l') in the new system, and the transformation formulæ are :

$$h' = u_1 h + v_1 k + w_1 l$$
$$k' = u_2 h + v_2 k + w_2 l$$
$$l' = u_3 h + v_3 k + w_3 l.$$

If a crystal is to be described, for instance, with the aid of the new axes [100], [010] and [112], then the following values are obtained for the new indices (h'k'l') of a plane (hkl) :

$$h' = h, \; k' = k, \; l' = h + k + 2l.$$

(e) Spacing of lattice planes.

In view of the fundamental importance of the distance between equivalent lattice planes for the diffraction of X-rays in crystals, the general expression for this distance d may be given here. It is a function of the indices (hkl) of the plane : the coefficients are functions of the axial intercepts a, b, c and the axial angles α, β, γ.

The formula applicable to the general case (triclinic crystal) is

$$\frac{1}{d^2} = \frac{1}{V^2}\{g_{11}h^2 + g_{22}k^2 + g_{33}l^2 + 2g_{12}hk + 2g_{23}kl + 2g_{13}hl\}$$

where
$$g_{11} = b^2c^2 \sin^2 \alpha \qquad g_{12} = abc^2(\cos \alpha \,.\, \cos \beta - \cos \gamma)$$
$$g_{22} = a^2c^2 \sin^2 \beta \qquad g_{23} = a^2bc(\cos \beta \,.\, \cos \gamma - \cos \alpha)$$
$$g_{33} = a^2b^2 \sin^2 \gamma \qquad g_{13} = ab^2c(\cos \gamma \,.\, \cos \alpha - \cos \beta)$$
$$V^2 = a^2b^2c^2 (1 - \cos^2\alpha - \cos^2\beta - \cos^2\gamma + 2 \cos \alpha \cos \beta \cos \gamma)$$

V represents the volume of the elementary parallelopiped.

Special cases of higher symmetry:

Orthorhombic $\dfrac{1}{d^2} = \left(\dfrac{h}{a}\right)^2 + \left(\dfrac{k}{b}\right)^2 + \left(\dfrac{l}{c}\right)^2$

Tetragonal $\qquad \dfrac{1}{d^2} = \dfrac{h^2 + k^2}{a^2} + \dfrac{l^2}{c^2}$

Cubic $\qquad\qquad \dfrac{1}{d^2} = \dfrac{h^2 + k^2 + l^2}{a^2}$

Hexagonal $\qquad \dfrac{1}{d^2} = \dfrac{4}{3} \cdot \dfrac{h^2 + k^2 + hk}{a^2} + \dfrac{l^2}{c^2}$ (valid for four-number indices with $i = h + k$).

It may be said in general that the simpler the indices of a plane the greater is the interplanar spacing, and consequently the greater the number of lattice points per unit area of the plane (density of distribution). Thus, for example, for the cube face (100) of a cubic crystal, $d = a$, for the dodecahedral face (110),

$$d = \frac{a\sqrt{2}}{2},$$

while for the octahedral face (111),

$$d = \frac{a\sqrt{3}}{3}.$$

ELASTICITY OF CRYSTALS

7. *Hooke's Law*

If a solid body is subjected to mechanical stresses, elastic deformations will both precede and accompany plastic strain, *i.e.*, there will be changes in shape which disappear when the state of stress ceases. In this reversible process the deformation (which in any case is usually only very small) is determined solely by the prevailing stress, from which it can be calculated. The relationship between stress and strain is linear. This linear characteristic (Hooke's law), which is based on a wide experience, can now be deduced theoretically from Born's lattice theory, by assuming that the atoms in the crystal lattice are in positions of stable equilibrium relative to the lattice forces. The assumption is justified in so far as it has hitherto proved impossible to destroy, or even deform to any perceptible extent, a crystal by the application of infinitesimally small forces. It is assumed that the forces acting between the particles of the lattice are central forces; no assumptions are necessary regarding the law of inter-atomic forces itself when studying elastic behaviour. The distortion of a lattice has two components : the lattice is deformed as a whole; and, in addition, the simple lattices of which a crystal is generally composed can, as a whole, be displaced with reference to each other. This latter type of macroscopically invisible distortion is a peculiarity of the lattice structure of crystals.

The effect of external forces on a lattice is to displace the lattice points from equilibrium until the opposing forces set up by distortion re-establish equilibrium with the external forces. In order to calculate this behaviour we develop the energy density, whose derivatives with respect to the strain components are the stresses, in a power series of the strain components. The linear terms disappear owing to the assumption of the stability of the initial position : the components of a third and higher order are neglected. In this way the six equations of Hooke's generalized law are obtained :

$$
\left.
\begin{aligned}
\sigma_x &= c_{11}\varepsilon_x + c_{12}\varepsilon_y + c_{13}\varepsilon_z + c_{14}\gamma_{yz} + c_{15}\gamma_{zx} + c_{16}\gamma_{xy} \\
\sigma_y &= c_{12}\varepsilon_x + c_{22}\varepsilon_y + c_{23}\varepsilon_z + c_{24}\gamma_{yz} + c_{25}\gamma_{zx} + c_{26}\gamma_{xy} \\
\sigma_z &= c_{13}\varepsilon_x + c_{23}\varepsilon_y + c_{33}\varepsilon_z + c_{34}\gamma_{yz} + c_{35}\gamma_{zx} + c_{36}\gamma_{xy} \\
\tau_{yz} &= c_{14}\varepsilon_x + c_{24}\varepsilon_y + c_{34}\varepsilon_z + c_{44}\gamma_{yz} + c_{45}\gamma_{zx} + c_{46}\gamma_{xy} \\
\tau_{zx} &= c_{15}\varepsilon_x + c_{25}\varepsilon_y + c_{35}\varepsilon_z + c_{45}\gamma_{yz} + c_{55}\gamma_{zx} + c_{56}\gamma_{xy} \\
\tau_{yx} &= c_{16}\varepsilon_x + c_{26}\varepsilon_y + c_{36}\varepsilon_z + c_{46}\gamma_{yz} + c_{56}\gamma_{zx} + c_{66}\gamma_{xy}
\end{aligned}
\right\} \quad (7/1)
$$

In these equations $\sigma_x(\sigma_y, \sigma_z)$ is the normal stress acting on a plane perpendicular to the x-(y-,z-) axis, ε_x is the normal dilatation in the direction of x, *i.e.*, the change of the spacing of two planes perpendicular to the x-axis, initially·of unit spacing. τ_{yz} is the shear stress in the direction of y in a plane perpendicular to the z axis; it is equal to the shear stress in the z direction in a plane perpendicular to the y-axis ($\tau_{yz} = \tau_{zy}$). γ_{yz} is the displacement in the direction of y of two planes of unit distance normal to the z-axis; it is equal to the (relative) displacement in the z direction of two planes perpendicular to the y-axis ($\gamma_{yz} = \gamma_{zy}$).

If the equations are resolved with respect to the strain components ε and γ, six corresponding equations are obtained :

$$
\left.
\begin{aligned}
\varepsilon_x &= s_{11}\sigma_x + s_{12}\sigma_y + s_{13}\sigma_z + s_{14}\tau_{yz} + s_{15}\tau_{zx} + s_{16}\tau_{xy} \\
&\vdots \\
\gamma_{yz} &= s_{14}\sigma_x + s_{24}\sigma_y + s_{34}\sigma_z + s_{44}\tau_{yz} + s_{45}\tau_{zx} + s_{46}\tau_{xy} \\
&\vdots
\end{aligned}
\right\} \quad (7/2)
$$

The parameters c_{ik} are designated as moduli, the parameters s_{ik} as coefficients of elasticity.

8. Simplification of the Equations of Hooke's Law as a Consequence of Crystal Symmetry

The equations expressing Hooke's law can be greatly simplified if the symmetry of crystals is taken into account. Nine different groups are obtained, for which the matrices of the moduli of elasticity are shown in Table I. The table also indicates the distribution of the thirty-two crystal classes over these nine groups.

TABLE I

Matrices of the Moduli of Elasticity of Crystals Corresponding to Symmetry

Group 1.   Class * C_1, S_2, triclinic system (twenty-one constants).

c_{11}	c_{12}	c_{13}	c_{14}	c_{15}	c_{16}
c_{12}	c_{22}	c_{23}	c_{24}	c_{25}	c_{26}
c_{13}	c_{23}	c_{33}	c_{34}	c_{35}	c_{36}
c_{14}	c_{24}	c_{34}	c_{44}	c_{45}	c_{46}
c_{15}	c_{25}	c_{35}	c_{45}	c_{55}	c_{56}
c_{16}	c_{26}	c_{36}	c_{46}	c_{56}	c_{66}

Group 2.   Class C_s, C_2, C_{2h}, monoclinic system (thirteen constants).

c_{11}	c_{12}	c_{13}	0	0	c_{16}
c_{12}	c_{22}	c_{23}	0	0	c_{26}
c_{13}	c_{23}	c_{33}	0	0	c_{36}
0	0	0	c_{44}	c_{45}	0
0	0	0	c_{45}	c_{55}	0
c_{16}	c_{26}	c_{36}	0	0	c_{66}

* For the symbols of the various crystal classes cf. detailed treatises on crystallography.

TABLE I—*continued*

Group 3. Class C_{2v}, V, V_h, rhombic system (nine constants).

$$
\begin{matrix}
c_{11} & c_{12} & c_{13} & 0 & 0 & 0 \\
c_{12} & c_{22} & c_{23} & 0 & 0 & 0 \\
c_{13} & c_{23} & c_{33} & 0 & 0 & 0 \\
0 & 0 & 0 & c_{44} & 0 & 0 \\
0 & 0 & 0 & 0 & c_{55} & 0 \\
0 & 0 & 0 & 0 & 0 & c_{66}
\end{matrix}
$$

Group 4. Class C_3, C_{3i}, hexagonal system (trigonal sub-group) (seven constants).

$$
\begin{matrix}
c_{11} & c_{12} & c_{13} & c_{14} & -c_{25} & 0 \\
c_{12} & c_{11} & c_{13} & -c_{14} & c_{25} & 0 \\
c_{13} & c_{13} & c_{33} & 0 & 0 & 0 \\
c_{14} & -c_{14} & 0 & c_{44} & 0 & c_{25} \\
-c_{25} & c_{25} & 0 & 0 & c_{44} & c_{14} \\
0 & 0 & 0 & c_{25} & c_{14} & \tfrac{1}{2}(c_{11}-c_{12})
\end{matrix}
$$

Group 5. Class C_{3v}, D_3, D_{3d}, hexagonal system (trigonal sub-group) (six constants).

$$
\begin{matrix}
c_{11} & c_{12} & c_{13} & c_{14} & 0 & 0 \\
c_{12} & c_{11} & c_{13} & -c_{14} & 0 & 0 \\
c_{13} & c_{13} & c_{33} & 0 & 0 & 0 \\
c_{14} & -c_{14} & 0 & c_{44} & 0 & 0 \\
0 & 0 & 0 & 0 & c_{44} & c_{14} \\
0 & 0 & 0 & 0 & c_{14} & \tfrac{1}{2}(c_{11}-c_{12})
\end{matrix}
$$

Group 6. Class C_{3h}, D_{3h}, C_6, C_{6h}, C_{6v}, D_6, D_{6h}; hexagonal system (five constants).

$$
\begin{matrix}
c_{11} & c_{12} & c_{13} & 0 & 0 & 0 \\
c_{12} & c_{11} & c_{13} & 0 & 0 & 0 \\
c_{13} & c_{13} & c_{33} & 0 & 0 & 0 \\
0 & 0 & 0 & c_{44} & 0 & 0 \\
0 & 0 & 0 & 0 & c_{44} & 0 \\
0 & 0 & 0 & 0 & 0 & \tfrac{1}{2}(c_{11}-c_{12})
\end{matrix}
$$

Group 7. Class C_4, S_4, C_{4h}, tetragonal system (seven constants).

$$
\begin{matrix}
c_{11} & c_{12} & c_{13} & 0 & 0 & c_{16} \\
c_{12} & c_{11} & c_{13} & 0 & 0 & -c_{16} \\
c_{13} & c_{13} & c_{33} & 0 & 0 & 0 \\
0 & 0 & 0 & c_{44} & 0 & 0 \\
0 & 0 & 0 & 0 & c_{44} & 0 \\
c_{16} & -c_{16} & 0 & 0 & 0 & c_{66}
\end{matrix}
$$

Group 8. Class C_{4v}, V_d, D_4, D_{4h}, tetragonal system (six constants).

$$
\begin{matrix}
c_{11} & c_{12} & c_{13} & 0 & 0 & 0 \\
c_{12} & c_{11} & c_{13} & 0 & 0 & 0 \\
c_{13} & c_{13} & c_{33} & 0 & 0 & 0 \\
0 & 0 & 0 & c_{44} & 0 & 0 \\
0 & 0 & 0 & 0 & c_{44} & 0 \\
0 & 0 & 0 & 0 & 0 & c_{66}
\end{matrix}
$$

Group 9. Class T, T_h, T_d, 0, 0_h, regular system (three constants).

$$
\begin{matrix}
c_{11} & c_{12} & c_{12} & 0 & 0 & 0 \\
c_{12} & c_{11} & c_{12} & 0 & 0 & 0 \\
c_{12} & c_{12} & c_{11} & 0 & 0 & 0 \\
0 & 0 & 0 & c_{44} & 0 & 0 \\
0 & 0 & 0 & 0 & c_{44} & 0 \\
0 & 0 & 0 & 0 & 0 & c_{44}
\end{matrix}
$$

This arrangement can also be applied to s_{ik} with the following slight modifications :

In the Groups 4, 5 and 6 the relationship $s_{66} = 2(s_{11} - s_{12})$ replaces $c_{66} = \frac{1}{2}(c_{11} - c_{12})$; in Group 4, $s_{46} = 2s_{25}$ replaces $c_{46} = c_{25}$; in Groups 4 and 5, $s_{56} = 2s_{14}$ replaces $c_{56} = c_{14}$.

In each case the crystal is set up in such a way that any single axis of highest symmetry is the z axis, and a digonal axis, if any, the y axis of the system of co-ordinates. From the elastic parameters relative to this co-ordinate system it is possible to calculate directly the stresses (or deformations) in the co-ordinate axes and planes. If, on the other hand, it is desired to calculate the deformations from the stresses (or the reverse) for any directions or planes, then a new system of co-ordinates is based on these directions or planes, and the elastic parameters with reference to the principal system of co-ordinates must be transformed to this new system. The transformation formulæ are generally complex, and for further particulars the reader is advised to consult the detailed discussions of the subject referred to in the bibliography.

The following table of moduli of elasticity is obtained for an isotropic solid :

$$\begin{matrix} c_{11} & c_{12} & c_{12} & 0 & 0 & 0 \\ c_{12} & c_{11} & c_{12} & 0 & 0 & 0 \\ c_{12} & c_{12} & c_{11} & 0 & 0 & 0 \\ 0 & 0 & 0 & c_{44} & 0 & 0 \\ 0 & 0 & 0 & 0 & c_{44} & 0 \\ 0 & 0 & 0 & 0 & 0 & c_{44} \end{matrix}$$

in which $c_{44} = \frac{1}{2}(c_{11} - c_{12})$. The same arrangement is also valid for the coefficients s_{ik} with $s_{44} = 2(s_{11} - s_{12})$.

This matrix, which is independent of the choice of the co-ordinate system, contains only two independent parameters. The relationship between these quantities and the constants used in the literature of the strength of materials, namely, Young's modulus E, modulus of shear G, and Poisson's ratio μ, is given by the equations

$$E = \frac{1}{s_{11}}, \ G = \frac{1}{s_{44}} \ \mu = \frac{s_{12}}{s_{11}}.$$

From this it follows that $\mu = \frac{E}{2G} - 1$.

9. Cauchy's Relations

The equations of Hooke's law for the triclinic crystal contain twenty-one constants. The number is reduced to fifteen, however, if the internal displacements of the constituent simple lattices are disregarded. This introduces six new relationships, known as the Cauchy relations, which formerly were obtained on the assumption that the central forces depended solely on the distance between the particles. The six equations are as follows :

$$c_{23} = c_{44},\ c_{56} = c_{14},\ c_{64} = c_{25},\ c_{31} = c_{55},\ c_{12} = c_{66},\ c_{45} = c_{36}$$

According to Born's theory these equations are valid if a crystal is so constituted that each particle is a centre of symmetry. Since this condition must persist for any distortion, relative displacements of the constituent simple lattices are excluded by virtue of the structure of the crystal.

Tables II and III contain the moduli of elasticity of various materials. It will be seen that the behaviour of the ionic crystals is in accordance with what would have been expected from their structure.

TABLE II

Validity of the Cauchy Relations for the Cubic Ionic Crystals

Material.	Moduli of elasticity in 10^{11} dyn/cm.^{2}.			References.
	c_{11}.	c_{12}.	c_{44}.	
(a) Cauchy relation $c_{12} = c_{44}$ demanded by theory.				
Sodium chloride	4·94	1·37	1·28	(2)
,, ,,	3·30	1·31	1·33	(2)
Potassium chloride	$\begin{cases}3\cdot70\\3\cdot88\end{cases}$	$\begin{cases}0\cdot81\\0\cdot64\end{cases}$	$\begin{cases}0\cdot79\\0\cdot65\end{cases}$	(2) (3)
,, bromide	3·33	0·58	0·62	(2)
,, iodide	2·67	0·43	0·42	(2)
(b) Cauchy relation $c_{12} = c_{44}$ not demanded by theory.				
Fluorspar	16·4	4·48	3·38	(4)
Sodium chlorate	6·5	−2·10	1·20	(5)
Pyrite	36·1	−4·74	10·55	(4)

With metals, on the other hand, there is a very marked discrepancy between c_{12} and c_{44} even in cases where the validity of the Cauchy relation is demanded (Table IIIa) [cf. especially (1)]. Particular attention is drawn to this failure of the lattice theory in the case of metals. It may be that, owing to the ease with which they can be displaced, the valency electrons should be regarded as independent constituents of the lattice, since it is difficult to doubt Born's second assumption relating to the stability of the lattice when deriving the Cauchy relations.[1]

For isotropic solids the six Cauchy equations are reduced to one : Poisson's equation $c_{11} = 3c_{12}$, so that in this case only a single constant survives. For Poisson's ratio the general value obtained

TABLE III

Validity of the Cauchy Relations for Metal Crystals

Material.	Moduli of elasticity in 10^{11} dyn/cm.^{2}.			References.	Poisson's ratio, μ.
	c_{11}.	c_{12}.	$c_{\cdot 44}$.		

(a) Cubic metals. Cauchy relation $c_{12} = c_{44}$ demanded if central forces assumed.

Material.	c_{11}.	c_{12}.	$c_{\cdot 44}$.	References.	Poisson's ratio, μ.
Copper	17·0	12·3	7·52	(7)	0·34
Silver	12·0	8·97	4·36	(8)	0·37
Gold	{19·4 / 18·7}	{16·6 / 15·7}	{4·00 / 4·36}	(6) / (8)	} 0·420
Aluminium	10.8_2	6.2_2	2.8_4	(6)	0·343
a-Brass (72% Cu)	14·7	11·1	7·2	(9)	—
a-Iron	23·7	14·1	11·6	(10)	0·280
Tungsten	{51·3 / 50·1}	{20·6 / 19·8}	{15·3 / 15·1}	(11) / (12)	} 0·17

Material.	Moduli of elasticity in 10^{11} dyn/cm.^{2}.					References.	Poisson's ratio, μ.
	c_{11}.	$3c_{12}$.	C_{44}.	C_{13}.	C_{33}.		

(b) Hexagonal metals. Cauchy relations $c_{11} = 3c_{12}$; $c_{44} = c_{13}$ not demanded.

Material.	c_{11}.	$3c_{12}$.	C_{44}.	C_{13}.	C_{33}.	References.	Poisson's ratio, μ.
Magnesium	{5·65 . 10^{11} / 5·94}	{6·96 / 6·09}	{1·68 / 1·14}	{1·81 / 2·03}	{5·87 / 5·94}	(13) / (14)	—
Zinc	{16·3 / 15·9}	{7·65 / 9·69}	{3·79 / 4·00}	{5·08 / 4·82}	{6·23 / 6·21}	(15) / (11)	} 0·33
Cadmium	{12·1 / 10·9}	{14·43 / 11·94}	{1·85 / 1·56}	{4·42 / 3·75}	{5·13 / 4·60}	(15) / (11)	} 0·30

is then $\mu = \frac{1}{4}$. With certain plausible assumptions, however, the constants of a quasi-isotropic crystal aggregate can be derived as mean values from the elastic parameters of the single crystal (cf.

[1] *Translator's footnote.*—According to the modern theory of metals (cf., e.g., Mott and Jones, *Theory of Metals and Alloys*, or A. H. Wilson, *Theory of Metals*), the cohesive forces in metals are far from being central forces, and there is no reason to expect the validity of the Cauchy relations.

Section 81). If the Cauchy relations apply in this case, Poisson's equation is applicable to the crystal aggregate as a first approximation. Consequently, when measuring quasi-isotropic fine crystalline material, the deviation which is observed from the value for Poisson's ratio μ 1/4, provides a criterion for the validity of the Cauchy relations in the case of the single crystal. For this reason Table III contains the values for Poisson's ratio, determined on polycrystals, together with the parameters of the single crystal; the parallelism between deviation from Poisson's ratio of 1/4, and non-validity of the Cauchy relations, will be immediately apparent.

10. *Determination of the Elastic Parameters*

The elastic parameters are determined with the aid of the elastic constants of single crystals of various orientations. These elastic constants are, as for any solid : Young's modulus (E) and the modulus of shear (G), the definition of which applies without modification to crystals. [E = the tensile stress that would be necessary to double the length of the specimen, G = the shear stress that would develop on the periphery of a cylindrical specimen having a length and diameter = 1 when twisted round an angle of 1 radian (57·3°).] The elastic moduli of single-crystal specimens are obtained experimentally by the same methods as are used for testing isotropic solids. The determination of characteristic acoustic frequencies—a method of testing which recently has been much in use—offers particular advantages [transverse, longitudinal, torsional vibrations (16), (17); cf. (18) and (19) for the corrections which have to be applied in this case].

The theory of crystal elasticity yields two equations for $\frac{1}{E}$ and $\frac{1}{G}$ as functions of the angles formed by the axis of the bar with the axes of the crystal (orientation); the coefficients of these equations are the elastic constants s_{ik}. The problem consists in representing, as far as possible, the observed dependence of $\frac{1}{E}$ and $\frac{1}{G}$ upon the orientation by a suitable choice of the s_{ik}. A check for the resulting s_{ik} values is provided by their connection with compressibility. For the cubic compressibility K, which is independent of orientation, this relationship is as follows for the triclinic crystal : $K = s_{11} + s_{22} + s_{33} + 2(s_{12} + s_{23} + s_{31})$. It is customary, however, to measure the *linear* compressibility S (the change of length in certain directions under hydrostatic pressure), which is usually dependent on direction. Only in the case of cubic crystals does a crystal sphere remain a sphere under hydrostatic pressure.

The following expressions give the theoretical dependence of E, G and S upon the orientation for cylindrical specimens of cubic and hexagonal crystals :

Cubic crystals :

$$\left.\begin{aligned} \frac{1}{E} &= s'_{33} = s_{11} - 2[(s_{11} - s_{12}) - \tfrac{1}{2}s_{44}] \\ &\qquad\qquad (\gamma_1{}^2\gamma_2{}^2 + \gamma_2{}^2\gamma_3{}^2 + \gamma_3{}^2\gamma_1{}^2) \\ \frac{1}{G} &= \tfrac{1}{2}(s'_{44} + s'_{55}) = s_{44} + 4[(s_{11} - s_{12}) - \tfrac{1}{2}s_{44}] \\ &\qquad\qquad (\gamma_1{}^2\gamma_2{}^2 + \gamma_2{}^2\gamma_3{}^2 + \gamma_3{}^2\gamma_1{}^2) \\ S &= s_{11} + 2s_{12} \end{aligned}\right\} \quad 10/1\ \ldots\ 3$$

Hexagonal crystals :

$$\left.\begin{aligned} \frac{1}{E} &= s'_{33} = s_{11}(1 - \gamma_3{}^2)^2 + s_{33}\gamma_3{}^4 + (2s_{13} + \\ &\qquad\qquad s_{44})\gamma_3{}^2(1 - \gamma_3{}^2) \\ \frac{1}{G} &= \tfrac{1}{2}(s'_{44} + s'_{55}) = s_{44} + [(s_{11} - s_{12}) - \tfrac{1}{2}s_{44}](1 - \\ &\quad - \gamma_3{}^2) + 2(s_{11} + s_{33} - 2s_{13} - s_{44})\gamma_3{}^2(1 - \gamma_3{}^2) \\ S &= s_{11} + s_{12} + s_{13} - \gamma_3{}^2(s_{11} - s_{33} + s_{12} - s_{13}) \end{aligned}\right\} \quad 10/4\ \ldots\ 6$$

In the case of the cubic crystals, $\gamma_1\gamma_2\gamma_3$ represent the cosines of the angles formed by the axis of the specimen with the three edges of the cube $(\gamma_1{}^2 + \gamma_2{}^2 + \gamma_3{}^2 = 1)$; in the case of the hexagonal crystal only the direction-cosine γ_3 of the angle formed with the hexagonal axis appears, since the elastic properties have rotational symmetry with respect to the six-fold axis.

CHAPTER III

PRODUCTION OF CRYSTALS

Methods of obtaining large crystals have been greatly improved in the past twenty years, and we are to-day in a position to produce crystals of many metals and alloys in almost any size and form by a great variety of methods. In the following discussion of the new methods of growing crystals the processes have been grouped according to the state of aggregation from which crystallization takes place.

A. PRODUCTION OF CRYSTALS FROM THE SOLID STATE: THE RECRYSTALLIZATION METHOD

The term recrystallization usually means the renewed formation, generally at elevated temperatures, of the crystal structure of crystalline materials. Numerous experiments have shown that this renewal of the crystal structure does not occur with cast metals that are completely free from internal stresses (20, 21). If, however, a specimen which cannot recrystallize is plastically deformed, it acquires the capacity for renewing its texture. This recurs as a result of nucleus formation followed by the consumption of the old grains by the new ones.

In addition to this recrystallization due to deformation (" work recrystallization "), there is a phenomenon known as " grain growth " which leads to renewal of the texture, and which is exhibited by fine-grained recrystallization-structures or by finely powdered metal compressed at high temperatures. This process is not initiated by the formation of new crystal nuclei, but consists instead of the preferential growth of individual grains, in certain directions, at the expense of the others. It results from the instability caused by the higher surface energy of the polycrystal as compared with the single crystal.

As regards their structural mechanisms, both types of recrystallization are phenomena of atomic rearrangement.

11. *Recrystallization after Critical Plastic Deformation*

If recrystallization, after cold working, is used as a method for the production of large crystals, it will, of course, be necessary to ensure that the number of nuclei is kept to a minimum. Since this number increases with the degree of strain, only small percentages of cold work should precede the annealing. A further necessary condition is that the deformation should be as homogeneous as possible so that a uniform capacity for recrystallization will result. To achieve this, a uniformly fine-grained structure is indispensable.

In order, therefore, to produce single crystals by recrystallization after cold working, it is necessary initially to obtain in the specimen, whether sheet, wire or tensile test bar, as fine and uniform a grain size as possible. If the required texture is not present in the initial material, it must be produced by previous recrystallization. A short annealing for about $\frac{1}{2}$ hour at a fairly low temperature after substantial cold working, or heating above a transformation temperature usually suffices for this purpose. After the specimens have been prepared in this way the most suitable percentage of working must be determined by subjecting them to small deformations of varying magnitude (usually elongations between $\frac{1}{2}$ and 4 per cent. are necessary), the utmost care being taken to exclude any additional strain, such as bending. The ensuing annealing treatment must also be carried out under very careful control with a view to reducing the number of nuclei to a minimum. For instance, annealing should begin at temperatures below that at which recrystallization starts, although for solid solutions they should be above the solubility limit. Metals that tend to oxidize easily should be annealed in a current of H_2 or *in vacuo*. By increasing the temperature very slowly (20–$50°$ per day) one of the first nuclei formed may be induced to grow through the entire specimen, at the same time avoiding the formation of further nuclei. It is good practice to maintain a slight temperature gradient in the furnace (eccentric location of the specimens; if heating takes place under a stream of gas, the resultant temperature gradient suffices), since in this way the formation of further nuclei is avoided. Towards the end of the annealing period, which usually lasts for several days, the temperature can be increased more rapidly, and the operation is concluded by heating for a brief period just below the melting point (or a transformation temperature, a solidus line or a eutectic temperature) in order that small grains which generally are still present may be consumed by grain growth. Cooling should usually take place *inside* the furnace in order not to damage the crystals, which are particularly sensitive at high temperatures. Subsequent etching develops the boundaries of the newly formed crystals and ensures the removal of any small grains that may still be present on the surface. By comparing the specimens which have been subjected to varying degrees of working, the critical working percentage at which crystals of maximum size can be obtained is determined (see Fig. 14).

In general, little of a positive nature can be said regarding the

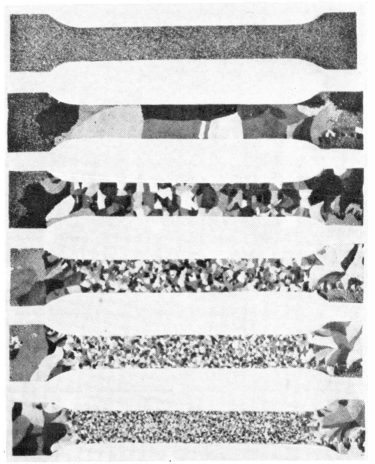

FIG. 14.—Recrystallization of Aluminium; Relationship between Grain Size and Plastic Strain.

(From top to bottom : 0, 2, 4, 6, 8 and 10% extension.)   (22.)

yield of this process, since the growth of the crystals depends to a very large extent upon the composition and purity of the material used. In the most favourable cases, however, it may amount to almost 100 per cent. Arbitrary control of the orientation of the crystals, *i.e.*, of the position of the lattice in the specimens, is impossible. As a rule the whole field of orientation can only be

TABLE IV

Crystal Production by Recrystallization after Critical Cold Working

Metal or alloy.	Pre-treatment, initial grain size.	Critical working, % elongation.	Annealing conditions.	References.
Mg	~120 grains/mm.²	0·2	From 300° to 600° C. in 6 days	(28)
Mg-solid solutions with Al, Zn, Mn, Al and Zn	Lengthy annealing just below eutectic temperature; 7–10% strain, followed by brief annealing	0·2–0·3	Anneal only in the range of homogeneous solid solution; daily increase in temperature 20–50° C.	(29)
Al~(99·5) *	~100 grains/mm.²	1·6	Start at 450° C.; increase temperature by 25°/day to 500° C. Then 1 hour at 600° C.	(22), (23)
Al-solid solutions with Zn (up to 18·6% Zn) and Cu	—	1–2	500–550° C.	(30), (31)
	—	1·5	In 6 days from 450° to 515° C.	(32)
Fe~0·13% C 0·4% Mn 0·02% Si 0·03% S 0·02% P	Pre-heat for protracted period at 950° C in H₂. About 120 grains/ mm.²	3·25	A few days at 880° C.	(33)
Fe~0·10% C 0·4% Mn traces Si 0·021% S 0·021% P	—	2·75	4 days at 880° C.	(34)
Fe (Armco) 0·03% C 0·025% Mn traces Si 0·025% S 0·01% P 0·06% Cu	—	2·75	Increase from 530° to 880° in 4 days, then 2 days at 880° C.	(35) cf. also (36)

* The production of crystals of very pure aluminium (99·99 per cent.) has hitherto been attended by difficulties.

covered by increasing the number of specimens. In certain cases, however, this method fails, and the crystals obtained have similar orientations. This is due to the difficulty, and sometimes even the impossibility, of obtaining a *fine-grained structure of random orientation* before the start of critical straining, and so avoiding the consequences of a preferred orientation.

The quality of the crystals obtained in this way is greatly influenced by that of the original material. Since all stresses are removed by the protracted anneal, it should be possible to satisfy requirements in regard to physical perfection. Owing to lack of experimental data it is not yet feasible to compare such crystals with those obtained by other methods.

The difficulties due to segregation, which occurs when crystals of solid solutions with a wide melting range are produced from the molten state, do not arise in the recrystallization method. Finally, the freedom which this method allows in the shaping of the specimen can be of great advantage in certain cases. It is a fundamental disadvantage of the work-recrystallization method that it involves plastic deformation, since brittle material is excluded from such treatment.

The main development of this method of growing crystals originates from (23) and (22) (cf. also 24). Details of the conditions employed by various authors will be found in Table IV. Apart from the metals mentioned in the table, the method has been adopted also with copper (25, 26) and β-brass; however, large crystals completely free from twins could not be produced in copper.

Using the same treatment it is possible to obtain from drawn tungsten wire those industrially important lamp filaments which consist of large crystals whose grain boundaries lie at small angles to the wire axis (27). The same texture can also be achieved without straining before the final anneal, if certain small additions are made to the base material. This method, however, should not be included among those described in the present section. It represents the transition to the methods of grain growth discussed in the following section.

12. *Crystal Production by Grain Growth*

The method of producing crystals by grain growth has been employed mainly in the case of tungsten. The so-called " Pintsch wires " are produced from a uniform and fine-grained tungsten powder which has been mixed with thorium oxide (approximately 2 per cent.) of maximum fineness; a bonding substance is added, and the mixture is then extruded through diamond dies to fine threads [(37), (38)]. After drying, these are moved at the rate of up to 1 mm. per second through a very hot and narrow zone (2500° C.) as a result of which the crystal growth starts at the end of the wire and then proceeds continuously. The furnace consists of a few coils of a spiral of tungsten wire, heated electrically in a hydrogen atmosphere. A second method of effecting continuous local heating consists in passing a current through the sintered wire which is to be transformed into a crystal by moving it slowly over two adjacent contacts. There is still some doubt as to the function of the thorium oxide which must be added if the method is to succeed. The thickness of the crystals obtained in this way amounts to as much as 0·1 mm.

By another method (39), which can also be regarded as a typical grain-growth process, tungsten powder is pressed into bars at a pressure of 4000 kg./cm.^{2}, sintered and then heated for about an hour just below the melting point (3268° C.) in a moist hydrogen atmosphere. By this method it is sometimes possible to convert the entire bar into a single crystal. Although thermal instability is the cause of crystal growth, the water-vapour content of the protective gas also performs a very important function, the significance of which, however, is not yet fully understood. Grain growth can also be materially assisted by additions to the base material (*e.g.*, ThO_2) [(40), (41)].

B. PRODUCTION OF CRYSTALS FROM THE MOLTEN STATE

For the purpose of description, methods of growing large crystals from the molten state will be divided into two groups : processes in which solidification of the whole melt is suitably induced *in* the melting crucible, and those in which parts of the melt are made to solidify *outside* the crucible.

13. *Crystallization in the Crucible*

Tammann observed that if a melt of bismuth was cooled slowly in a glass tube, a single crystal of about 20 cm. length could be obtained (42). In succeeding years this process was perfected by a number of investigators. Glass vessels which had been drawn out at one end to a capillary tube were used. Solidification took place in the interior of a vertical tubular furnace containing two heating coils which could be switched on independently, thereby enabling solidification to start at the thinned-down end of the tube (43) : see also (44) to (47). The sketch of another melting vessel of glass which has been frequently used is shown in Fig. 15. The cavity *A* is charged with the metal to be melted and then drawn out at its front end *D* into a tube to which an air-pump is connected. After the parts *B* and *C* have been heated in an electric furnace to about 50–100° C. above the melting point of the metal, and the air has been exhausted from the entire vessel, the metal contained in *A*, which projects from the horizontal furnace, is melted in a gas flame. The occluded gases are then driven off by shaking the molten mass, an operation which, for bismuth, must be prolonged for 1 hour. The furnace is

FIG. 15.—Sketch of Vessel for the Production of Metal Crystals [see (48)].

now brought into a vertical position so that the metal flows down to the lower end, its quantity being so determined that finally a small amount of the melt remains in A, while all impurities, such as oxide skins, etc., are kept away from the actual melting chamber B. The vessel is now sealed at D, and crystallization is initiated by slowly lowering it (rate of lowering 4–60 mm. per hour). The purpose of the capillary between C and B, as in the case of the melting chamber

mentioned above, is to allow only one of the initially formed crystals to penetrate to the chamber B in which crystallization actually occurs. The capillary works selectively, so that of the nuclei which first appear the one which ultimately develops is that for which the direction of maximum speed of growth encloses the smallest angle with the axis of the tube, as is shown in Fig. 16. It is seen clearly how crystals with small axial rates of growth are eliminated.

With a view to securing the maximum liberation of gas when producing crystals of metals and alloys of high melting point (copper, cobalt, nickel, the precious metals), vacuum furnaces were used repeatedly in which the melt, contained in a crucible made of carbon or alumina, was moved slowly through a graphite heating tube [(50), (51)].

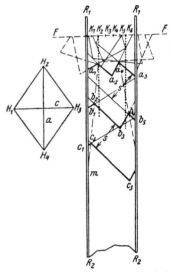

FIG. 16.—The Growth of Crystals in Tubes; Selection Based on Speed of Growth (49).

K_1–K_6: variously orientated crystal nuclei on the surface of the melt.

The charge of the whole crucible can solidify in vacuum to large crystals (52) if provision is made for solidification to begin exclusively at one point, and for the heat to be withdrawn at a suitable speed mainly from this point, which may be, e.g., a conical depression in the crucible. In this case the size of the crystals is limited solely by the quantity of the material (size of the furnace). A copper crystal weighing 6 kg. was obtained by these means. Later the process was considerably improved [(53), (54)]. A high-frequency furnace was used, care being taken to avoid turbulence in the melt; the vacuum was improved to 1/100 mm. Hg; and, in particular, the orientation of the growing crystal was influenced. Indeed, whereas control of the orientation of the crystal is practically impossible with the crucibles already described (although it might conceivably be effected by tilting, at various angles, the capillary tubes at the lower end of the melting vessel, thereby exploiting the selective growth), the crucible design in use at present is such that the orientation of the crystal can be determined in advance by inoculation. For this purpose a small hole is drilled into the base of the graphite crucible, and into this hole a piece of crystal of the desired orientation is inserted. Heating is so regulated that only the upper portion of this crystal melts, so that crystallization proceeds from the seed crystal which serves as an artificial nucleus. It is, of course, always necessary to adjust the rate of cooling, or the speed at which the crucible is lowered through the furnace, to the speed of crystallization.

By this method single crystals have been produced from a large number of pure metals [Cu, Ni, Ag, Au, Mg, Zn, Cd, Hg (55, 56), Sn, Sb, Bi, Te] and alloys (α- and β-brass, Cu–Al, Cu–Sn, Au–Ag, Au–Cu, Au–Sn, Fe–Ni, Sb–Bi, Austenite).

If satisfactory crystals are to be obtained by this method care must be taken, by degassing the melts thoroughly, to avoid microporosity during crystallization. However, the segregation which occurs when producing the crystals of alloys cannot be avoided when crystallization proceeds from the melt. The material of the crucible must be chosen with the utmost care, both with a view to restricting contamination of the metal, and ensuring that the crystal will not be damaged when it is removed from the crucible. The production of completely undamaged, stress-free crystals by this method is in any case very difficult, and there has been no lack of experiments to ensure a still more careful handling of the crystal in the course of its production.

Of the methods evolved with this object in view there are four types [(57), (58)]. In the first method, the rod of metal which is to be converted into a single crystal is placed on a copper plate in which a temperature gradient has been produced by one-sided heating. First of all the rod is melted by adequate heating of the plate; owing to its surface tension and the presence of a thin oxide skin it will retain its cylindrical shape. Crystallization is initiated at the cooler end of the plate by gradually decreasing the current. The orientation of the growing crystal can be influenced at will by contacting the melt with a " seed crystal " which has been placed at a suitable angle and which, of course, must not be completely melted down. By the second method the metal rod, which has been placed in a wide silica tube, is moved slowly through a narrow heating coil in a manner similar to that described in Section 12, melting and crystallization taking place consecutively. According to the third and simplest method the metal rod is put in a narrow cylindrical cavity bored into a copper block which is heated on one side only and which can be cooled under control. In both these cases it is possible to influence the orientation by the use of a suitable seed crystal. The fourth variant of the process (58) purports to improve upon the other three by eliminating also the stresses which emanate from the oxide skin. To this end the metal is melted down and solidified in a graphite channel placed in an evacuated silica tube which is moved slowly through the furnace. A current of purified hydrogen is passed through the tube. Each specimen is remelted several times before being brought into contact with the seed crystal. In this case the shape of the crystal is determined by the shape of the graphite channel. It is remarkable that hitherto this treatment should have been used only for bismuth, and (recently) tin (59).

So far we have described methods employed mainly for producing crystals from metals. Naturally they can also be adapted to the production of *salt crystals*. For instance, a furnace has been described (60) in which the melt is shaped like a plano-convex lens, in which the isotherms run parallel to the surface of the melt. Solidification proceeds uniformly from the bottom of the lens upwards. Large crystals have been produced in this way from sodium nitrate, potassium nitrate and sodium chloride, as well as from bismuth and zinc. By a further development of this method the internal stresses which lead to a tearing of the crystal can be eliminated (61). The following particulars are available of a method which has already been adopted for a large number of ionic crystals (sodium chloride, sodium bromide, potassium iodide, rubidium chloride and lithium fluoride) (62). A platinum tube, closed at the lower end and cooled internally by air, is immersed in the melt to the depth of a few millimetres. If the temperature has fallen to about 70° above the melting point of the salt, intensified cooling will cause crystallization to start on the platinum tube. When the diameter of the hemisphere which has crystallized out (Fig. 17) amounts to about four times that of the platinum tube, the cooler is raised with the aid of a micrometer screw until the surface of contact between the crystallized spherolite (I) and the melt is composed of one crystal only. After this the greater part of the melt is made to crystallize by increased cooling (II). The operation is broken off before the crystal reaches the wall of the crucible. If this method is to be successful a period of growth of several hours will be needed for crystals averaging 3 cm. in size, and the temperature of the furnace (supply by accumulator battery) and of the cooling stream must be kept constant.

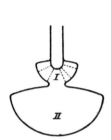

FIG. 17.—Illustrating the Production of Salt Crystals According to (62). Shape of the Crystal " Pears " Obtained in this Way.

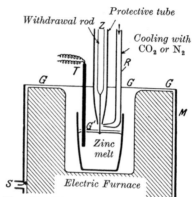

FIG. 18.—Diagram showing the Method of Producing Crystals by Drawing from the Melt. According to Czochralski [see (65)].

14. *Production of Crystals by Drawing from the Melt*

Czochralski (63) initiated a process which, although originally devised for the determination of the speed at which metals crystallize, was later much used for growing crystals. In this method a small rod (glass, clay or, better still, a seed crystal), immersed in the melt and serving as a centre of crystallization, is slowly and steadily raised. If the temperature of the melt is only slightly above the melting point, and if the rod is lifted slowly enough, there will adhere to it a small molten thread which crystallizes above the surface of the melt. The length of the crystals obtained in this way is limited by the dimensions of the apparatus : in shape they are approximately cylindrical.

Fig. 18 illustrates the type of equipment subsequently used

[(64), (65)] for producing crystal wires of a number of metals of low melting point. Through a small disc of mica (G'), perforated in the centre and weighted, a thread of molten metal is drawn, which is cooled just above the disc by a current of inert gas. In order to obtain crystals of maximum uniformity great care must be taken not only to maintain a constant melt temperature and constant speed of drawing and cooling but also to avoid vibrations. The diameter of crystal wires so obtained can be varied between about 0·2 and 5 mm., according to the conditions of heat removal. The requisite crystal orientations are obtained by using seed crystals with suitable lattice orientations. If intimate contact is maintained between melt and seed crystal (removal of the oxide skin), then the seed crystal will in most cases continue to grow. The role of the speed of drawing is still not fully explained. A speed of about 1 cm. per minute would appear to be smaller than the minimum rate of crystallization of the metals employed. This represents, of course, the natural upper limit of the speed of drawing.

By this method abundant material has been obtained for experiments, including crystals of Zn, Cd, Sn, Bi, Zn–Cd and Zn–Sn alloys.

The orientation of crystals produced in this way is always subject to slight but constant variations along the wire (66), caused, it is believed, by variations in the temperature gradient above the melt [cf. also (67)]. The maximum differences in orientation observed in zinc crystals 30 cm. long amounted to approximately 10°; however, if the experimental conditions were held constant they did not exceed about 2°.

The drawing method is also frequently used to-day for the production of *salt* crystals (68). In their case special cooling is usually unnecessary. The orientation of the growing crystal bars or " pears " can be materially influenced by inoculation.

The drawing process as described above cannot be used for reactive melts. For magnesium crystals, therefore, it is customary to employ a carbon tube which is immersed deep in the metal (the surface of which is covered with a protective layer) and is then slowly withdrawn (69). For protective purposes, this tube is enclosed in an iron tube, open at the bottom. This is provided at the top end with a cock which must be closed before withdrawal from the melt.

C. OTHER METHODS OF GROWING CRYSTALS

15. *Crystal Growth by Precipitation from Vapour*

When producing crystals from vapour it is customary to start with an existing seed crystal upon which the material is deposited.

In the case of tungsten the decomposition of the WCl_6 vapour is used. By one process (70) the tungsten hexachloride is reduced by hydrogen according to the equation

$$WCl_6 + 3H_2 = W + 6HCl.$$

Another method (71) makes use of the thermal decomposition of WCl_6 into its two components, which takes place at temperatures above 1500° C. In both cases a thin tungsten crystal (Pintsch wire) is used as a seed crystal, which, according to its orientation, grows into a 4-, 6- or many-sided crystal bar with a thickness of up to about 10 mm. By this method it is possible to deposit alternating layers of tungsten and molybdenum. In a similar manner crystals of Ta, Fe, Zr and Ti have also been precipitated from vapour (72).

The production of *zinc* and *cadmium* crystals by sublimation is described in (73). By this method six-cornered crystals with regular faces can be obtained without the use of seed crystals.

16. *Crystal Growth by Electrolytic Deposition*

This process, which is applied to tungsten (74), is similarly based on the use of a " Pintsch " wire as a seed crystal, the wire serving as a cathode in the axis of the cylindrical sheet of tungsten which constitutes the anode. Sodium tungstate forms the electrolyte, its decomposition being effected at 900° C. with a current density of 150 $mA/cm.^2$.

This concludes the description of methods for the production of crystals. Processes for growing salt crystals from the *solution* have not been included in this review; for information on this subject see the summary given in (75).

CHAPTER IV

DETERMINATION OF THE ORIENTATION OF CRYSTALS

The orientation of a crystal is the position of the lattice with reference to directions conveniently defined by the form of the crystal specimen. Whereas with crystals bounded by natural faces the position of the lattice is immediately obvious from the face angles (goniometrically measurable), with crystals of irregular shape it must be determined by special methods. Often it is only a question of determining the orientation of the lattice with reference to a *single* prominent direction of the crystal specimen (*e.g.*, the longitudinal direction of a cylindrical crystal rod) which is fixed by the angle it makes with the principal crystallographic axes. In the following pages we shall describe methods for determining the orientation of opaque crystals, with special reference to the investigation of metal crystals.

A. MECHANICAL AND OPTICAL METHODS

17. *Symmetry of Percussion-figures. Investigation of Light Reflected from Crystal Surfaces*

Orientation can be determined by mechanical tests only if a plane face is present on the specimen under investigation. The percussion-figure resulting from plastic deformation reveals the crystallographic nature of the plane normal (76).

Orientation can be determined much more precisely by investigating a crystal surface which, by etching, has been covered with crystallographically regular etch pits. These pits can also be replaced by a surface relief, consisting of microscopic or submicroscopic negative crystals, such as occurs, for instance, when metal crystals solidify from the melt. The first method of examining such a surface, *i.e.*, the determination by microscopical means of the shape of the etch pits on plane surfaces (77), is no more powerful than the percussion-figure method. The same is true of the " maximum lustre " method (78, 79), which uses the intensity of the light reflected from the etched surface. The specimen under investigation, exposed to oblique parallel light, is rotated around the axis of the microscope. From the number of alternations in intensity in the course of a complete revolution, the symmetry properties of the plane normal are deduced.

Whereas the above methods are restricted to certain simple cases and are dependent for their success on the presence of plane faces on the crystals, orientation can also be determined by investigating reflected light in the more general case [(80), (81)]. Since the principle is the same for both methods, the variant referred to in (81), which is the simpler and more commonly used of the two, will be described in detail. The crystal under investigation (we assume this to be a cylindrical rod) is placed in a hole drilled radially in a wooden sphere, and illuminated with parallel light (sunlight). The crystal is turned, together with the sphere, until a reflexion of maximum intensity is obtained. This direction is indicated by means of an adjustable mirror which is placed tangentially on the sphere, and which is also brought into the reflecting position, its point of contact with the sphere being marked. As a result of reflexions from several different faces a pole figure is obtained on the sphere, from which, assuming the interfacial angles of the crystal to be known, the position of the lattice in relation to the longitudinal axis of the crystal can be deduced. The only precaution needed is in regard to double reflections, which, however, can be easily recognized. Direct readings of the desired angles can be obtained if the surface of the sphere bears a suitable network of meridians and parallel circles. It is obvious that this method can also be used to determine the crystallographic orientation of a polished surface; in this case the specimen must be attached tangentially to a pole of the sphere. A small systematic error arises in using this method owing to the fact that the normal of the reflecting crystal face does not pass through the centre of the sphere. The correction to be applied is discussed in (82), where particulars are also given of an extension of the method, using polarized light.

B. X-RAY METHODS

Very great possibilities for the investigation of orientation are opened up by the use of X-rays.

18. *The Diffraction of X-rays by Crystal Lattices*

A detailed account of the diffraction of X-rays by crystal lattices will be found in the relevant literature. Here we shall deal only with the features important to our special application—the determination of crystal orientations. The main progress which has resulted from the application of X-ray methods to the examination of solids has been in the determination of their internal structure. The distribution of the various structures among the elements of the periodic system, together with numerical values of the lattice dimensions of the more important metals and of some ionic crystals,

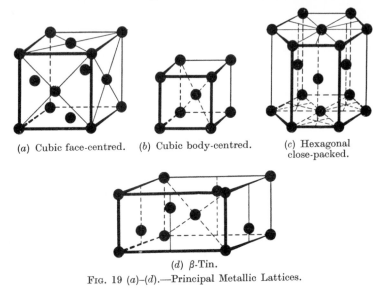

(a) Cubic face-centred.   (b) Cubic body-centred.   (c) Hexagonal close-packed.

(d) β-Tin.

FIG. 19 (a)–(d).—Principal Metallic Lattices.

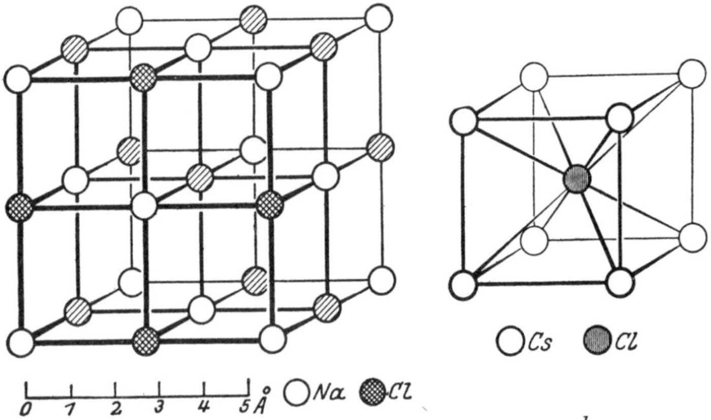

○ Cs   ● Cl

○ Na   ● Cl

FIG. 20.—Lattices of (a) Sodium Chloride and (b) Cæsium Chloride.

will be found in Tables XL–XLII at the end of the book. A few of the principal lattice types are shown in Figs. 19 and 20.

The diffraction of a beam of X-rays in a crystal can be regarded, according to Bragg, as reflexion of the rays from the different lattice planes of the crystal. The reflexion of X-rays differs from ordinary optical reflexion (in which the angles of incidence and of reflexion are the same and lie in the same plane) in that it does not occur at *all* angles of incidence. It can occur only when the wavelength (λ) of the incident beam, the spacing (d) of the reflecting lattice plane, and the glancing angle [1] ($\theta/2$) satisfy the condition.

$$n \cdot \lambda = 2d \cdot \sin \theta/2 \text{ (Bragg's equation; } n = 1, 2, 3 \ldots)$$

Consequently a crystal at rest, if exposed to X-rays of a *single* wavelength, will not normally yield any reflexion, since none of the lattice planes will be at an angle to the ray which corresponds to d. There are two ways by which the requirements of Bragg's equation can be satisfied. (1) By using " white " X-rays the wavelength can be varied (Laue photographs); and (2) where monochromatic radiation is employed, by rotating the crystal (Bragg) the lattice planes can be made to pass through the reflecting position (rotation photographs).   On the other hand, on irradiating a fine-grained crystal powder (polycrystalline material) the conditions for reflexion are satisfied even when the specimen is at rest, owing to the completely random orientation of the individual grains (Debye–Scherrer, Hull photographs).

(a) THE ROTATING-CRYSTAL METHOD

19. *Basic Formulæ*

The use of the rotating-crystal method for determining orientation is illustrated in Fig. 21 in terms of the geometry of the reference sphere [(83), (84) and (85)]. We observe the reflexion of the incident beam by the lattice plane (hkl) from which it is reflected at the glancing angle $\theta/2$. The plane normal, therefore, makes an angle of ($90-\theta/2$) with the incident beam at the moment of reflexion, and the geometrical locus of the normal for all possible reflecting positions of the plane (hkl) is the circle of reflexion which is shown in the illustration as R_k. In the course of the rotation of the crystal about a geometrically important direction (longitudinal axis), which in Fig. 21 is chosen at right angles to the incident beam, the normal describes about the axis of rotation a double cone, whose intersection with the reference sphere is represented by the two parallel circles at polar distance ρ. The points at which the circle of reflexion intersects these two circles indicate the positions of the plane normal

---

[1] The glancing angle is the angle between the incident beam and the reflecting plane; consequently it is complementary to the angle of incidence of 90° used in optics.

at the moment of reflexion. Since the plane normal bisects the angle between reflected and incident beams, and is co-planar with them, the positions of the reflected rays are unequivocally determined. They make with the incident beam the angle of deviation θ, and therefore lie on a cone with vertical angle 2θ, having the incident beam as axis. The intersection of this cone with the photographic film gives the " Debye–Scherrer ring " of this plane, which represents the geometrical locus of the reflexion for all possible positions of the plane. This takes the form of a circle if a plate is used at right angles to the incident beam. However, if a cylindrical film is

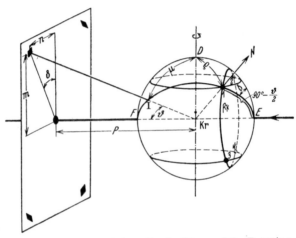

FIG. 21.—Determination of Orientation by Means of the Rotating-crystal Method.

employed, the axis of which would normally also be placed at right angles to the incident beam, then a curve of the fourth order is obtained. When the axis of the cylinder and the incident beam coincide, the curve degenerates into a circle; this in turn becomes a straight line when the film is laid out flat. Corresponding to the reflexion positions (1–4) of the reflecting plane, the reflected beams are also symmetrical in relation to the plane of the incident beam and axis of rotation, and to the " equatorial plane ", which is perpendicular to the plane of incidence.

If the film cylinder is perpendicular to the beam, the diagram contains the Debye–Scherrer rings of all the reflecting planes. The magnitude of the angle of deviation of X-rays on reflexion at a plane (hkl) is found by inserting the lattice-plane spacing (d) into

Bragg's equation (Section 6). The higher the indices and, consequently, the smaller the spacing of the lattice plane, the greater is the angle of deviation (θ) of the beam, which may be as much as $180°$.

The intensity of the Debye–Scherrer rings depends on many factors, including primarily the structure-factor, which, however, will not be further discussed. Table V contains the indices of the lattice planes of the more important lattice types, arranged in order of increasing θ angles of the corresponding Debye–Scherrer rings.

TABLE V

Sequence of Debye–Scherrer Rings for some of the Metallic Lattices

Cubic.		Close-packed hexagonal.	
Face-centred.	Body-centred.	$c/a = 1{\cdot}633$ $(=2\,\sqrt{\tfrac{2}{3}}).$	$c/a = 1{\cdot}86$ (Zn).
111	110	$10\bar{1}0$	0002
200	200	0002	$10\bar{1}0$
022	112	$10\bar{1}1$	$10\bar{1}1$
113	022	$10\bar{1}2$	$10\bar{1}2$
222	013	$11\bar{2}0$	$10\bar{1}3$
004	222	$10\bar{1}3$	$11\bar{2}0$
313	213	$20\bar{2}0$	0004
024	004	$11\bar{2}2$	$11\bar{2}2$

The problem of orientation now consists in determining, by measurement of the positions of the reflexions on the diagram, the angle between the normal to the reflecting planes and the axis of rotation.

From the triangle $E1D$ (Fig. 21) we obtain first for the unknown angle ρ between face normal and axis of rotation the equation

$$\cos \rho = \cos \theta/2 \cos \delta \quad . \quad . \quad . \quad . \quad (19/1)$$

in which δ represents the angle between the plane of the incident beam and the axis of rotation, and that of the incident beam, normal, and reflected beam. The angle δ can be determined in various ways, according to whether the photograph has been recorded on a plate or cylindrical film. In the case of the plane shown in Fig. 21, the determination is very easy. δ can be measured either directly with a protractor, or calculated by measuring the distance between diffraction spots (reflexions). Let P be the distance of the plate from the specimen, and n and m the rectangular co-ordinates of a reflexion (diffraction spot) with reference to the vertical and horizontal lines of symmetry of the diagram; it then follows that $tg\delta = \dfrac{2n}{2m}$ or, if the radius $r\ (= Ptg\theta)$ of the Debye–Scherrer ring is introduced, on which the four reflexions lie, then $\cos\delta = \dfrac{m}{r}$ and finally

$$\cos\rho = \frac{\cos\theta/2}{2P\cdot tg\theta}\cdot 2m \quad . \quad . \quad . \quad . \quad (19/2)$$

Consequently, if the distance of the plate and the indices of the reflecting lattice plane (θ) are known, measurement of the distance $2m$ between two reflexions parallel to the axis of rotation will give the desired angle between plane normal and axis of rotation. Should it be necessary to determine orientations in large numbers, it is recommended that the relationship between ρ and $2m$ be represented in graphic or tabular form.

If the photograph is recorded on a cylindrical film of radius R, the axis of which is parallel to the axis of rotation of the crystal (perpendicular to the incident beam), then, if $2m$ represents the distance between reflexions (parallel to the axis of the cylinder), the following equation is obtained :

$$\cos\rho = \frac{1}{2\sin\theta/2}\cdot\frac{m}{\sqrt{m^2 + R^2}} \quad . \quad . \quad . \quad (19/3)$$

Before we proceed to illustrate, by means of examples, the practical application of X-ray technique to the determination of orientation, it will be useful to offer some further observations on the method under discussion.

20. Oblique Photographs

In the first place, as may be seen from Fig. 21, it may happen that the plane normal no longer intersects the reflexion circle when the crystal is rotated. This will occur when the angle ρ is smaller than $\theta/2$. For $\rho = \theta/2$ the two parallel circles make contact with the reflexion circle at two points, and two reflexions are obtained on the vertical axis of the diffraction pattern, lying symmetrically above and below the equator. In the most frequent case of $\theta/2 < \rho < 90°$, the four reflexions referred to above will appear. As ρ increases, they will approach the equator of the Debye–Scherrer ring belonging to the plane, where ultimately for $\rho = 90°$, each pair of reflexions combines to form a single reflexion. Consequently the reflexions for all those planes which belong to the zone of the axis of rotation will lie on the equator. In order to ensure that reflexions shall occur in all cases (including cases where $\rho < \theta/2$) the axis of rotation of the crystal is fixed at an angle $(90° - \theta/2)$ to the incident beam (Fig. 22). Symmetry relative to the equator disappears in these " oblique photographs " (83). The required angle ρ is obtained from the triangle $E1D$ by means of the relation

$$\cos \rho = \sin^2 \theta/2 + \cos^2 \theta/2 \cos \delta \quad . \quad . \quad (20/1)$$

δ can again be measured directly on the plate, or it can be calculated,

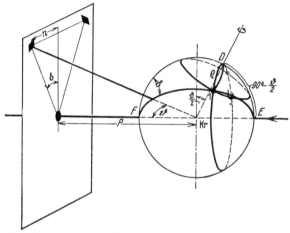

FIG. 22.—Determination of Orientation with the Aid of " Oblique " Photographs.

with the aid of the formula $\sin \delta = \dfrac{n}{P \cdot tg\theta}$, from the distance $2n$ between the reflexions and the plate distance P.

If the angle between the axis of rotation and the incident beam is not $(90 - \theta/2)$, but β, then the equation which determines ρ is as follows :

$$\cos \rho = \cos \beta \sin \theta/2 + \sin \beta \cos \theta/2 \cos \delta \quad . \quad (20/2)$$

Oblique photographs are mainly important for use with crystals with unique planes (e.g., hexagonal and tetragonal) whose position is important for the determination of the orientation of the crystal.

21. Layer-line Diagrams

A further observation concerns the special case in which the axis of rotation coincides with a crystal direction defined by simple indices.   The diagrams then possess certain specially simple features [namely, the arrangement of the reflexions on Polanyi " layer lines ",

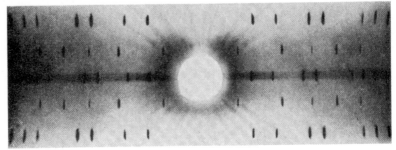

FIG. 23.—KBr Crystal : Layer-line Diagram Recorded on Cylindrical Film.

which appear as parallel straight lines when photographed on a cylindrical film, and as a set of hyperbolæ on a plate (Figs. 23 and 24)].   In discussing these diagrams we refer again to Fig. 21, in

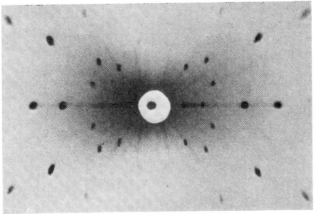

FIG. 24.—The Same Crystal : Layer-line Diagram Recorded on Flat Plate.

which the polar distance of the reflected beam is the angle μ.   From the triangle FID we first obtain

$$\cos \mu = \sin \theta \cos \delta$$

from which, by substituting for $\cos \delta$ from equation 19/1, the following equation is obtained :

$$\cos \mu = 2 \sin \theta/2 \cos \rho \quad . \quad . \quad . \quad (21/1)$$

We next assume that the co-ordinates have been so transformed that the axis of rotation coincides with the new c-axis. The transformed index l' of the plane in question, relative to the new c-axis, indicates that the intercept on the axis amounts to $\frac{1}{l'}$ on this axis.

Consequently l' equivalent lattice planes lie on the c-axis (axis of rotation) between two successive lattice points with the identity period J. For cos ρ Fig. 25 gives the expression

$$\cos \rho = \frac{l'd}{J}.$$

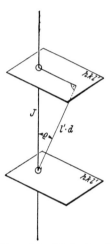

It then follows that for cos μ the equation is

$$\cos \mu = 2 \sin \theta/2 \frac{l'd}{J} = \frac{l'n\lambda}{J} \quad . \quad . \quad (21/2)$$

It will be seen from this equation that for a given value of l' all reflected rays intersect the reference sphere at points on a parallel circle, at a distance μ from the pole, and that consequently the reflected rays lie on the surface of a cone with semi-vertical angle μ about the axis of rotation. The parallel circles corresponding to successive values of l' make equal intercepts on the axis of rotation

Fig. 25.—Method of Indexing   Layer-line Diagrams (84).

so that the surface of the reference sphere carries a series of equidistant parallel circles. The index l' remains constant along each circle and increases by 1 on passing from one circle to the next.[1]

The maximum index $l'_{max.}$ is derived from the condition $l'_{max.} \leq \frac{J}{\lambda}$ since cos μ ≤ 1. The index O corresponds to the equator of the reference sphere.

The intercepts of the cones of reflected rays belonging to the individual parallel circles with the film constitute the characteristic " layer lines ". If the photograph is recorded on a cylindrical film with an axis which coincides with the axis of rotation, the film will be divided by the cones into circles, which will appear as straight lines when the film is laid out flat. If a plate is used, a set of hyperbolæ will appear with vertices on the axis of symmetry of the diagram parallel to the axis of rotation. Finally, if a cylindrical

---

[1] Provided that all reflexions on a layer line are not absent.

film is used with axis along the incident beam, the layer lines will
be represented by an array of curves of the fourth order. This last
type of photograph contains all the possible layer lines.

The significance of layer-line diagrams for the determination of
crystal orientation lies in the fact that, from the spacing between
the layer lines, or the vertices of the hyperbolae, the identity period
parallel to the axis of rotation, and hence its crystallographic nature
(when the lattice dimensions are known), can be readily ascertained
(84). If for instance $2e_l$ is the space between the lth layer lines, or
between their vertices, then if R is again the radius of the cylinder
of the film (or P the distance of the plate) :

$$\cot \mu_l = \frac{2e_l}{2R} \left(= \frac{2e_l}{2P} \right) \quad . \quad . \quad . \quad . \quad (21/3)$$

from which μ_l is calculated and can be inserted in

$$J = \frac{l \cdot n \cdot \lambda}{\cos \mu_l} \quad . \quad . \quad . \quad . \quad . \quad (21/2a)$$

The shortest identity period is obtained according to this formula
if the order of the reflexion (n) in the Bragg equation is put equal to 1.

On the other hand, should the axis of rotation, as is more usual,
not be a simple lattice direction, and thus be characterized by a
large identity period, then the layer lines will fall so close together
that it will no longer be possible to assign individual reflexions to
separate layer lines.

22. X-Ray Goniometer

We conclude with a final observation on the subject of the complete determination of the orientation of the lattice in the specimen.
The methods hitherto described yield the orientation in relation to
one direction only.   If the lattice orientation of the specimen is to
be fully known, then either two such photographs must be obtained
in different directions, or a photograph must be taken in an " X-ray
goniometer " [(86), (87), (88)].   In this case rotation of the crystal
is synchronized with movement of the film holder so that for each
reflexion obtained the position of the specimen is known.

23. Examples of the Determination of Orientation by the Rotating-crystal Method

As a first example of the determination of orientation by X-ray
methods we will work out the orientation of the axis of a rod-

shaped, cubic face-centred crystal with reference to the three axes of the cube. Fig. 26a shows the rotation photograph (flat plate, copper radiation $\lambda_{\kappa\alpha} = 1.54$ Å.) of an aluminium crystal, rotated about its longitudinal axis which is perpendicular to the beam. Owing to the distance from the plate (45.8 mm.) only the two innermost Debye–Scherrer rings ({111} and {200}) with the θ angles 38° 30' and 44° 50' are present. With the aid of the formula (19/2) or the

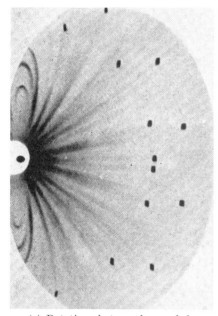

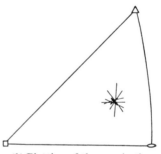

(a) Rotation photograph recorded on plate.

(b) Plotting of the arcs in the stereographic diagram.

FIG. 26.—Determination of the Orientation of an Aluminium Crystal.

equivalent graphical representation the corresponding ρ angles are now determined from the measured distances ($2m$) between reflexions. The above diagram gives :

$$2m_{(111)} = 2.9,\ 21.0,\ 52.0\ \text{and}\ 70.0\ \text{mm.}$$

from which it follows that

$$\rho_{(111)}\quad = 87°\ 45',\ 74°\ 15',\ 47°\ 45'\ \text{and}\ 24°\ 45'$$

and $2m_{(200)} = 20.2$ and 54.0 mm.

i.e., $\rho_{(200)}\quad = 78°\ 15'$ and $56°\ 45'$.

Although two angles suffice to determine the orientation of the longitudinal axis, there is close agreement (between the different values), and, as will be apparent from the stereographic diagram shown in Fig. 26b, this increases the degree of accuracy of the final result. The arcs in Fig. 26b are plotted in the first place with reference to the normals of two reflecting planes (e.g., in the present case two cubic axes); arcs having the known ρ angles are then described round these. The crystallographically equivalent normals from which the remaining angles are to be plotted (cf. Fig. 11)—in this case the four angles with the cube body diagonals—will appear from the initial rough determination of orientation (intersection of the first two circles to be described). By taking advantage of the known symmetry of the crystal it is always possible to complete the plotting of the orientation in the same basic triangle. The mean values for

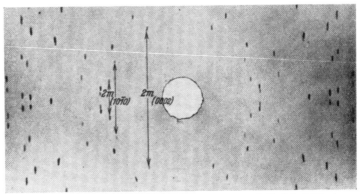

FIG. 27.—Rotation Photograph of a Zn Crystal (Cylindrical Film).

the angles between the axis of the rod and the cubic main axes in the present example are found ultimately to be 78°, 56° 30′ and 36°.

In our next example we will determine the orientation of a hexagonal crystal (zinc) in the form of a wire with the aid of a rotating-crystal photograph on a cylindrical film perpendicular to the beam (Fig. 27). In this case the orientation will be specified by the angle χ between the wire axis and basal plane, and the angle λ between wire axis and the nearest digonal axis type I. The reflexion from the basal plane appears in the case of zinc on the innermost Debye–Scherrer ring, that of the prism planes of type I on the next ring (Table V); the corresponding θ angles are 36° 20′ and 39° 10′ for copper radiation.

It follows from $2m_{(0002)} = 36\cdot4$ mm., with the aid of equation (19/3) ($2R = 57\cdot3$ mm.) that the angle between wire axis and hexagonal axis is $\rho_{(0002)} = 30° 40′$. From $2m_{(10\bar{1}0)} = 9\cdot4$ and $19\cdot8$ mm. for the reflexions from the prism planes type I we obtain the two angles $\rho_{(10\bar{1}0)} = 76°$ and $60° \ 20'$ between wire axis and two digonal axes type II (normals to the prism planes type I). Once again there is agreement between the different values of the orientation deduced from these figures. By forming an average in the stereographic net we find that

$$\chi = 60°$$
$$\lambda = 63° \ 30'.$$

Finally, from the two layer-line diagrams in Figs. 23 and 24 we will determine the crystallographic nature of the axis of rotation of a rod of potassium bromide.

From the distances between the layer lines

$$2e_1 = 13\cdot7 \text{ and } 2e_2 = 30\cdot0 \text{ mm.}$$

the diameter of the camera being $57\cdot3$ mm., it follows from the formula (21/3) that the angles

$$\mu_1 = 76° \ 30' \text{ and } \mu_2 = 62° \ 20'.$$

Further, we obtain from (21/2a), with $n = 1$, a value of $6\cdot63$ for J from the first layer line, and of $6\cdot65$ for J from the second layer line; giving a mean value for $J = 6\cdot64$ Å. Comparison with the measurements for the KBr lattice (Table XLII) shows that the rotation axis was parallel to the cubic axis of the crystal.

The same result is arrived at from an evaluation of Fig. 24. The distances between the vertices of pairs of hyperbolae are $2e_1 = 7\cdot8$ and $2e_2 = 16\cdot6$ mm., which, for a plate distance of $P = 16\cdot0$ mm., gives a value of $76° \ 20'$ for μ_1 and of $62° \ 30'$ for μ_2, and so similarly indicates the cubic axis as the axis of rotation.

(b) LAUE METHOD

Compared with the methods described above for determining the orientation of crystals by the use of monochromatic radiation, the method based on Laue photographs is relatively unimportant. It can, however, be used with advantage to determine the crystallographic orientation of the face normals of plate-shaped crystals.

There is a general similarity in the appearance of Laue photographs, in so far as the individual reflexions corresponding to the different wavelengths lie on curves which are conic sections, and always include, as vertex, the point of incidence of the direct beam, which is perpendicular to the plate. All beams reflected from the planes of a zone constitute the generators of a circular cone about the zone axis, whose intercept with the photographic plate gives the conic sections known as " zone circles ".

As a rule the specimen is held between the plate, which is perpendicular to the incident beam, and the X-ray tube. So long as the angle between the zone axis and the direct beam is less than 45° the zone circles appear on this plate as ellipses, which become parabolas when the zone axis is inclined at 45°. If the axis of the zone is more steeply inclined to the incident beam, then the zone circle becomes one branch of a hyperbola. In this case the zone also appears as one branch of a hyperbola on a photographic plate placed *between* X-ray tube and specimen, but the vertex does not coincide with the point of incidence of the direct beam, as in the case of the first plate. Finally, if the zone axis is perpendicular to the incident beam, the cone degenerates into a plane which intersects both plates in straight lines passing through the point of incidence of the direct beam. Both transmission (89) and backreflexion Laue photographs can now be used to determine crystal orientation. As already mentioned, by this method (in which the specimen is not rotated), the position of the lattice is determined in relation to a system of co-ordinates in space and not in relation to a direction only (namely the direction of rotation) as is the case when using the rotating-crystal method.

24. *Example of the Determination of Orientation by Laue Photographs*

The method can be illustrated by two examples, the examination of a magnesium crystal by means of a transmission photograph [according to (90)] and of an aluminium crystal by means of a backreflexion photograph. The two diagrams are shown in Figs. 28a and 29a. The zone circles are easily recognized in both pictures. Since the individual reflexions are due to various unknown wavelengths, it is not at first possible to assign them to definite lattice planes. On the other hand, it is easy to ascertain the position of the reflecting lattice plane with reference to the direction of the incident beam (*i.e.*, the direction of the normal to the crystal plate).

Let S_1 or S_2 be the distance of each reflection from the point of incidence of the primary beam, then the angle of deviation θ_1 of the X-ray beam is obtained for the transmission photograph from $tg\theta_1 = \dfrac{S_1}{P_1}$, the deviation θ_2 for the back-reflexion photograph from $tg\,(180° - \theta_2) = \dfrac{S_2}{P_2}$, which in P_1 or P_2 represent the distances from

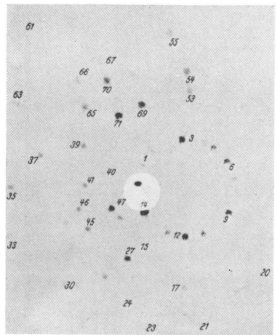

(a) Laue transmission photograph.

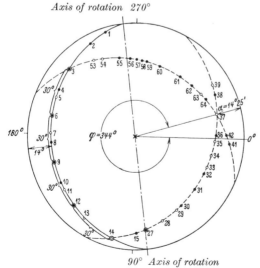

(b) Pole figure of the reflecting lattice planes (in stereographic projection).

FIG. 28 (a) and (b).—Determination of the Orientation of a Magnesium Crystal.

crystal to plates (Fig. 30).   The normal N_1 or N_2 to the reflecting
lattice planes lies in each case in the plane defined by the incident

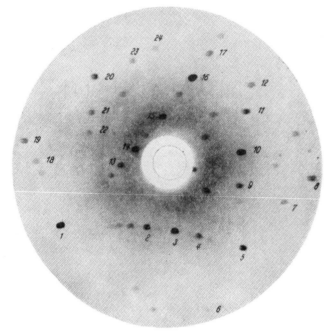

(a) Laue back-reflexion photograph.

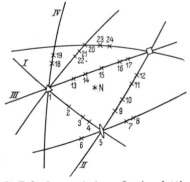

Fig. 29 (a) and (b).—Determination of the Orientation
of an Aluminium Crystal.

(b) Pole figure of the reflecting lattice
planes (in stereographic projection).

and diffracted beams and makes with the direct beam the angle
$(90° - \theta_1/2)$ or $(90° - \theta_2/2)$.   If now the position of the plane of

the two beams is defined by the angle, ϕ_1 or ϕ_2, which it makes with an arbitrarily selected plane of reference, then the pole of the

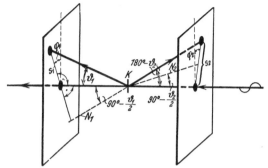

FIG. 30.—Analysis of Laue Photographs.

reflecting plane can be shown on a stereographic projection whose equatorial plane is the plane of the crystal plate.

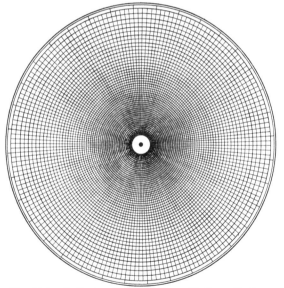

FIG. 31.—Polar Reflexion Chart for Use with Laue Photographs.

In order to simplify the determination of the θ-angles for the various reflexions it is convenient to employ a polar reflexion chart (Fig. 31) which indicates the θ-angles corresponding to the various S-values for a given plate distance (91). By superimposing Laue photograph and chart, the θ and φ angles for all reflexions can be read off directly. If these are plotted in the stereographic projection, then the pole figure for all reflecting-lattice planes is obtained (Figs. 28b and 29b). As will be seen from Fig. 30 the angle (90° − θ/2) from the centre must be plotted in the direction of the corresponding reflexion for back-reflexion photographs, and in the *opposite* direction for transmission photographs.

The orientation can now be determined from these pole figures in two ways. By the first method the pole figure, in which the zones

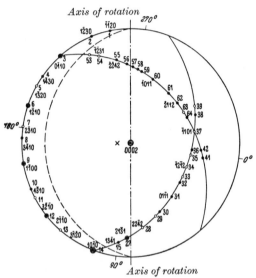

FIG. 32.—Determination of the Orientation of a Magnesium Crystal. Pole Figure of Fig. 28b, Rotated through 14°.

appear clearly as great circles passing through the reflexions, can be transformed into a pole figure corresponding to a simple setting of the crystal by rotation about an axis lying in the equatorial plane (the rotation method). By the second method the crystallographic description of the zones and plane normals can be derived directly from the relationship between the angles which appear on the stereographic pole figure.

For the first method we employ the pole figure of the transmission photograph (Fig. 28b) in which the intensity of the reflexions is also indicated. By rotating in such a way that the zone 1, 2 . . . 14 comes into coincidence with the basic circle, a pole figure (Fig. 32) is obtained which coincides with the pole figure of the magnesium crystal about the hexagonal axis (Fig. 33).[1] As a consequence of this rotation the plate normal moves to X; its orientation, *i.e.*, the angles which it makes with the hexagonal and digonal axes, can be

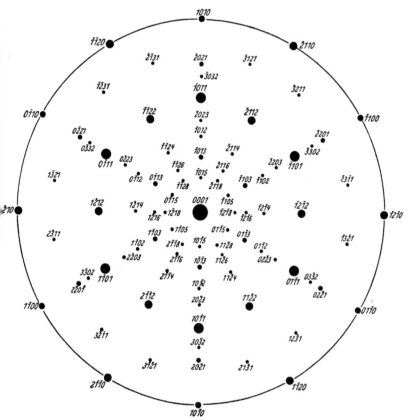

Fig. 33.—Pole Figure of Magnesium about the Hexagonal Axis (90).

seen from Fig. 32; they are $\rho = 14°$ with the hexagonal axis, and $77°$ and $76°$ with the nearest digonal axes of types I or II. If the longitudinal and transverse directions were also entered on the pole figure, the crystallographic orientations of these two directions would be immediately apparent.

[1] In this case the zone 1, 2 . . . 14 is seen to be a zone of the hexagonal axis from the fact that it exhibits reflexion angles of 30°.

Applying the second method to the stereographic figure in Fig. 29b we obtain the following angles between the important zones :

$$(I-II) = 90°; \quad (I-III) = 60°; \quad (I-IV) = 60°$$
$$(II-III) = 45°; \quad (II-IV) = 45°; \quad (III-IV) = 60°.$$

From this it follows that I, III, IV are zones of cube-face diagonals, while II is a cubic-axis zone. Zone intersections are represented by a cube edge (II–III), a face diagonal (I–II) and a body diagonal (I–III–IV). In this way the orientation of the normal (N) of the specimen which emerges from the centre of the net is fixed. It makes the angles 34°, 61° and 74° with the three cubic axes.

If the direction of incidence of the X-ray beam coincides with a symmetry axis of the crystal, the Laue photograph will exhibit a correspondingly symmetrical pattern. As a rule the nature of the symmetry of the Laue photograph suffices in this case to determine the crystallographic direction of the incident beam. Photographs obtained by the back-reflexion method are specially suitable for this purpose, besides having the additional advantage of being practicable with thick specimens [cf., *e.g.*, (92)].

Finally, it should be mentioned that atlases of the Laue photographs, obtained with face-centred cubic and body-centred cubic crystals by systematically varying the orientation, have been prepared (93).

CHAPTER V

GEOMETRY OF THE MECHANISMS OF CRYSTAL DEFORMATION

Before discussing in the second part of this book the results of investigations into crystal plasticity, we propose in the present chapter to examine the geometry of the various mechanisms of deformation in so far as it serves to explain the processes to be described.

It has long been known that crystals can be plastically deformed. The mechanisms which accompany plastic flow are based on the phenomena of glide (translation) and mechanical twinning (simple shear), which were discovered by Reusch on rock salt and calcite (1867). In both cases the deformation is plane and homogeneous, straight lines and planes remaining straight and plane; a sphere is transformed into an ellipsoid. The crystalline structure is also retained in the deformed part, the lattice being transformed into itself by the deformation.

A. GLIDE [1]

25. *Model of Gliding*

Glide consists in the slipping of portions of the crystal along crystallographic planes of low indices in the direction of densely packed atomic rows. Consequently neither the plane nor the direction of slip is determined by the loading (for instance, by a maximum of the shear stress); both these lattice elements are fixed by the structure of the lattice.

Fig. 34 shows the model of a cylindrical crystal which has been extended by glide (94). The model represents the basal glide of a hexagonal crystal under uni-axial tension. The upper and lower boundaries of the cylindrical specimen are made up of basal planes. The direction of glide is chosen to be a digonal axis of type I, and the hexagon sketched on the basal plane indicates that in general this glide direction does not coincide with the large elliptical axis. Glide along the crystallographically preferred glide system [2] leads

---

[1] Throughout this book the word "glide" has been used in preference to "slip". The terms are, however, synonymous and in the literature either may be used.

[2] The glide system that is mechanically preferred by a maximum of the shear stress consists, for uni-axial tension, of the plane that makes 45° with the axis of tension, and the major axis of the ellipse as the glide direction [cf., *e.g.* (40/1)].

55

to the configuration shown in Fig. 34, c and d. It is obvious that
the change in shape produced in this way is quite specific (band
formation) : the originally cylindrical crystal contracts considerably
in one direction, while it expands somewhat in the direction perpendicular to it, owing to the divergence of the glide direction from
the major axis of the translation ellipses. A comparison between
Fig. 34, b and d, shows also that the extension has been accompanied

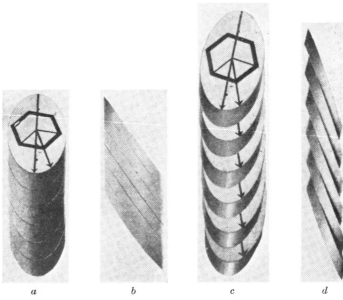

a   b   c   d

FIG. 34 (a)–(d).—Models of Gliding.

(a) and (b), initial state; (c) and (d), after extension.

by a very marked rotation of the lattice with respect to the longitudinal direction (direction of tension).

How closely this model corresponds to reality is shown by the
photographs of elongated metal crystals reproduced in Fig. 35.[1]
The traces of the glide planes appear as sharply defined elliptical
bands on the surface of the strip. The position of the apex of the
ellipses outside the central plane of the crystal band reveals clearly
the divergence between glide direction and the major axis of the
ellipses.

[1] For very early observations of glide bands and band formation resulting
from the elongation of metal crystals see (95) and (96).

26. *Geometrical Treatment of Simple Glide*

The formulæ expressing the relationship between deformation and lattice rotation when a single-glide system is operative will be examined in the first place for *extension* with the aid of Fig. 36. The initially cylindrical crystal will be extended between the two glide planes that go through the points A and B respectively.   Glide

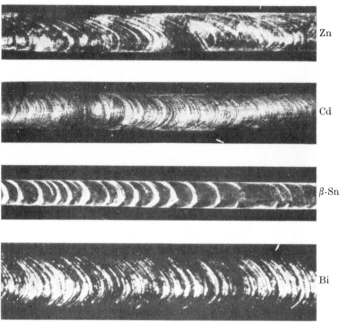

Fig. 35.—Glide of Metal Crystals, Viewed Perpendicularly to the Plane of the Band.

takes place parallel to the glide plane T in the glide direction t. The resultant configuration is drawn with thick lines; it has a double kink.   It will again be seen from the diagram that the glide has been accompanied by lattice rotation relative to the longitudinal axis, although in this case it is the lattice position (position of the glide elements) and not, as in Fig. 34, the longitudinal axis which has been kept fixed.   The angle between longitudinal axis and glide direction diminishes with increasing extension.   It will now be seen from Fig. 36 that the lattice rotation consists in a movement of the longitudinal axis towards the glide direction, during which the longitudinal axis always remains in the plane determined by its original position and the direction of glide. The relationship between the amount of extension: $\left(\delta = \dfrac{l_1 - l_0}{l_0} \right.$; l_0 and l_1 representing the length before and after extension) and the rotation of the lattice is obtained directly from the triangles ABB', ABN and $AB'N$ (94). The triangle ABB' gives

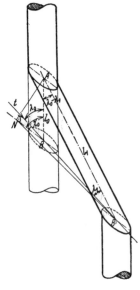

$$\frac{l_1}{l_0} = 1 + \delta = \frac{\sin \lambda_0}{\sin \lambda_1} \quad (26/1)$$

where λ_0 and λ_1 represent the angles between the direction of tension and the glide *directions* before and after extension.

From the two right-angled triangles ABN and $AB'N$ (AN is the normal on the glide plane) we obtain for the small side AN common to both triangles the expressions

$$AN = l_0 \sin \chi_0 = l_1 \sin \chi_1$$

and further

$$\frac{l_1}{l_0} = 1 + \delta = \frac{\sin \chi_0}{\sin \chi_1} \quad (26/2)$$

in which χ_0 and χ_1 are the angles between glide *plane* and directions of tension before and after extension. It is seen from these formulæ that the amount of extension by glide can be very considerable; as a rule it will increase with the

FIG. 36.—Diagram Illustrating the Formula for Extension by Gliding.

original obliquity of the glide elements to the direction of deformation.

The introduction of the *plastic shear strain* instead of the tensile strain has proved very convenient in the crystallographic analysis of stress–strain curves of crystals. This quantity, also known as the crystallographic glide strain (a), is the relative displacement of two glide planes of unit distance from each other [(97), (98)]. It is therefore given by the quotient $\dfrac{BB'}{AN}$; in which the numerator represents the total amount of slip and the denominator the thickness of the deformed glide packet. The connection between glide strain and extension can be easily derived from Fig. 36. For BB' we obtain from the triangle ABB'

$$BB' = \frac{l_1 \sin (\lambda_0 - \lambda_1)}{\sin \lambda_0}.$$

Using the first of the above expressions for AN, it follows that

$$a = \frac{BB'}{AN} = \frac{l_1}{l_0 \sin \chi_0} \cdot \frac{\sin (\lambda_0 - \lambda_1)}{\sin \lambda_0}.$$

In order to simplify the formulæ we denote $\frac{l_1}{l_0} = 1 + \delta$ by d, and after eliminating λ_1, we finally obtain, with the aid of the extension formula (26/1)

$$a = \frac{1}{\sin \chi_0} (\sqrt{d^2 - \sin^2\lambda_0} - \cos \lambda_0) \quad . \quad . \quad (26/3)$$

In this formula the glide strain is expressed by the *initial position*

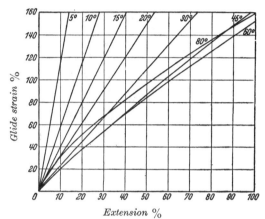

FIG. 37.—Glide Strain as a Function of Extension for Various Initial Angles of the Glide Plane.

of the glide elements and by the amount of extension. If, instead of the extension, the final position of the glide elements is introduced, the following formulæ, which are particularly convenient for numerical calculations, are obtained :

$$a = \frac{\cos \lambda}{\sin \chi} - \frac{\cos \lambda_0}{\sin \chi_0} \quad . \quad . \quad . \quad (26/4)$$

or

$$a = (\cot \lambda - \cot \lambda_0) \frac{\sin \lambda_0}{\sin \chi_0} \quad . \quad . \quad (26/4a)$$

The formula (26/3) shows that equal extensions may be accompanied by different amounts of glide strain, dependent upon the initial position of the glide elements. How strong the effect of the orientation is, can be seen from Fig. 37, in which the glide strain is expressed as a function of the extension for various initial angles of the glide plane (for the sake of simplicity $\lambda_0 = \chi_0$ has been chosen). For *every* position of the glide elements the glide strain corresponding to a small extension can be calculated according to the formula

$$\Delta a = \frac{\Delta d}{\sin \chi_0 \cos \lambda}$$

which is obtained by differentiation from the equation (26/3).

Another formal description of glide (by means of a transformation of co-ordinates) which we shall use in what follows, is contained in

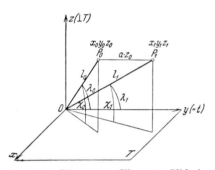

FIG. 38.—Diagram to Illustrate Glide by Means of a Transformation of Co-ordinates.

Fig. 38. The xy plane represents the glide plane and the y axis the glide direction. Let OP_0 be the length of the crystal to be stretched; the direction of tension is given with reference to the glide elements by the co-ordinates $(x_0 y_0 z_0)$. When glide occurs there is a gliding parallel to T in the direction t, as a result of which P_0 arrives at P_1. Let a represent the crystallographic glide strain, the glide displacement is then given by the product $a \cdot z_0$. Hence, the transformation of the co-ordinates is:

$$x_1 = x_0$$
$$y_1 = y_0 + z_0 \cdot a$$
$$z_1 = z_0$$

and for

$$\frac{l_1}{l_0} = \sqrt{\frac{x_1{}^2 + y_1{}^2 + z_1{}^2}{x_0{}^2 + y_0{}^2 + z_0{}^2}}$$

we obtain

$$\frac{l_0}{l_1} = \sqrt{1 + \frac{2a y_0 z_0 + a^2 z_0{}^2}{l_0{}^2}}.$$

Further, since $y_0 = l_0 \cos \lambda_0$ and $z_0 = l_0 \sin \chi_0$
we have

$$\frac{l_1}{l_0} = d = 1 + \delta = \sqrt{1 + 2a \sin \chi_0 \cos \lambda_0 + a^2 \sin^2 \chi_0} \qquad (26/5)$$

In this formula the extension is expressed by the glide strain and the initial orientation of the glide elements; it is also obtained directly by solving, *e.g.*, (26/3), with respect to d.

By means of this representation of glide it is now quite easy to follow the change of the cross-section which occurs during extension of cylindrical crystals. It has already been pointed out that the elliptical cylinder (the ribbon), which develops during glide, is usually broader than the original crystal. In what follows, the expressions for the two semi-axes of the elliptic cross-section will be given as functions of the original orientation and the orientation at any given moment in the course of extension (99). These formulæ are obtained as follows :

We start by finding the equation of the original cylinder in the co-ordinate system shown in Fig. 38. By introducing the transformed co-ordinates which correspond to the glide we obtain the equation of an elliptical cylinder, which is compared with the normal form. In this way the following expression is obtained :

$$2 + a^2 \sin^2 \lambda_0 + 2a \sin \chi_0 \cos \lambda_0 = R^2 . \left(\frac{1}{A^2} + \frac{1}{B^2} \right),$$

in which R denotes the radius of the initial circular section, A and B the half-axes of the sectional ellipse after extension. If now we bear in mind that the volume remains constant during extension by glide [1] ($R^2 \pi l_0 = AB\pi l_1$) it follows that after some calculation

$$p^4 - p^2 \left(\frac{2 \cos (\lambda_0 - \lambda) . \sin \lambda}{\sin \lambda_0} + \frac{\sin^2 (\lambda_0 - \lambda)}{\sin^2 \chi_0} \right) + \frac{\sin^2 \lambda}{\sin^2 \lambda_0} = 0 \quad . \quad (26/6)$$

The four roots p_i of this equation, of which two each differ only by the sign, are the desired ratios $\frac{A}{R}$ and $\frac{B}{R}$.

Two special cases must be briefly mentioned. For $\lambda_0 = \chi_0$, *i.e.*, for glide along the great axis of the translation ellipses, there are two roots of the equation equal to 1. The width of the band always coincides with the diameter of the initial section. The second special case relates to the maximum width at infinite extension (94). This case corresponds to the convergence of the longitudinal direction towards the operative glide direction. $\lambda = 0$ must therefore be inserted in the above formula, and the limiting value for increase in width by simple glide is then found to be

$$\left(\frac{A}{R} \right)_{max.} = \frac{\sin \lambda_0}{\sin \chi_0} \quad . \quad . \quad . \quad (26/6a)$$

[1] How exactly this condition is fulfilled will be shown in Section 60.

If for λ_0 we substitute the angle κ_0 between the direction of glide and the great axis of the glide ellipses, the maximum increase in width can also be expressed by the formula

$$\left(\frac{A}{R}\right)_{\max} = \sqrt{\frac{\sin^2 \kappa_0}{\sin^2 \chi_0} + \cos^2 \kappa_0} \qquad . \quad . \quad (26/6b)$$

since χ_0, λ_0 and κ_0 are connected by the relationship $\cos \lambda_0 = \cos \chi_0 . \cos \kappa_0$ (cf. Fig. 39).

Simple glide in *compression* can be described much more briefly

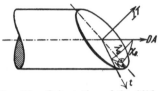

FIG. 39.—Orientation of the Glide Elements in the Crystal. Diagram Illustrating an Analysis of the Components of Tensile Stress.

[(100), (101)]. Fig. 40 shows diagrammatically a crystal specimen before and after compression. Contrary to expectation the lattice rotation which accompanies compression is not simply the opposite of that which occurs in tension.

If, in the compression test, the longitudinal direction is regarded as that of the normal to the compression plates, then on the great circle this will not move away from the operative glide direction. Lattice rotation takes the form rather of an approach of the longitudinal direction to the pole of the

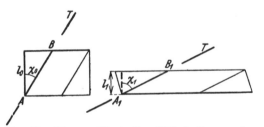

FIG. 40.—Diagram Illustrating Glide in Compression.

glide plane, which follows directly from the fact that the line of intersection of the glide plane with the compression plates retains the same inclination to the crystal axes throughout the test.

The formula connecting compression with glide strain can also be found directly from Fig. 40. Let χ_0 and χ_1 be again the angles between the glide plane and the longitudinal direction before and after compression; then

$$e = \left(\frac{l_1}{l_0}\right) = \frac{\cos \chi_1}{\cos \chi_0} . \qquad . \quad . \quad . \quad . \quad (26/7)$$

and since the eccentricity (κ) of the glide direction, corresponding to the rotation of the lattice, remains constant, this formula can be extended to include also λ_0 and λ_1.

The relationship between glide strain and compression is then

$$\frac{1}{e^2} = \left(\frac{l_0}{l_1}\right)^2 = 1 + 2a \sin \chi_0 \cos \lambda_0 + a^2 \cos^2 \lambda_0 \quad (26/8)$$

In the case of plastic *bending* by glide, the exact geometrical treatment of which has still to be formulated, parallel epipedic plates acquire as a rule a saddle shape [aluminium crystals (102)]. When the operative glide elements are in a certain special position relative to the axis of bending, cylindrical bending can also occur.

27. *Transformation of Indices in Glide*

In Section 26 we discussed the relation of glide to lattice rotation and deformation. In the present section we will deal briefly with its effect on the crystallographic symbols of directions and planes. From the rotation during glide of the longitudinal axis of the specimen relative to the crystallographic axes, it is apparent that the indices are not retained. In general, the atomic array constituting a direction or a plane changes its crystallographic nature continuously during glide.

Let (HKL) be the indices of the glide plane, $[UVW]$ those of the glide direction, then the following transformation formulæ will apply according to (103):

The direction $[uvw]$ is transformed into $[u'v'w']$ by

$$\left.\begin{array}{l} u' = u + U(uH + vK + wL)N \\ v' = v + V(uH + vK + wL)N \\ w' = w + W(uH + vK + wL)N \end{array}\right\} \quad . \quad . \quad . \quad (27/1)$$

The plane (hkl) is transformed into $(h'k'l')$ by

$$\left.\begin{array}{l} h' = h + V(hK - kH)N + W(hL - lH)N \\ k' = k + U(kH - hK)N + W(kL - lK)N \\ l' = l + U(lH - hL)N + V(lK - kL)N \end{array}\right\} \quad . \quad (27/2)$$

in which N is a whole number and signifies the number of interatomic distances by which two neighbouring glide planes are displaced relative to one another.

The directions and planes for which the indices are retained during glide can be readily recognized from these equations. The glide plane is the geometrical location of all directions which remain invariable relative to glide, the zone of the glide direction contains all planes whose symbol remains unchanged during glide.

28. *Double Glide*

Owing to the symmetry of crystals, the plane and direction of glide are not usually unique elements of the lattice. It may often happen, therefore, that two or more crystallographically equivalent glide systems will be equivalent geometrically, and so become operative in plastic deformation. So far only the simultaneous activity of two crystallographically equivalent systems has been considered in detail.

The simplest case is that of glide in two directions in the same plane. Double glide in this instance is gliding on the same plane, with the bisector of the angle of the two glide directions serving as direction of glide. The lattice rotation which accompanies extension consists in a movement of the longitudinal direction towards the line of symmetry of the two glide directions. The only extension formula to remain valid is (26/2), with which (26/1) conforms, provided that the angles formed with the " resulting glide direction " are now designated λ.

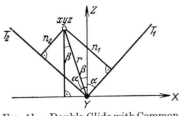

FIG. 41.—Double Glide with Common Direction of Glide.

The next case concerns the simultaneous operation of two glide planes with a common glide direction (their line of intersection). In the diagram shown in Fig. 41 the common glide direction coincides with the y-axis (which is perpendicular to the plane of the drawing) of the system of co-ordinates that is symmetrical to the two glide systems. The resulting displacement (Δy) of a point P is composed of the two separate displacements. Let a again represent the glide strain which is equal for the two systems; the glide displacements are then given by the products an_1 and an_2; and $\Delta y = an_1 + an_2$ represents the total glide strain. If now the two expressions which follow directly from Fig. 41 :

$$n_1 = r \sin (\alpha + \beta)$$
$$n_2 = r \sin (\alpha - \beta)$$

are inserted for n_1 and n_2, we obtain finally

$$\Delta y = 2a \sin \alpha . z,$$

i.e., the resulting movement is equivalent to a simple glide along the x–y plane in the common glide direction, that of the y-axis.

The yz-plane, too, contains the common glide direction and lies symmetrically to the two planes T_1 and T_2. Which of these two planes of symmetry is to be regarded as the resultant glide plane depends upon the direction of the imposed deformation. Deformation is accompanied by lattice rotation in the same way as in *simple glide*; the extension formula (26/1), relating to the common glide direction applies.

The third case to be discussed arises when neither glide plane nor glide direction is common to the two glide systems (Fig. 42). In order to superpose the two glides in the case of an extension [(104), (105)] we represent glide by means of a co-ordinate transformation (106) which was given in Section 26. We transform the glides which occur in the systems xyz and $x'y'z'$ to a new rectangular

FIG. 42.—a-Brass Crystal Showing Double Glide (103a).

system $\xi\eta\zeta$. The position of the new axes in relation to the old ones is given by the two matrices of the direction cosines :

	x	y	z			x'	y'	z'
ξ	a_1	b_1	c_1		ξ	a'_1	b'_1	c'_1
η	a_2	b_2	c_2	and	η	a'_2	b'_2	c'_2
ζ	a_3	b_3	c_3		ζ	a'_3	b'_3	c'_3

The changes in the co-ordinates due to glide were, in view of the special position of the co-ordinate system,

$$\begin{cases} \Delta x = 0 \\ \Delta y = az_0 \\ \Delta z = 0 \end{cases} \quad \text{and} \quad \begin{cases} \Delta x' = 0 \\ \Delta y' = a'z_0 \\ \Delta z' = 0 \end{cases}$$

These have now to be transformed into the new common system of co-ordinates $\xi\eta\zeta$. The changes $\Delta\xi$, $\Delta\eta$, $\Delta\zeta$ amount to :

$$\begin{cases} \Delta\xi = a_1\Delta x + b_1\Delta y + c_1\Delta z + a'_1\Delta x' + b'_1\Delta y' + c'_1\Delta z' \\ \Delta\eta = a_2\Delta x + \ldots \qquad\qquad + a'_2\Delta x' + \ldots \\ \Delta\zeta = a_3\Delta x + \ldots \qquad\qquad + a'_3\Delta x' + \ldots \end{cases}$$

Since Δx, $\Delta x'$, Δz, and $\Delta z' = 0$, the equations can be simplified to :

$$\begin{cases} \Delta\xi = b_1\Delta y + b'_1\Delta y' \\ \Delta\eta = b_2\Delta y + b'_2\Delta y' \\ \Delta\zeta = b_3\Delta y + b'_3\Delta y'. \end{cases}$$

The equations can be further simplified if the new system of co-ordinates is symmetrical to the two glide systems. Let the ζ-**axis** be perpendicular both to the z and z'-axes; it is then the line of intersection of the two glide planes (of the xy plane with the $x'y'$ plane). Let the $\eta\zeta$ plane bisect the angle between the two glide planes. The z-, z'-, η- and ξ-axes then lie in a plane perpendicular to ζ (Fig. 43). From these conditions there follow for the direction cosines the relationships :

$$\begin{aligned} c_3 = c'_3 = 0 \qquad & (\zeta\perp z,\ \zeta\perp z') \\ c_2 = c'_2 \qquad & (\sphericalangle\eta z = \sphericalangle\eta z') \\ c_1 = -c'_1 \qquad & (\sphericalangle\xi z = \sphericalangle\xi z') \end{aligned}$$

$b_1 = -b'_1$, $b_2 = b'_2$, $b_3 = b'_3$ (y-axes symmetrical to the η-ζ plane). In this way

$$\begin{cases} \Delta\xi = b_1(\Delta y - \Delta y') = b_1(az_0 - a'z'_0) \\ \Delta\eta = b_2(\Delta y + \Delta y') = b_2(az_0 + a'z'_0) \\ \Delta\zeta = b_3(\Delta y + \Delta y') = b_3(az_0 + a'z'_0). \end{cases}$$

Assuming that the glide strains in both systems are of the same magnitude ($a = a'$),

$$\begin{cases} \Delta\xi = ab_1(z_0 - z'_0) \\ \Delta\eta = ab_2(z_0 + z'_0) \\ \Delta\zeta = ab_3(z_0 + z'_0) \end{cases}$$

in which z_0 and z'_0 are still to be expressed by ξ, η, ζ. From the above equations ($c_3 = c'_3 = 0$) we obtain $z_0 = c_1\xi + c_2\eta$,

hence $\qquad\qquad z'_0 = -c_1\xi + c_2\eta$

so that for simultaneous glide of magnitude a on each of the systems we finally obtain the following formulæ :

$$\left.\begin{aligned} \Delta\xi &= 2ab_1c_1 \cdot \xi \\ \Delta\eta &= 2ab_2c_2 \cdot \eta \\ \Delta\zeta &= 2ab_3c_2 \cdot \eta \end{aligned}\right\} \quad \cdots \cdots \quad (28/1)$$

If, when double glide begins, the wire axis is in the plane $\eta\zeta$, which is symmetrical to the two glide systems, then $\xi = 0$ and

$\Delta \xi = 0$. This means, however, that in the case of equal glide strain in both systems the wire axis will remain always in the $\eta\zeta$ plane.[1]

The resulting movement can now be described in the same way as for simple glide by introducing the resulting glide direction τ (Fig. 44), whose angle ψ to the η-axis is given by the expression

$$tg \, \psi = \frac{\Delta \zeta}{\Delta \eta} = \frac{b_3}{b_2}$$

If ε is the angle between the axis of the bar and the resultant glide direction τ then it follows from Fig. 44 :

$$\frac{l_1}{l_0} = \frac{\sin \varepsilon_0}{\sin \varepsilon_1} \qquad \cdots \qquad (28/2)$$

FIG. 43. Orientation of the New System of Co-ordinates in the Case of Double Glide.

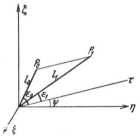

FIG. 44.—Double Glide in the New System of Co-ordinates.

From this extension formula for double glide it is again apparent that, so long as the wire axis remains in the same plane during rotation, the ensuing lattice rotation will follow the same law as applies to simple glide, with the resulting glide direction assuming the role of the glide direction. Just as the glide direction is attained by simple glide only after infinite extension, so, too, in double glide the resultant glide direction represents the final position of the wire axis for infinite extension. However, the purely formal nature of this analogy can be seen from the fact that a crystal whose longitudinal axis is parallel to the glide direction cannot be extended; on the other hand, if the longitudinal axis coincides with the resultant glide direction, it can be extensible by any amount by double glide, without change of orientation.

It is unnecessary to examine the formula by which glide strain

[1] But if the wire axis is not originally in the plane of symmetry, then it will move in a plane which is inclined towards it.

and extension are connected [cf. (99) and (107)]. For particulars of double glide in *compression*, readers are referred to the literature [(100), (101)].

29. *Formation of Spurious Glide Bands*

On the surface of extended crystals, in addition to the typical glide bands, other elliptical markings whose plane is usually inclined to the wire axis in a direction opposite to that of the glide bands (Fig. 45) are sometimes observed. That this second set of markings does not represent traces of crystallographic planes is shown by the fact that, in contrast to the glide ellipses, the eccentricity of these ellipses increases with increasing extension. A geometrical consideration of

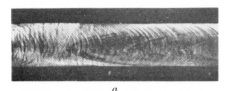

a

b

Fig. 45 (*a*) and (*b*).—Extended Zinc Crystals with Markings.

these markings reveals that they are obviously due to the distortion of grooves which originally ran circumferentially on the crystal (growth defects) (94).[1]

The special case of $\chi = \lambda$ ($\varkappa = 0$), which does not restrict the generality of the conclusions arrived at, is illustrated in Fig. 46. This shows the plane through the longitudinal axis of the crystal (l_0 or l_1), and the slip direction (AB). The plane BC of the groove which ran originally perpendicularly

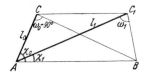

Fig. 46.—Origin of the Additional Markings.

to the longitudinal direction arrives by glide in the position BC_1, which makes the angle ω_1 with the longitudinal axis of the extended crystal.

From the triangles ABC and ABC_1 we obtain

$$\frac{l_1}{l_0} = d = \frac{\sin (\chi_1 + \omega_1)}{\cos \chi_0} \cdot \frac{1}{\sin \omega_1}$$

---

[1] These defects are, in general, found only on crystals drawn from the melt (Czochralski method).

Using the formula for the extension it follows that either

$$\cot \omega_1 = \frac{\sin^2 \chi_0 - \sin^2 \chi_1}{2 \sin^2 \chi_1} \quad \cdot \quad \cdot \quad \cdot \quad \cdot \quad (29/1)$$

or

$$\cot \omega_1 = \frac{d^2 \cos \chi_0 - \sqrt{d^2 - \sin^2 \chi_0}}{\sin \chi_0} \quad \cdot \quad \cdot \quad (29/1a)$$

These two formulæ give the angle between the plane of the groove and the longitudinal axis of the crystal as a function of the original position of the glide elements and their final position, or the amount of extension. It will be seen from formula (29/1) and from Fig. 46 that for values of χ_0 in excess of $45°$ the plane of the groove will come to lie once again perpendicularly to the longitudinal direction of the crystal, namely, when χ_1 (which is always $<\chi_0$) equals $(90° - \chi_0)$. Consequently in the initial positions $\chi_0 > 45°$ the traces of groove plane and glide ellipses coincide at the start of extension; whereas, however, in the course of further extension the glide ellipses remain inclined in the same direction, the plane of the grooves regains its upright position, reverts again to the transverse position, and then inclines in the opposite direction. If for the initial positions $\chi_0 < 45°$, then from the start of extension the plane of the groove, like the glide plane, approaches to the longitudinal axis but in the opposite direction.

B. MECHANICAL TWINNING

30. *Model of Twinning*

Viewed externally, mechanical twinning represents a type of plastic deformation entirely different from that of glide. It is a process by which portions of the stressed crystal discontinuously take up positions which are symmetrical to a plane or direction in the rest of the crystal. The symmetrical position of the two lattices in the undeformed and deformed parts is essential to this process, but not the symmetry of the external form. The latter occurs only when the crystal is bounded by certain crystallographic planes.

The best-known example of mechanical twinning is provided by the compression twins of calcite (Fig. 47). This type of deformation also occurs extensively with metal crystals (Fig. 48).

Macroscopically, deformation twins are formed by simple shear : plane layers glide by an amount proportional to their distance from the plane separating the undeformed from the deformed part, *i.e.*, the twinning plane. Fig. 49 illustrates this process by means of a wooden model.

The difference between twinning and glide appears clearly from Figs. 50 and 51. The illustrations show the lattice plane that passes

FIG. 47.—Calcite Twin, Produced by Insertion
of a Knife (Baumhauer).

through the direction of shear and is perpendicular to the plane of shear. Whereas in glide the shear cannot amount to less than an

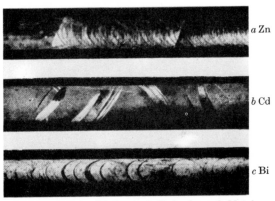

a Zn

b Cd

c Bi

FIG. 48 (*a*)–(*c*).—Mechanical Twinning of Metal
Crystals.

interatomic distance in the direction of glide (identity translation), the magnitude of displacement in twinning is usually no more than a small fraction of a lattice spacing. As is apparent from glide bands, the shear strain is by no means uniform over the whole length of the crystal in the case of glide. In mechanical twinning,

on the other hand, all planes in the deformed part which lie parallel
to the shear plane are identically displaced with reference to the
neighbouring plane. Consequently bands do not occur within a
twin lamella. In the case of glide, the shear can take place in both

a   b

Fig. 49.—Model of a Crystal (a) before and (b) after Simple Shear, showing
the Mechanism of Twinning (According to Mügge).

directions (corresponding to the extension or compression of the
crystal), whereas in twinning the shear direction is polar. This
deformation process leads to a change in shape of definite type and

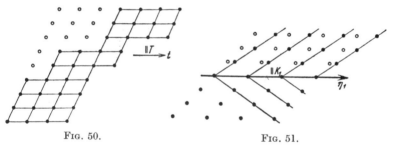

Fig. 50.   Fig. 51.

Movement of the Lattice Points in Glide and Mechanical Twinning (103).

magnitude, which is determined by the crystallographic nature of
the twinning elements.

31. *Geometrical Treatment of Mechanical Twinning*

The geometrical relationships between the deformation occurring
in twinning and the twinning elements will now be discussed with
the aid of Fig. 52. The plane of the drawing is again the plane of
displacement, and the perpendicular twinning plane (K_1) intersects
it in the shear direction η_1. The intersection of the unit sphere
with the plane of shear is also included in the form of a circle.

Twinning distorts the unit sphere into an ellipsoid of equal volume, and the intersection of this ellipsoid with the plane of shear is shown also.   For the two extreme semi-axes of the ellipsoid, a and c, which lie in the plane of shear, the relationship $a \cdot c = 1$ follows from the constancy of volume.

The shear plane (K_1) changes neither its shape ("1st undistorted plane") nor its position in the course of deformation.   On the other hand, all other planes, together with the directions not contained in K_1, become tilted.   One of the planes perpendicular to the plane of shear remains, similar to K_1, as a circle ("2nd undistorted plane K_2"). η_2 represents the intersect of this plane on the shear plane.   The amount by which a point at unit distance from the twinning plane is displaced is termed the shear strain s.   It is related to the angle 2ϕ between the two undistorted planes.   It follows from the triangle OAC and from the proportionality between gliding and the distance from $K_1 (AA_1 = s \cdot OC)$ that

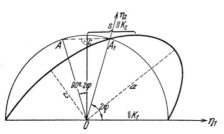

Fig. 52.—Diagram Illustrating the Geometrical Treatment of Mechanical Twinning.

$$tg\,2\phi = \frac{2}{s} \quad . \quad . \quad . \quad . \quad . \quad (31/1)$$

The amount of shear is therefore fixed by the crystallographic nature of the two undistorted planes.

Since mechanical twinning is a simple shear, the resulting changes of length are again given by formula [(26/5) (108), (109)]. χ^* and λ^* now denote the angles between the longitudinal direction and the directions of K_1 and η_1, and the glide strain a should be replaced by the shear strain s.   The formula for twinning is therefore :

$$\frac{l_1}{l_0} = d = \sqrt{1 + 2s \cdot \sin \chi^* \cos \lambda^* + s^2 \cdot \sin^2 \chi^*} \; . \quad (31/2)$$

The condition $2s \sin \chi^* \times \cos \lambda^* + s^2 \sin^2 \chi^* = 0$ gives the geometrical position of the directions whose length remains unchanged in the course of twinning.   For these directions we first obtain $\sin \chi^* = 0$, i.e., the 1st undistorted plane.   The second solution yields $-\dfrac{\sin \lambda^*}{\cos \lambda^*} = \dfrac{2}{s}$, or, employing equation (31/1) $-\dfrac{\sin \chi^*}{\cos \chi^*} = tg\,2\phi$.

However, as can be easily ascertained from Fig. 53, this is the equation of the 2nd undistorted plane in its initial position.

With the aid of these two undistorted planes all the orientations on the pole sphere are divided into four regions, of which each opposite pair on twinning leads to either extension or compression. Extension occurs for all the directions contained in the obtuse angle between K_1 and the original position of K_2 (K°_2), and compression within those contained in the acute angle between the two planes (cf. also Fig. 52).

Extreme values for changes in length are obtained for orientations in the plane of shear $(\chi^* = \lambda^*)$, since outside this plane λ^* always exceeds χ^*. From the differ-

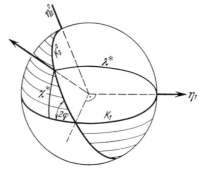

FIG. 53.—Change in Length in Mechanical Twinning : Extension in the Unshaded, Compression in the Shaded Portion.

entiation of the equation (31/2) in the case of $\chi^* = \lambda^*$, we obtain the orientation which leads to maximum extension or compression :

$$tg\chi^*_{1,\,2} = \frac{s}{2} \pm \sqrt{\frac{s^2}{4} + 1} \quad . \quad . \quad . \quad (31/3)$$

The amount of maximum extension or compression is obtained by inserting (31/3) in (31/2). For this purpose, equation (31/2) is transformed into

$$d = \frac{1}{\sqrt{1 + tg^2\chi^*}} \sqrt{1 + 2s \cdot tg\chi^* + tg^2\chi^* + s^2 tg^2\chi^*}$$

It follows from (31/3) that $s \cdot tg\chi^* = tg^2\chi^* - 1$ so that finally

$$d_{1,\,2} = \pm\, tg\chi^* \quad . \quad . \quad . \quad . \quad (31/4)$$

The choice of sign is fixed by the condition that $d = \dfrac{l_1}{l_0}$ must always remain positive.

The maximum extension is therefore

$$d_{\max.} \text{ extension} = \frac{s}{2} + \sqrt{\frac{s^2}{4} + 1} \quad . \quad . \quad (31/4a)$$

and the maximum compression

$$d_{\max.} \text{ compression} = -\frac{s}{2} + \sqrt{\frac{s^2}{4} + 1} \quad . \quad . \quad (31/4b)$$

For $l_0 = 1$ these two values represent the axes a and c of the deformation ellipsoid. Their product is unity, as is required by the constancy of volume.

The above formulæ show that the values of the shear strain (s) are small, and in contrast to glide the amount of deformation by twinning is therefore usually slight.

32. *Possibility of Twinning and Transformation of Indices*

Whereas a plane lattice represented by a net of parallelograms can always undergo twinning, this is not always so with space lattices. The condition which must be satisfied if mechanical twinning is to occur in a space lattice can be obtained from the following considerations [(110), (111)]. If, after twinning, each lattice point of the deformed portion is to have a corresponding point of the original lattice mirror reflected along the gliding surface (HKL), it should be possible before shear to join each pair of points in the lattice by lines which are parallel to the direction $\eta_2[UVW]$ and which are halved by the glide plane. Let $[[mnp]]$ and $[[m_1n_1p_1]]$ be two points in a simple primitive space lattice, then the condition which must be fulfilled if their connecting line is to run parallel to $[UVW]$ is :

$$(m_1 - m) : (n_1 - n) : (p_2 - p) = U : V : W.$$

In addition the equation

$$Hm_1 + Kn_1 + Lp_1 = -(Hm + Kn + Lp)$$

indicates that the points on different sides are at the same distance from K_1. The solution of both equations according to m, n, and p leads to

$$m = m_1 - 2U\frac{Hm_1 + Kn_1 + Lp_1}{HU + KV + LW}$$

and two analogous equations for n and p.

Since mnp and $m_1n_1p_1$ must, according to definition, be whole numbers, while UVW as indices of a direction have no common factor, the denominator $HU + KV + LW$ must be either ± 1 or ± 2. Therefore this relationship between the indices of the shear plane and the direction η_2 furnishes the condition for the possibility of twinning in a simple primitive lattice ; it must be correspondingly modified for multiple centred lattices (111).

In conclusion, the following particulars are given of the transformation of indices for directions and planes which occurs with mechanical twinning (112). Let $[UVW]$ and (HKL) be the inaices of η_2 and K_1 or η_1 and K_2; then, if ρ is a factor of proportionality the following transformation formulæ apply :

The direction $[uvw]$ is transformed into $[u'v'w']$ by

$$
\left.
\begin{aligned}
\rho u' &= u - 2U\,\frac{Hu + Kv + Lw}{HU + KV + LW} \\[4pt]
\rho v' &= v - 2V\,\frac{Hu + Kv + Lw}{HU + KV + LW} \\[4pt]
\rho w' &= w - 2W\,\frac{Hu + Kv + Lw}{HU + KV + LW}
\end{aligned}
\right\} \qquad . \;\; . \;\;(32/2)
$$

The plane (hkl) is transformed into $(h'k'l')$ by

$$
\left.
\begin{aligned}
\rho h' &= h(UH + VK + WL) - 2H(Uh + Vk + Wl) \\
\rho k' &= k(UH + VK + WL) - 2K(Uh + Vk + Wl) \\
\rho l' &= l(UH + VK + WL) - 2L(Uh + Vk + Wl)
\end{aligned}
\right\} \quad . \;\;(32/3)
$$

From these equations it will be apparent that, generally speaking, directions and planes change their indices consequent upon twinning. The indices remain unaltered for those directions which lie in the twin plane ($Hu + Kv + Lw = 0$) and for those planes which belong to the zone of η_2 ($Uh + Vk + Wl = 0$). In shear of type 2 (cf. Section 33) the directions in K_2 and the planes of the zone of η_1 retain their indices.

33. *Empirical Crystallographic Rules*

Hitherto we have discussed the model and the geometry of mechanical twinning. We will now examine briefly the crystallo-graphic aspects of mechanical twinning and empirical data concerning observed twinning elements.

Glide plane K_1 or glide direction η_1 (or both together in cases of higher symmetry) are simple, rational lattice elements. Polyhedra with rational faces are also bounded after twinning by planes with rational indices.

If K_1 is rational, it is customary to speak of twinning of type 1. The deformed part of the crystal is the reflexion of the undeformed part with respect to the twinning plane. In this case we find that not only K_1 but also η_2, which is the intersection of the second undistorted plane with the plane of twinning, is also a rational lattice element. K_1 and η_2 serve to characterize twinning crystallo-graphically. They define the plane of twinning, the direction of shear, the magnitude of shear and the second undistorted plane.

If η_1 is rational, the twinning is said to be of type 2. The deformed part of the crystal, compared with the original crystal, has now rotated 180° relative to the direction of shear. In this case K_2 is rational as well as η_1. Consequently both these elements are used to characterize a twinning of type 2, which in any case is observed only with crystals of low symmetry. The other elements of shear are then determined in the same way as described above.

The choice of twinning elements is also restricted in so far as K_1 cannot be a symmetry plane in twinning of type 1 (neither can it be perpendicular to an even-numbered axis of rotational symmetry); while in shear of type 2, η_1 cannot be a digonal axis, since it is only as a result of twinning that this lattice element acquires the character of a symmetry element.

PLASTICITY AND STRENGTH OF METAL CRYSTALS

Chapter V contained a description of the general principles governing the change in shape of crystals during plastic deformation, together with the resultant changes in orientation. Both mechanisms of deformation—gliding and mechanical twinning—were discussed. We now turn to the technique and the results of experiments on the plastic deformation of crystals; for the present, we shall confine ourselves to crystals which, owing to their extreme ductility, provide excellent experimental material. The result of investigations on the plasticity and strength of salt crystals will be given in the next chapter.

A. ELEMENTS OF GLIDING AND TWINNING

34. *Determination of the Elements of Gliding*

The elements used in the characterization of gliding—the glide plane (T) and the glide direction (t)—can be determined in various ways.

It is not proposed to discuss here the goniometric and microscopic methods applicable to polyhedrons with plane faces. Instead, a description will be given of the new methods developed for opaque crystals which are not bounded by plane faces. These methods can be considered under several headings according to the conditions governing the experiment. When applying them, however, certain modifications or combinations are often necessary.

To take the simplest case first, let us consider a crystalline material which admits of substantial plastic extension, accompanied by the appearance of glide bands or striations indicative of the operation of a single glide system (cf. Fig. 35). In this case the glide plane (T) can often be found by means of a Laue photograph (with the incident beam perpendicular to the plane of the striation). If the glide plane is perpendicular to a symmetry axis of the crystal, the resulting symmetry of the Laue photograph will usually suffice to index the glide plane (Fig. 54). T is more difficult to determine when its normal does not coincide with a symmetry axis, since an exact evaluation of the Laue photographs is often greatly hampered by the pronounced asterism caused by previous extension (cf. Section 59).

In crystals which have been substantially stretched, the direction of slip (t) can be usually determined by means of a rotation photo-

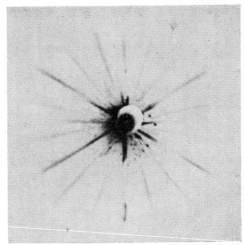

(a) Zn crystal: 6-fold picture: $T = (0001)$; (113).

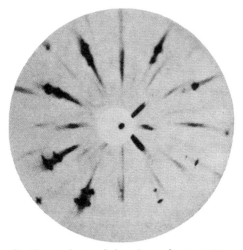

(b) Al crystal extended at elevated temperature;
4-fold picture: $T = (001)$; $(11\bar{7})$.

FIG. 54 (a) and (b).—Determination of the Glide
Plane (T) from the Symmetry of the Laue
Photographs.

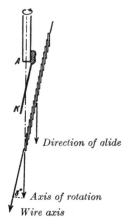

Direction of glide

Axis of rotation
Wire axis

FIG. 55.—Adjustment of
the Crystal in Order to
Determine the Direction of
Glide from Rotation Photographs.

graph (132). The position of t is indicated approximately by the
intersection of T with the symmetry plane of the crystal band which
is normal to the plane of the band and which contains its longi-

tudinal axis. Consequently the crystal is adjusted for the rotating-crystal photograph so as to ensure that the intersection of the glide plane with the median plane of the band to which reference has

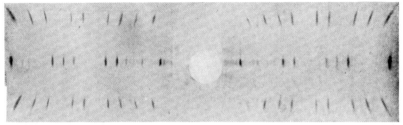

(a) Sn crystal; $t = [001]$; (114).

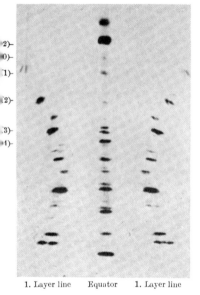

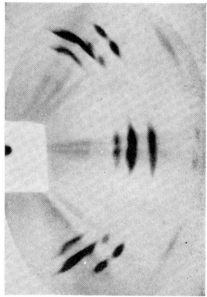

2)-
0)-
1)-
2)-
3)-
1)-

1. Layer line   Equator   1. Layer line

(b) Cd crystal; film axis = incidental beam; $t = [11\bar{2}0]$; (131).

(c) Al crystal; extended hot; $t = [110]$; (117).

FIG. 56 (a)–(c).—Determination of the Direction of Glide (t) from Rotation Photographs.

already been made, forms the axis of rotation (Fig. 55). Thus the crystal wire describes a cone round the direction of glide, with a cone angle that corresponds to the small angle of inclination between T and the wire axis. Fig. 56 contains a few rotation photographs which have been obtained in this way, and which, evaluated by means of the Polanyi layer-line relationship (formula 21/2), plainly reveal the crystallographic nature of the glide direction.

So much for the X-ray method, which supplies direct evidence of the elements of glide. We shall now consider a more indirect method for the determination of the glide *direction*. In this connection reference will be made to the change in orientation caused by glide during extension, which was discussed in Section 26. The change was found to be a movement of the longitudinal axis of the crystal towards the operative direction of glide. Consequently the method of obtaining t consists in determining that direction to which the longitudinal axis approaches as extension increases—having first ascertained the crystal orientation at various stages of extension. The resultant direction of glide can be checked by applying the formula for extension (26/1), which relates the amount of extension to the change of the angle between the longitudinal axis and the direction of glide. Fig. 57 shows

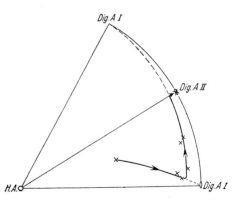

FIG. 57.—Determination of the Direction of Glide from the Lattice Rotation in the Course of Extension; Mg Crystal at >225° C. [see (129)].

the application of this method to a magnesium crystal which has been extended at elevated temperature. A digonal axis of type 1 was found to be the effective glide direction both for the initial glide in the direction of the basal plane and for the subsequent glide in the direction of a pyramidal plane.

Similarly, it is possible to determine the glide *plane* from the lattice rotation during compression, since in this case the change of orientation consists in a movement of the longitudinal direction towards the normal of the glide plane (checking the compressive strain by the formula 26/7). Tensile tests afford no direct indication of T apart from the fact that T contains the glide direction and must therefore belong to the zone of t. If, however, we make the obvious physical assumption that, where glide systems are crystallographically equivalent, that system will operate in which the resolved shear stress is a maximum [1] it will be seen that in certain circumstances the glide planes can be determined from study of the lattice rotation alone. With the further assumption that the glide plane will have low indices, the whole range of orientations can be subdivided in such a way that the shear stress within each division is greater in one glide system than in the remaining crystallographically equivalent systems. If, on this assumption, the glide system which invariably operates is that in which the shear stress is a maximum, then the lattice rotation which accompanies extension in each division, and consequently in the total range of orientation, is established. A comparison of the experimentally ascertained lattice rotation with the changes of orientation which have been calculated on various assumptions about T, may serve to determine the glide plane.

An example of the application of this method is shown in Fig. 58, in which the lattice rotation for the extension of close-packed hexagonal crystals is given $\left(\dfrac{c}{a} = 2\sqrt{\dfrac{2}{3}} = 1\cdot633\right)$.

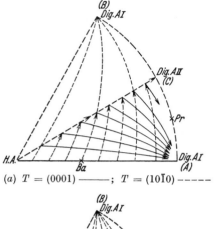

(a) $T = (0001)$ ——— ; $T = (10\bar{1}0)$ - - - - -

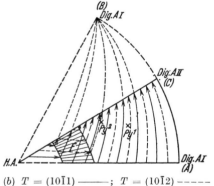

(b) $T = (10\bar{1}1)$ ——— ; $T = (10\bar{1}2)$ - - - - -

Fig. 58 (a) and (b).—Indirect Determination of the Glide Elements of Hexagonal Crystals from the Lattice Rotation in the Course of Extension.

The diagram is based on a direction of glide parallel to the digonal axis, type I, and on glide planes parallel to the basal plane (0001), prism type I (10$\bar{1}$0), pyramidal plane I, types 1 and 2 (10$\bar{1}$1) and (10$\bar{1}$2). Basal glide can be readily distinguished from the other three types by the fact that in the whole of the orientation triangle the wire axis moves

[1] It will be shown in Section 40 that this assumption is entirely justified.

towards the digonal axis type 1 (A) in a corner of the triangle (Fig. 58a). On the other hand, if one of the other three glide planes becomes operative the wire axis tends towards a digonal axis type 1 (B) lying outside the triangle for most initial orientations. With prism glide this occurs in the whole of the orientation triangle, whereas with pyramidal glide (Fig. 58b), sections adjacent to the hexagonal axis involve an initial rotation towards (A). By examination of the behaviour of crystals in the area F it is possible to distinguish between the two pyramidal glide planes. Whereas with basal glide a considerable degree of extension will result in an approach to a digonal axis type I (A) in the final stage, with prism and pyramidal glide a digonal axis type II (C) represents the final orientation of the longitudinal axis of extended crystals. It is achieved in this case by double glide on two equivalent systems.

In Section 40 we shall indicate a further possible method for determining the glide plane, by examining the dependence of the yield point upon orientation. The points which refer to this are shown in Fig. 58 as Ba, Pr, Py^1 and Py^2, which indicate the initial orientation for minimum values of the yield point.

G. I. Taylor and his associates in their investigations of the plastic deformation of metal crystals (116) adopted a method entirely different from that described above.

In this method it is not assumed that deformation is due to crystallographic glide; instead, the mechanism of deformation is deduced from changes in the shape of the test piece, or of the systems of lines drawn upon it. This can be illustrated by a compression test carried out on a cylindrical disc (115).[1] Two groups of lines are scratched on the compression surface of the plate, and their distortion under compression is observed. The axes of co-ordinates are such that the x-axis runs parallel to the lines, the y-axis passes at right angles to them through the compression surface, while the z-axis lies normal to the compression surface (parallel to the direction of compression). Let the co-ordinates of a given portion before and after deformation be ($x_0 y_0 z_0$) and ($x_1 y_1 z_1$). The transformation equations will then be [2]

$$x_1 = \alpha x_0 + l y_0 + \mu z_0$$
$$y_1 = m y_0 + \nu z_0$$
$$z_1 = \gamma z_0$$

[1] Application of this principle to the tensile test is discussed in (121).
[2] Owing to the particular choice of the direction x, y_1 is independent of x_0. Planes parallel to the compression surface preserve this parallel position; consequently z_1 depends only on z_0 and the degree of compression.

α, l, μ, m, ν and γ are constants which can be calculated from the change in shape of the test piece and the distortion of the lines.

It is now necessary to establish the geometrical locus of all those directions which have not altered their length during deformation. This is governed by $x_0{}^2 + y_0{}^2 + z_0{}^2 = x_1{}^2 + y_1{}^2 + z_1{}^2$.

Adoption of the above transformation formulæ gives the equation of the " unstretched cone " in the initial position (cone-surface type 2). After introducing polar co-ordinates (θ = the angle between the observed direction and the z-axis, ϕ = the angle between the projection of this direction on the xy-plane and the x-axis), the equation of the unstretched cone surface is obtained in the form of $f(\theta, \phi) = 0$. With the help of this formula the θ values appropriate to the individual ϕ values can be

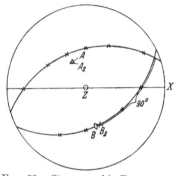

calculated, and the unstretched cone surface expressed in a stereographic projection by using the xy-plane as projection plane. It is usually found that the unstretched cone surface has degenerated into two planes and that the deformation is therefore a plane one.[1]

Which of the two planes is the glide *plane* (T) can be determined in various ways. 1. A determination of the orientation of the crystal lattice with respect to the two planes before and after deforma-

Fɪɢ. 59.—Stereographic Representation of the Unstretched Cone Surface of a Compressed Al Crystal (115).

tion reveals that only one of them (the one corresponding to T) retains its orientation relative to the crystal axes. 2. If glide bands occur during deformation these will usually suffice, even if only indistinct, to distinguish between the two planes. 3. The initial state and *two* stages of deformation are examined. The planes that remain undeformed are determined after the first and second deformation. One of them (T) will have retained its crystallographic character.

The *direction* of glide (t) is shown by this method to be that direction which is perpendicular to the intersection of the two planes in T (cf. the analogous geometrical discussion in Section 31 of the two undistorted planes in mechanical twinning).

As an example of the use of this method, Fig. 59 shows a stereo-

---

[1] If x_0, y_0 and z_0 are eliminated from the above condition, in place of x_1, y_1 and z_1, then we obtain the equation of the unstretched cone surface after deformation.

graphic projection of the unstretched cone surface of a compressed aluminium crystal, which clearly reveals degeneration into two planes. One of them corresponds to a (111) plane of the crystal, as will be seen from the coincidence of its normal (A_2) with a [111] direction (A). The direction (B_2) contained in this plane, and separated by 90° from the intersection of the two planes, very nearly coincides with a [101] direction (B). Consequently these two lattice elements, the octahedral plane (111) and the plane diagonal [10$\bar{1}$], have been recognized as the glide elements of the aluminium crystal.

By means of the unstretched-cone method it is also possible to analyse that type of crystal deformation (already discussed in Section 28) in which the two glide systems are equally favourably placed in relation to the direction of applied stress, although in this case the cone surface normally will not have degenerated into two planes. It is found that the deformation takes the form of double glide along the two equally favoured glide systems.

35. *The Glide Elements of Metal Crystals*

The results of existing determinations of the glide elements of metal crystals are given in Table VI. The last column contains the cleavage or rupture planes which have been observed in the crystals.

With cubic face-centred metals only octahedral glide occurs at room temperature. The position with regard to cubic body-centred metals is not yet clear. In the case of α-iron in particular, the existence of a " pencil "-glide in which only the *direction* of glide is supposed to be crystallographically fixed (121) [1] has been indicated. However, later work [(122), (123)] makes it seem highly probable that in this case too the selection of *both* glide elements proceeds strictly along crystallographic lines. In *hexagonal* metals the basal plane has hitherto always been found to be a unique glide plane with the three digonal axes type I as glide directions.

In the case of the tetragonal β-tin crystal, crystallographically non-equivalent glide systems are already found at room temperature. However, glide $T = (110)$, $t = [001]$ is greatly preferred. Among the rhombohedral metals only bismuth exhibits extensive glide. The best glide plane in this case is the basal plane, although the three equivalent (11$\bar{1}$)-planes appear also to occur as glide planes. It is not yet known for certain whether, in addition to the [101] directions, the [101] directions are also possible glide directions. The

---

[1] This mechanism is said to apply, within a certain range of orientation, to cubic body-centred β-brass also. The other orientations give glide with $T = (101)$, $t = [11\bar{1}]$ (125).

Glide Elements and Cleavage Planes of Metal Crystals

Metal.	Lattice type, crystal class.	Glide elements. At 20°. T.	At 20°. t.	Additional at elevated temp. T.	Additional at elevated temp. t.	Close packed. Lattice planes.	Lattice directions.	Literature.	Cleavage planes.
Aluminium. Copper. Silver. Gold. Lead.	Cubic face-centred O_h	(111) (111)	[10Ī]	from 400° C. (100)	[011]	1. (111) 2. (100) 3. (110)	1. [10Ī] 2. [100] 3. [112]	(116), (117) (118), (119) (120)	— — —
α-Iron	Cubic body-centred O_h	(101) (112) (123) (123) (112)	[111] [111]	—	—	1. (101) 2. (100) 3. (111)	1. [111] 2. [100] 3. [110]	(121) (122) (123), (124) (126)	(001) —
Tungsten			[11Ī] [11Ī]						· (001)
Magnesium	Hexagonal close packed D_{6h}	(0001)	[11Ž0]	from 225° C. (10Ī1) or (10Ī2)	[11Ž0]	(0001)	[11Ž0]	(127), (128) (129)	(0001), (10Ī1) (10Ī2), (10Ī0)
Zinc. Cadmium.						— —	— —	(130) (131)	(0001) —
β-Tin (white)	Tetragonal D_{4h}	(110) (100) (101) (121)	[001] [001] [10Ī]	from 150° C. (110)	[11Ī]	1. (100) 2. (110) 3. (101)	1. [001] 2. [111] 3. [100] 4. [101]	(132), (133) —	— —
Arsenic. Antimony. Bismuth.	Rhombohedral D_{3d}	{ (111) (111)	[10Ī] also [101]	—	—	1. (11Ž0) 2. (11Ī)	1. [10Ī] 2. [101]	(134) (135)	(111), 110) (111), (110), (11Ī) (111), (11Ī), (110)
Tellurium	Hexagonal D_3	(10Ī0)	[11Ž0] ?	—	—	(10Ī1)	—	(136)	(10Ī0)

plasticity of tellurium crystals merely takes the form of a slight flexibility, due to glide along the (11$\bar{2}$) plane, which is also a cleavage plane (prism plane type I in the hexagonal representation).

A study of Table VI reveals a clear connection between capacity for glide and density of packing—a connection which has long been recognized (137). It is invariably the most dense atomic row that serves as a direction of glide, and usually the planes of highest atomic density are the glide planes. Attempts to explain the glide elements in terms of lattice theory have so far been unsuccessful : considerations of glide in a packing of spheres with cubic face-centred and cubic body-centred symmetry threw no light on the glide elements which had been observed in these crystals (138); determination of the modulus of shear of various planes revealed that the operative glide elements are by no means characterized by a minimum value (139).

In no case did an increase in the temperature of the test lead to a disappearance of the glide elements which are operative at room temperature; on the contrary, in many cases new glide systems appeared. It is quite likely that a detailed investigation of plastic deformation at temperatures near the melting point would reveal new glide systems in many other cases.

Although most of the glide elements of metal crystals have been determined from static tensile tests, it should be specially noted that with all other types of stressing as well [compression, torsion, dynamic and alternating stressing, cf. (140)] the same glide elements, arising from the nature of the lattice, become effective.

In conclusion, it should be mentioned that the glide elements observed in the crystals of pure metals are always observed in the α solid solutions of the respective metals.

36. *Manifestations of Glide*

Having enumerated the glide elements observed in metal crystals we will now describe the occurrence of these elements under different types of stress, and the resulting lattice rotations.

The lattice rotation which occurs in an extended crystal where a single glide plane is present (hexagonal metals) is shown in Fig. 58a. If the initial orientation is near a prism plane of type II, a second and only slightly less favourable digonal axis type I becomes operative as a glide direction in the initial stages, so that glides tarts along the same plane in two directions (see Section 28). If the longitudinal axis lies exactly in a prism plane type II, both glides persist, and the longitudinal axis of the crystal approaches increasingly a digonal axis type II, to which the two operative directions of glide are symmetrical. On the other hand, where the two glide directions are not equally favourable, the one which is preferred geometrically soon predominates, and the subsequent change in orientation proceeds as shown in Fig. 58a (129). The change of orientation as a result of pyramidal glide which occurs with magnesium at high temperatures has already been discussed (Fig. 57). The surface bands on the crystal which result from the new glide and the

(a) In addition to the coarse markings resulting from basal glide, two additional systems of markings caused by pyramidal glide can be seen.

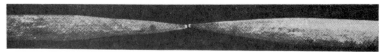

(b) Necking due to pyramidal glide.

FIG. 60 (a) and (b).—Mg Crystals Stretched at Elevated Temperatures (129).

change in the shape of the crystal (necking of the band) are shown in Fig. 60.

There are twelve equivalent glide systems available in the cubic face-centred metals (four octahedral planes, each with three face diagonals). The choice of system in the case of extension depends upon the direction of the tensile stress in relation to the crystallographic axes. As was explained at length for the case of aluminium (116), glide always occurs along the system subjected to maximum shear stress.[1] Fig. 61 indicates the choice of available glide systems for different directions of stress. Each triangle which, by the application of symmetry operations characteristic of the crystal, results in the whole of the orientation range being covered, is characterized by a definite most favourable glide system.

The lattice rotations which accompany plastic deformation in

[1] Calculation of the shear stress is dealt with in detail in Section 40.

extension can be illustrated by means of an experimental example [aluminium (Fig. 62)] and by Fig. 63 (which represents a section of

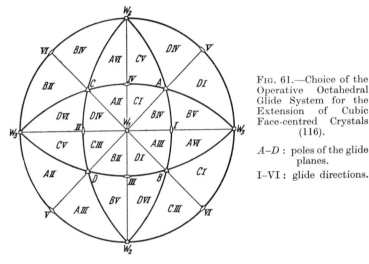

Fig. 61.—Choice of the Operative Octahedral Glide System for the Extension of Cubic Face-centred Crystals (116).

A–D : poles of the glide planes.

I–VI : glide directions.

Fig. 61).   In the heavily outlined orientation area W_1AI, the most favourable glide system is B IV.   Consequently, the lattice rotation

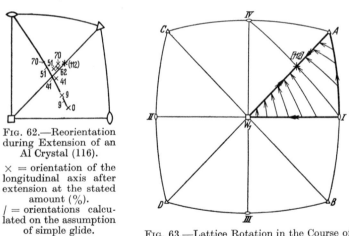

Fig. 62.—Reorientation during Extension of an Al Crystal (116).

× = orientation of the longitudinal axis after extension at the stated amount (%).

| = orientations calculated on the assumption of simple glide.

[Formula (26/1).]

Fig. 63.—Lattice Rotation in the Course of Extension by Octahedral Glide.

which accompanies extension is indicated by the arrows pointing to IV.   The glide direction IV, however, is not attained, for when the longitudinal axis enters the dodecahedral plane W_1A a second glide system (*CI*) becomes geometrically equally favourable. The double glide along two systems which now sets in leads to a movement of the longitudinal axis in the symmetry plane W_1A towards the bisector [112] of the operative glide directions. The longitudinal

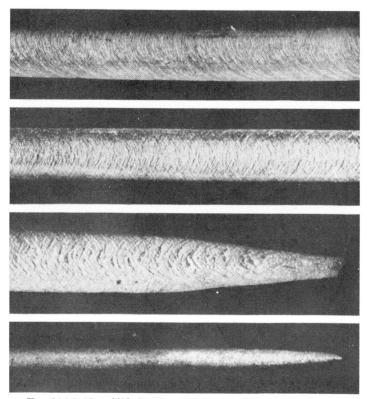

FIG. 64 (*a*)–(*d*).—Glide Bands on Al Crystals Extended at 400° C.

(*a*) and (*b*) simple and double octahedral glide.   (*c*) and (*d*) cubic glide; viewed perpendicular and parallel to the plane of the original band.

axis of crystals extended at room temperature tends towards this direction.

Deviations from the normal case described above occur in the first place when the longitudinal axis of the crystals lies originally in a symmetry plane, so that from the outset two or more systems are equivalent [(141), (142)]. Secondly, deviations are observed at elevated temperatures. In this case, for initial orientations in the area *AI* [112] of the orientation triangle (Fig. 63), an aluminium specimen, in addition to octahedral glide, will undergo cubic glide according to the most favourable system W_1V. Under the influence of these two glide systems the longitudinal axis tends towards a final position parallel to the body diagonal (A). The sequence of glide bands is shown in Fig. 64, a to d. Crystals in the orientation area W_1I [112] also exhibit irregular behaviour at elevated tempera-

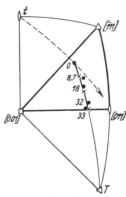

FIG. 65.—Reorientation of an Al Crystal as a Result of Compression (144).

● = as observed.
× = as calculated from formula (26/7) on the basis of the stated percentages of compression.

tures. In this instance several octahedral glides appear to become active even before the " symmetrals " are attained, thus resulting in a movement towards the cube edge W_1 [(117) cf. also (141) and (143)].

In the compression of aluminium crystals, too, that octahedral glide system in which the resolved shear stress is a maximum is the first to come into operation (144). The subdivision of the entire field of orientation into regions of equally favourable glide systems is the same as in the tensile test (Fig. 61), since it is only the sign of the stress that has changed. As already indicated in Section 26, however, the lattice rotation which accompanies the operation of an octahedral glide system is not that obtained by an inversion of the sign from the rotation in extension. It rather consists in a movement of the " longitudinal direction " towards the normal to the operative glide plane. This is illustrated in Fig. 65.

37. Determination of Twinning Elements

Like the glide elements, the twinning elements (twin plane K_1, and direction η_2 for twinning of type 1; 2nd undistorted plane K_2

FIG. 66.—Mg Crystals with Deformation Twins about $K_1 = (10\bar{1}1)$; (149).

and shear direction η_1 for twinning of type 2) can be determined by various methods. In principle, the problem is solved when the

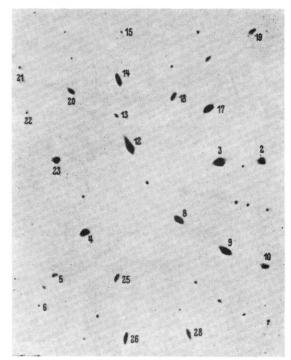

(a)—Laue Photograph Perpendicular to K_1.

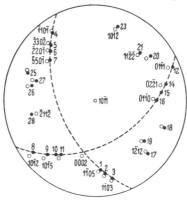

(b) Pole figure of the reflecting planes; close agreement with the pole figure about (10Ī1).

FIG. 67 (a) and (b).—Determination of Twin Planes from Laue Photographs : Mg Crystal (149).

indices for several (at least two) crystallographic planes are known, both before and after twinning. With plane crystal boundaries they can be determined by a microscopic and goniometric measurement of the twin lamellæ. The transformation formulæ (32/3), resolved with respect to (HKL) and $[UVW]$ then furnish the crystallographic indices of the twinning elements.[1]

However, in the case of the metal crystals with which we are mainly concerned there are as a rule no crystallographically bounded polyhedrons. We therefore propose to describe the methods to be applied in cases where the crystal " habit " is absent. These methods, however, only lead directly to the first undistorted plane K_1.

K_1 can be determined directly in all cases where a Laue photograph is obtainable at right angles to the twinning plane. For this it is necessary that the position of the plane should be adequately defined by twin bands. Figs. 66 and 67 illustrate the use of this method in establishing the $(10\bar{1}1)$ plane as the twinning plane of magnesium. Difficulties may arise owing to the pronounced distortion of the interference spots (asterism, see Section 59) which sometimes makes the evaluation of the diagrams impossible.

A second way of determining K_1 (or η_1) by X-rays is open in cases where it is possible to produce in the crystal a twin lamella so broad that not only the orientation of the initial crystal but also that of the twin can be determined, with respect to the same system of co-ordinates, by means of Laue photographs or photographs taken with an X-ray goniometer. The relative orientation of the two lattices is determined by the law of twinning : they arise from each other by reflexion on the twinning plane K_1 (or by rotation through 180° round the direction of glide η_1). If then the pole figures for the initial and the twinned positions are plotted in the same stereographic plot, it will be found that for twinning of the first type the position of *one* plane will be identical in both parts of the crystal, and this will correspond to K_1. Reflexion at its pole transforms corresponding directions into one another. In case of a twinning of the second type a rational crystal direction η_1 is common to both lattices which transform into each other by rotation through 180° round this direction. It is easier to find the twinning element in crystals containing unique sets of crystallographic planes ; in such cases the twinning element bisects the angle between the two positions of the unique plane. Although this method has not so far led to the discovery of new twinning elements, it has in many cases confirmed previous results [(145), (146), 147)].

Another method which has been successfully used for determining K_1 involves measurement of the angles of twinning lamellæ with respect to a crystallographically known system of co-ordinates (148). On a specimen of very coarse-grained material or a single crystal, two

---

[1] This is conditional on the second plane not being also in the zone which is determined by the first plane when in the original and twinning position.

faces are ground inclined to each other at a known angle (in the simplest case they may be perpendicular to each other), and then by a slight compression twinning is produced in individual grains. After the surfaces have been polished again the angles between the twin and the reference planes can be measured, as well as those between the twin planes, should several sets of twins appear. It is a condition of this method that the same twin lamella must be recognizable on both polished surfaces (see Fig. 68). If the positions of the reference planes with respect to the crystallographic axes are known, the positions of the twin planes with reference to these axes can also be determined. If there are several sets of equivalent twin planes, their crystallographic nature is already revealed by the angles between them.

The three methods described above for boundaries which are not well-defined crystal planes, lead solely to the first undistorted plane K_1. The second of these methods is the only one that is capable of indicating, in addition, the direction of glide η_1 and thus also the plane of the shear. But with crystals characterized by a high degree of symmetry, as in the case of metals, the following consideration will lead to the determination of η_1 in the other methods also. If there is a symmetry plane at right angles to K_1, this plane must coincide with the plane of shear, and so its intersection with K_1 must be η_1. Otherwise the symmetry plane which was originally present would lose this property in the twinned portion. This is immediately obvious from the fact that in planes at right angles to the plane of shear only those directions which form equal angles with the plane of shear suffer equal distortion during twinning. If there were in addition a symmetry plane at right angles to K_1 and *inclined* towards the plane of the shear, directions symmetrical in relation to this plane would also suffer equal distortion, an assumption which cannot be reconciled with the mechanism of twinning.

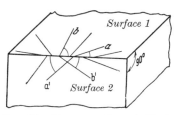

FIG. 68.—Determination of K_1 by Measuring the Traces of Twin Lamellæ on Two Polished Surfaces of the Crystal (148).   Be Crystal.

If in addition to K_1 the indices of a direction or a plane are known before and after twinning, the indices $[UVW]$ of the direction η_2, together with the second undistorted plane and the shear strain, can be calculated from the equations (32/2) or (32/3).

In very elementary cases it should be possible, if K_1 and η_1 are known, to determine the second undistorted plane K_2 and η_2 by observing the movement of the lattice points in the plane of shear (cf. the following section).

38. *The Twinning Elements of Metal Crystals*

A summary of the elements of shear which have so far been observed in the mechanical twinning of metal crystals is given in Table VII. In addition to the two undistorted planes the amount

TABLE VII

Shear Elements of Metallic Crystals

Metal.	Lattice type, crystal class.	1st undistorted plane, K_1.	2nd undistorted plane, K_2.	Amount of shear, s.	Literature.
a-Iron	Cubic space centred O_h	(112)	(11$\bar{2}$)	0·7071 $(= 1/2\sqrt{2})$	(150)
Beryllium		(10$\bar{1}$2)	(10$\bar{1}\bar{2}$) [1]	0·186	(148)
Magnesium	Hexagonal close packed D_{6h}	(10$\bar{1}$2) (10$\bar{1}$1)	(10$\bar{1}\bar{2}$) [1]	0·131	(148) (151)
Zinc		(10$\bar{1}$2)	(10$\bar{1}\bar{2}$)	0·143 $=\left(\dfrac{(c/a)^2 - 3}{c/a\sqrt{3}}\right)$ †	(148, (152)
Cadmium		(10$\bar{1}$2)	(10$\bar{1}\bar{2}$) *	0·175	(148)
β-Tin (white)	Tetragonal D_{4h}	(331)	(11$\bar{1}$)	0·120	(153)
Arsenic	Rhombo-	?(011)	(100)	0·256	(154)
Antimony	hedral	(011)	(100)	0·146	(155), (156)
Bismuth	D_{3d}	(011)	(100)	0·118 $=\left(\dfrac{2\cos a}{\sin a/2}\right)$ ‡	(156)

* In these cases K_2 has not yet been determined. In line with the behaviour of the Zn crystal, (10$\bar{1}$2) has been adopted as the 2nd reference circle and used for the calculation of s.

† c and a = lengths of axes.

‡ a = angle of the rhombohedron.

of the shear is also indicated. The table contains no particulars of the numerous instances of *recrystallization twins* in metals; these occur frequently during recrystallization after previous cold working.

The results of quantitative investigations on cubic metals are available so far for body-centred α-iron only. Here the two undistorted planes are crystallographically equivalent, and so, too, are the direction of shear and η_2 ([111]-directions). When K_1 and K_2 or η_2 and η_1 can be interchanged, the shears are said to be reciprocal. If twinning occurs on several planes, internal cavities are formed : Rose's channels [(157), (158)]. These cause an increase in volume which may be substantial, and which in the case of iron may amount theoretically to as much as 50 per cent. (159). Although mechanical twinning has been repeatedly observed with cubic face-centred metals [see for instance (159a)] it has not yet formed the subject of systematic investigations. However, it is possible that in this case K_1 is identical with the twinning plane (111), which is always observed with recrystalliza-tion twins. This assump-tion points to $(1\bar{1}0)$ as a plane of twinning, to $\eta_1 = [11\bar{2}]$ and, in accordance with the movement of the lattice points in the plane of shear as shown in Fig. 69, to $K_2 = (11\bar{1})$, $\eta_2 = [112]$ and $s = 1/2\sqrt{2} = 0\cdot7071$ [1] [cf. (138)].

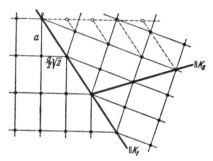

Fig. 69.—Simple Shear of Cubic Face-centred Crystals having $K_1 = (111)$ (Plane of Shear).

In the case of the hexagonal metals, twinning has hitherto been observed to occur on a pyramidal plane type I, order 2. Here we have a *single* glide plane in contrast to *six* planes of twinning. An exact determination of K_2 is available for the zinc crystal only, for which, apart from the direct determination of K_1, it was also possible to ascertain the indices of a plane (basal) before and after twinning. The particulars of K_2 and s in other cases have hitherto been inferred from analogy only. A consideration of the movements of the lattice points in the plane of displacement shows that in the case of the hexagonal metals, twinning is by no means a simple shear with the stated strain (s). Only a quarter of the lattice points are conveyed by this movement into the twin position. The remainder have still to carry out additional displacements towards or parallel to K_1 [(160), (152)]. With magnesium crystals in rare cases the pyramidal plane order 1 has been observed as a twinning plane, in

[1] These provisional shear elements of the cubic face-centred metal crystals fulfil the conditions of twinning for this type of lattice.

addition to the pyramidal plane of type 1, order 2. The *tetragonal* lattice of white tin is not capable of twinning with the elements mentioned above (161), so that here, too, the actual movement of the lattice points must differ substantially from a simple shear. Similarly the individual atoms of *rhombohedral* bismuth (and antimony) follow no straight line when twinning. On the other hand, the movement of the centres of gravity of pairs of co-ordinated atoms on the trigonal axis corresponds to simple shear (162).

Already this brief description confined to metal crystals shows how complicated are the circumstances in mechanical twinning — a fact which is abundantly confirmed by the numerous observations on salt crystals.

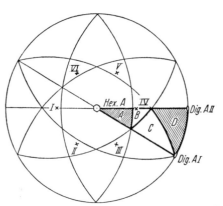

Fig. 70.—Change in Length as a Function of Orientation in the Case of Mechanical Twinning of Zn Crystals (163).

I–VI, poles of the twin planes.
A : I–VI, compression.
B : II, III, V and VI, compression; I and IV, extension.
C : II and V compression; I, III, IV and VI, extension.
D : I–VI, extension.

39. *Operation of Mechanical Twinning*

Now that particulars have been given of the elements of twinning observed with the various metals, we will proceed to a description of their operation under various types of loading.

First we must deal with the sign of the change in length due to twinning, since, owing to the polarity of the directions of glide, either only extension or only compression, according to the orientation, can occur in the direction considered (see Fig. 53). Fig. 70 represents a subdivision of the orientation field of the zinc crystal, based on its six twin planes $(10\bar{1}2)$. Directions at angles of between $0°$ and approximately $50°$ to the hexagonal axis are shortened by twinning on each of the six planes; while directions approximately at right angles to the hexagonal axis undergo extensions. A transition between the two is found in orientation regions in which some of the twinning planes produce extension and others compression. The same applies to cadmium crystals.

The change of shape is just the opposite for the important group of *hexagonal crystals* having an axial ratio c/a of approximately

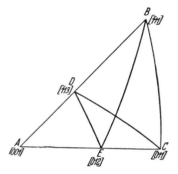

Operative K_1 plane :

1, 2, 7, 8 : extension occurs in the whole of the orientation region.

3, 4 : compression in the whole of the region.

5, 11 : in EBC extension, in ABE compression.

6, 12 : in $EBDC$ extension, in ADE compression.

9, 10 : in BCD extension, in ACD compression.

FIG. 71.—Mechanical Twinning of a-Fe. Orientation Regions in which Extension and Compression Occur. K_1 Planes Numbered 1–12 in Accordance with Fig. 11.

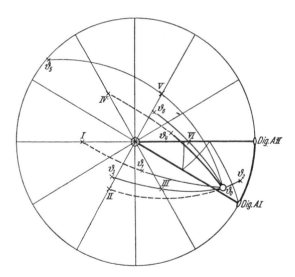

FIG. 72.—Change of Orientation with the Twinning of Zn.

(N.B.—In the text, $\vartheta = \theta$.)

1·63 (Mg, Be, etc.). In this case the hexagonal axis makes an angle of more than 45° with the twin planes (and, on the very probable

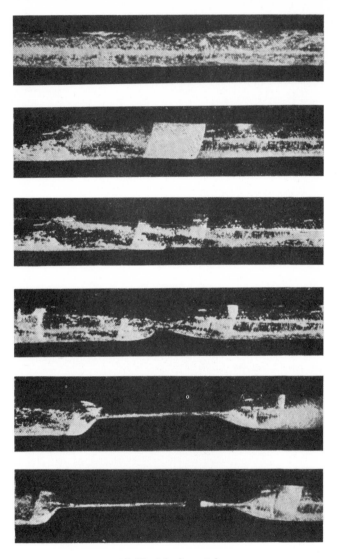

(*a*) Unetched crystal.

assumption of a reciprocal shear, also with the second undistorted planes), which means that the hexagonal axis and the adjacent

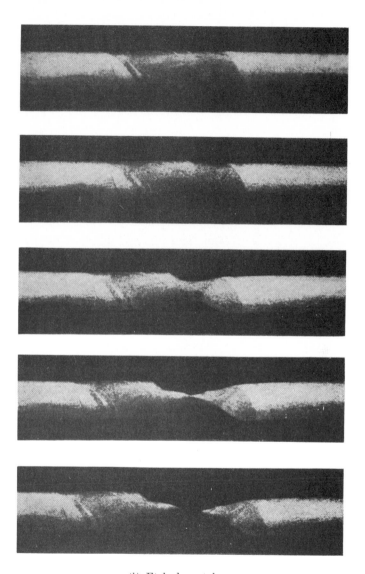

(b) Etched crystal.

FIG. 73 (a) and (b).—Secondary Basal Glide in the Twin Lamellæ of Zn
Crystals.   Viewed Perpendicularly to the Plane of the Band Resulting
from Primary Basal Glide.

directions are *extended* by twinning on ($10\bar{1}2$). As a rule, any twinning process that produces extension with zinc (and cadmium) leads to compression, and vice versa. Crystals with an axial ratio of $c/a = \sqrt{3}$ form the boundary between these two groups of crystals exhibiting opposite characteristics. Twinning on a ($10\bar{1}2$) plane is impossible in this case, since the intersection of the plane of displacement with the hexagonal cell would represent a *square* of the length of the side $a\sqrt{3} = c$, with a diagonal in the direction of movement.

With body-centred α-iron having $K_1 = (112)$ and $K_2 = (11\bar{2})$ we obtain, for the twelve available twin planes, the subdivision of the basic triangle shown in Fig. 71. In this case there are no areas in which the formation of twins on all twin planes results exclusively in either extension or compression. The directions [012] and [113] are the corners of the resulting subdivisions.

Contrary to glide, mechanical twinning produces *no continuous change* of orientation; it transfers the lattice discontinuously into a new position. Fig. 72 illustrates this re-orientation, again for the case of the zinc (cadmium) crystal. The crystallographic system of co-ordinates is retained in this diagram. If the orientation of a given direction in the initial state is θ_0, then its position *after* twinning (θ_1 . . . . . θ_6) is obtained by reflexion on the six ($10\bar{1}2$) planes. The choice of the initial position θ_0 results in a shortening of the direction considered in two of the twins (θ_2 and θ_5), and a lengthening in the other four.

Owing to the small extent of plastic deformation by twinning, its immediate importance for large-scale deformations of crystals is slight. Indirectly, however, and especially where hexagonal crystals are concerned, it can exert an important influence on plasticity. Owing to the characteristic swing-over of the lattice, a glide plane that is unfavourably situated to the direction of tension can suddenly adopt a position very favourable to further glide (*e.g.*, θ_1, θ_3, θ_4 and θ_6 in Fig. 72). Thus further appreciable increases of plastic deformation by glide, which otherwise would have been impossible, are facilitated by the conditions created by mechanical twinning. Zinc and cadmium crystals illustrate this very clearly [(148), (152)]. The "secondary extension", a new glide which occurs after the principal extension by basal glide has become exhausted, represents a basal glide in a twin band in which, owing to the swing-over of the lattice through an angle of about 60° to the direction of tension, the basal plane has taken up a position highly favourable to further extension (see Fig. 73, *a* and *b*).

B. THE DYNAMICS OF GLIDE

We shall now turn from the crystallographic to the mechanical aspect of the phenomenon of glide. We shall describe the laws governing the initiation and continuation of glide, with special

Fig. 74.—Schopper Tensile-testing Machine with Device for Automatically Recording the Stress–Strain Curve.

reference to the effect of alloying (the degree of purity), and to the time factor.

We shall confine our attention mainly to the common tensile test—a particularly simple type of loading—and shall deal only briefly with the behaviour of crystals subjected to more complicated types of stress.

Two types of apparatus which are much used for testing the tensile properties of wire-shaped metal crystals are illustrated in Figs. 74

and 75.   The Schopper instrument records the stress–strain diagram automatically.   With the filament-extension apparatus the load is determined from the deflection of a piece of steel sheet carrying the upper grip (mirror-reading), while the amount of extension is

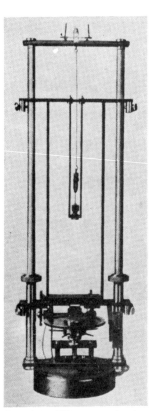

Fig. 75.—Filament-extension Apparatus—Polanyi (164).

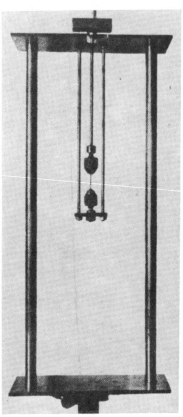

Fig. 76.—Adaptor for Carrying Out Extension Tests in the Schopper Tensometer in Baths of the Required Temperature (165).

measured by means of a micrometer screw attached to the lower grip.   As shown in Fig. 75, the apparatus is also suitable for carrying out tests at temperatures other than room temperature.   In the Schopper tensile machine by means of the adaptor illustrated in Fig. 76 the specimen can also be wholly immersed in a bath of the temperature required.

(a) FUNDAMENTAL LAWS

Fig. 77 represents a stress–strain curve with the usual co-ordinates —stress per square millimetre referred to the initial cross-section, and extension—such as is obtained when metal crystals are extended by glide. The process of deformation exhibits two clearly marked phases. In the first the stress increases sharply, while extension remains mainly within the elastic limit. Then the second phase begins, usually abruptly, and is characterized by *substantial* plastic deformation at only slightly increasing stress, until finally the crystal breaks.

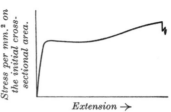

FIG. 77.—Normal Shape of the Stress–Strain Curve of Metal Crystals.

40. *Initiation of Glide in the Tensile Test (Yield Point).* *The Critical Shear Stress Law*

The initial parts of two stress–strain curves for cadmium crystals are shown in Fig. 78, in which the extension has been plotted on a very large scale. It is seen that a phase of very pronounced work hardening (steep increase in stress accompanied by slight increase in extension) is suddenly followed by a phase of appreciable extension with only a slight increase in stress. The transition is so abrupt that a physical significance may be justifiably attached to the stress at which this sudden glide occurs (166); it is termed "yield point of the crystal". It is seen that the yield point of a metal crystal, unlike that of the polycrystalline material, is not fixed by convention (0·2 per cent. plastic extension), but is determined by the nature of the deformation process itself. Nothing certain is yet known about

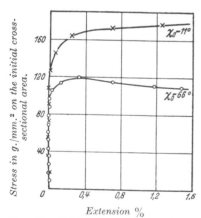

FIG. 78.—Initial Parts of the Stress–Strain Curves of Two Cd Crystals (170).

the plastic deformation which precedes the yield point. This is primarily due to the fact that strain does not take place uniformly over the whole length of the crystal under examination, but starts instead with local contractions, as shown in Fig. 79.

FIG. 79.—Start of Extension, by Necking, of a Cd Crystal.

The yield point of metal crystals is clearly marked in other ways too. Fig. 80 shows this by means of flow curves obtained with a cadmium crystal which was exposed to various initial stresses in the filament extension apparatus. It indicates that the speed of flow increases suddenly at a given stress, and that any substantial increase above this stress is quite impossible. Table VIII reveals that the yield point obtained in this way agrees closely with that which results from the stress–strain curve.

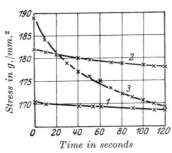

FIG. 80.—Flow Curves of a Cadmium Crystal. The Reduction in Stress is a Measure of the Increase in Length (170).

It has been found that this yield point is by no means constant for crystals of a given material, but strongly depends on the orientation of the crystal lattice to the direction of the applied stress. In the case of hexagonal magnesium the yield point of a crystal oriented to give maximum strength is forty times that of the weakest crystal. The differences with cubic metal crystals are much less pronounced, though still discernible.

If, in keeping with the process of deformation, the stress operative in yield-point tests (σ_0) is resolved into two components, a shear stress (S_0) in the direction of glide (t), on the glide plane (T), and a normal stress (N_0) on T, the following two expressions are obtained (cf. Fig. 39) :

$$S_0 = \sigma_0 \sin \chi_0 \cos \lambda_0 \quad . \quad . \quad . \quad . \quad . \quad (40/1)$$
$$N_0 = \sigma_0 \sin^2 \chi_0 \quad . \quad . \quad . \quad . \quad . \quad . \quad (40/2)$$

in which χ_0 or λ_0 represent the angles between the direction of the pull and the glide plane or direction respectively.
Experimental studies of the relationship between yield point and

TABLE VIII

Comparison of the Yield Points of Cadmium Crystals obtained from Flow Curves and Stress–Strain Curves

Angle between the glide plane and the direction of pull.	Yield point (g./mm.^{2}) obtained from	
	flow curve.	stress–strain curve.*
21·3°	155	159
23·5°	189	178
28·8°	158	136
43·3°	$\begin{cases} 106 \\ 114 \end{cases}$	115
44·8°	$\begin{cases} 87 \\ 83 \end{cases}$	99

* The figures given represent the mean of 2 to 4 determinations on portions of the same crystal.

the orientation of the glide elements have led to the conclusion that a *critical value of the resolved shear stress is required for the initiation of glide on a substantial scale* (Critical Shear Stress Law (166)). It has been found that the normal stress operating on the glide plane—a stress for which differences up to 1 : 2500 have been recorded—is of no importance ; this fact had also been ascertained by direct methods using tensile tests under hydrostatic pressure (up to 40 metric atmospheres) (167).

Fig. 81 gives the results of tests carried out to determine the influence of orientation on the yield point of hexagonal metal crystals. The differences in this case are very considerable, owing to the uniqueness of the basal glide plane. In these diagrams the yield point is plotted above the product $\sin \chi_0 \cos \lambda_0$, which gives the orientation of the glide elements. The left half of the diagrams refers to angles between the glide plane and the direction of tension,

of 0–45°; the right half to χ_0 values between 45° and 90°. The
minimum values for the yield point are obtained when the glide
elements are in positions of 45° to the direction of tension. The
variation of the yield point, calculated on the assumption of a
constant critical shear stress, is shown as an unbroken curve obtained
from formula (40/1). This theoretical curve is an equilateral hyperbola which, in accordance with our diagram, is reflected on the
ordinate which passes through the abscissa at 0·5. In all cases, the
agreement with observed results is satisfactory.

With cubic crystals the orientation possibilities of the operative
glide system are much more restricted. In this case there are
no " unique " planes; conse-
quently, if the geometrical
position of a glide system is
unfavourable a crystallographically equivalent system in
a more favourable position
always becomes available. The
choice of the operative glide
system for octahedral glide (T
$= (111)$, $t = [10\bar{1}]$) for different
orientations of the direction of
stress has already been shown
in Fig. 61. The experimental
results obtained with cubic
face-centred metals [a copper–
aluminium alloy (172), α-brass
(containing 72 per cent. copper)
(173), silver, gold, and their

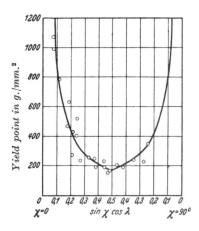

Fig. 81 (a) Magnesium (168).

alloys (174), and nickel and nickel–copper alloys (175)] agree very
well with the Critical Shear Stress Law. Deviations occur only
when the orientation is such that the simultaneous operation of
several glide systems disturbs the progress of simple glide. With
crystals of pure aluminium, permanent set was usually observed to
begin very gradually; recently, however, a clearly defined yield
point has been found in this case too, although it is confined to
crystals produced by recrystallization. With crystals of cast
aluminium a permanent set was observed as soon as any stress was
applied (171).

Fig. 82 shows the experimentally determined dependence of the
yield point on the orientation for *cubic body-centred α-iron* crystals.
The observed differences are compatible with the validity of the

Critical Shear Stress Law for the most probable glide system $(T = (123),\ t = [11\bar{1}])$.

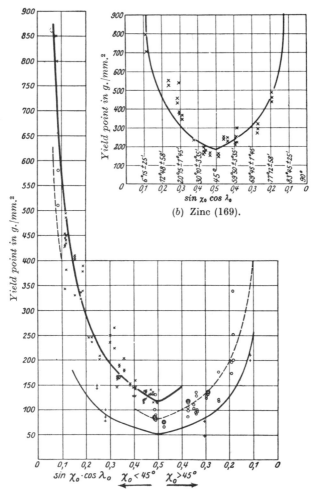

(b) Zinc (169).

(c) Cadmium (170). In this case the three groups of crystals investigated were drawn from the melt at various speeds (cf. Section 42).

Fig. 81 (a)–(c).—The Yield Point of Hexagonal Metal Crystals as a Function of Orientation.

Quantitative investigations on metal crystals of lower symmetry are available only for bismuth (177) and tin (178). These tests, too.

108   Plasticity and Strength of Metal Crystals

confirm the existence of a definite yield point characterized by a constant value of the critical shear stress in the operative glide system.

Since the Critical Shear Stress Law has so far been found consistently applicable, it is tempting to use it for the determination of slip elements that are still unknown. A comparison of the dependence of yield point on orientation, as calculated on the basis of various assumptions regarding the glide system, with that obtained experimentally, can in some circumstances reveal a preference for certain glide elements.

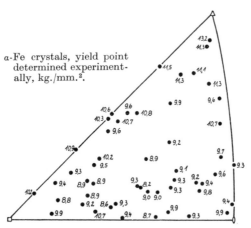

a-Fe crystals, yield point determined experimentally, kg./mm.².

FIG. 82.—The Yield Point of Fe Crystals in Relation to Orientation (176).

In (179) the plasticity condition for crystals is described in an entirely different way from that outlined above. The yield condition is represented by a quadratic function of the principal normal and shearing stresses, which would attain a characteristic constant value at the yield point. However, the results obtained with cubic metal crystals appear to us to dispose of this mathematically attractive attempt at a solution, and to confirm the superiority of the Critical Shear Stress Law (180).

As already explained, the yield point is characterized by the start of *substantial* plastic extension. It does not reveal the true start of permanent deformation. In fact, it is certain that plastic extension can be observed at much lower stresses, and that its mechanism is identical with the glide which subsequently causes large deformations. Owing to the minute stresses involved, as well as to the very gradual start of deformation, an experimental determination of the actual beginning of plasticity is very difficult. Nevertheless, tests carried out on zinc crystals with the filament-extension apparatus showed that even at a plasticity limit corresponding to a permanent set of approximately 0·002 per cent., an approximately constant shear stress was found whose value was less than half of that at the yield point (181). Moreover, tests carried out with the Martens apparatus on the elastic limit (0·001 per cent. permanent set) of aluminium crystals (182) indicate an approximately constant shear stress in the operative octahedral glide system (183). Furthermore, the recording of the first part of the stress–strain curve by precision measurement has shown that the section of the crystal is not without influence, inasmuch as with increasing diameter of the crystal the shear stress needed to produce a given amount of glide diminishes. This influence largely disappears within the limits of ordinary measurements (184).

41. *Torsion of Crystals*

Results are available of quantitative tests which have been carried out to determine the start of plastic torsion in crystals. A heat-treated copper–aluminium alloy (5 per cent. Cu) was used (185). In this case, too, a very marked dependence on orientation was observed. It would, therefore, seem advisable to resolve the applied stress into components with respect to the relevant octahedral glide systems.

In the first place it will be necessary to calculate, for a given torque (M), the shear and normal stresses $(S$ and $N)$ in a glide system (186). Let χ and λ again represent the angles between glide plane (T) and glide direction (t) and the longitudinal direction, while r represents the radius of the cylindrical crystal. The tangential shear stress $\tau_{max.}$, which acts at the surface in an element of the section, is represented by the equation :

$$\tau_{max.} = \frac{2M}{\pi r^3} \quad . \quad . \quad . \quad . \quad (41/1)$$

The point at the surface in which the stress is resolved into components is made the origin of a rectangular system of co-ordinates, the z-axis of which runs parallel to the cylindrical axis, while the x-axis points radially outwards. In this co-ordinate system the stress components are as follows :

$$\sigma_x = \sigma_y = \sigma_z = 0$$
$$\tau_{xy} = \tau_{zx} = 0$$
$$\tau_{yz} = \tau_{max.} = \frac{2M}{\pi r^3}.$$

The glide direction and the normal to the glide plane are in this system given by the angles λ or $90 - \chi$ with the z-axis and the angles ψ_t or ψ_T between the x-axis and the projection of t, or of the normal to T, on to the xy-plane. The condition for t to lie in T (*i.e.*, to be perpendicular to the normal of the glide plane) is expressed by the equation :

$$tg\chi = -\, tg\lambda \,.\, \cos(\psi_t - \psi_T) \quad . \quad . \quad . \quad (41/2)$$

In order to calculate the stress components with reference to the glide system, the system of co-ordinates is transformed into a new one, the z'-axis of which coincides with the normal of the glide plane,

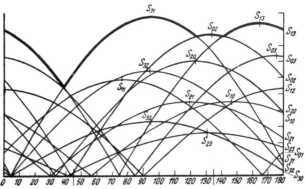

FIG. 83.—Torsion of an Aluminium Crystal. Relative Shear Stress in the 12 Octahedral Glide Systems along the Circumference of the Bar (186).

while the y'-axis coincides with the glide direction. $\sigma_{z'}$, and $\tau_{y'z'}$ are then the components required.   They are given by :

$$\sigma_{z'} = \tau_{\max.} \,.\, \sin^2 \chi \sin \psi_T \quad . \quad . \quad . \quad (41/3)$$

$$\tau_{y'z'} = \tau_{\max.} \,.\, (\cos \chi \cos \lambda \sin \psi_T + \sin \chi \sin \lambda \sin \psi_t) \,. \quad (41/4)$$

The normal stress varies according to a sine function along the circumference of the crystal cylinder.   A similar dependence holds also for the shear stress in the glide system, as is seen if ψ_t in formula (41/4) is eliminated by using equation (41/2).   Fig. 83 shows an example for the variation of the shear stress in the twelve glide systems of an aluminium crystal.

Several series of tests on crystals of aluminium (186), silver (187), iron (188) and zinc (189) have shown that under torsional stress (alternating torsion) the operative glide system is the one that is exposed to maximum shear stress.   In the work mentioned at the outset (185) the question has also been discussed whether, in addition, a torsional yield point characterized by a constant critical shear stress ($S_0 = \tau_{y'z'}$) in the operative system exists.   The dependence of the torque upon the orientation at such a yield point (186, 190) is given by :

$$M = \frac{\pi}{2} r^3 \tau_{\text{max.}} =$$

$$\frac{\pi}{2} r^3 \frac{S_0}{\sqrt{\cos^2 \chi \cos^2 \lambda + \sin^2 \chi \sin^2 \lambda - 2 \sin^2 \chi \cos^2 \lambda}} \cdot \quad (41/5)$$

In the case of cubic face-centred crystals (octahedral glide) the extreme values of M are as follows :

$$M_{\text{min.}} = \frac{\pi}{2} r^3 S_0$$

for $\chi = \lambda = 0$, *i.e.*, [110] is parallel to the longitudinal axis and $\chi = \lambda = 90°$, *i.e.*, [111] is parallel to the longitudinal axis,

$$M_{\text{max.}} = \frac{\pi}{2} r^3 S_0 \frac{1}{0.577} = 1.73 M_{\text{min.}}$$

for $\chi = 35° 16'$, $\lambda = 90°$ *i.e.*, [100] parallel to the longitudinal axis.

The maximum torque at the yield point should therefore occur in crystals whose longitudinal direction is parallel to the cube edge ; it should exceed by 73 per cent. the torque valid for orientations parallel to the face or body diagonal.

The behaviour thus calculated theoretically on the basis of a constant critical shear stress was, in fact, approximately confirmed by experiments (cf. Fig. 84, which represents $\tau_{\text{max.}}$ for a shear strain of 0·2 per

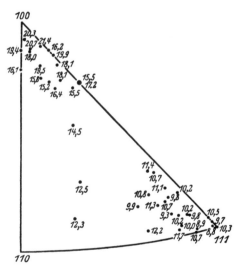

FIG. 84.—The Torsional Yield Point of Al Crystals (with 5% Cu) in Relation to Orientation (in kg./mm.²) (185).

cent.). It is true that the dependence on orientation revealed by the tests is more pronounced than that deduced from theory.

The torsional yield point for the cube-edge orientation is 2·2 times greater than for the body diagonal orientation. In order to explain this difference it was assumed that plastic torsion begins not only in those parts of the crystal whose orientation to the applied stress is the most favourable, but simultaneously in the whole surface layer. A diagram of the " mean plastic resistance " of a thin-walled crystal tube in relation to the orient-

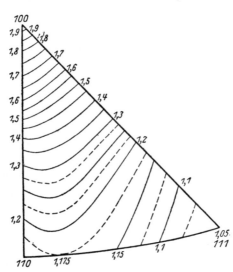

ation of its longitudinal axis is found in Fig. 85. It will be noticed that on this assumption a better agreement with the experimental result is obtained. Even so, the theoretical dependence on orientation still falls somewhat short of the observed dependence.

The value of the critical shear stress in the torsional test (9–11 kg./mm.²) approximates closely to the value of 9·0 kg./mm.² obtained with the same crystal material in the tensile test.

FIG. 85.—The " Mean Plastic Resistance ", against Torsion, of an Al Crystal Tube, in Relation to Orientation (185).

If crystals (aluminium) are subjected to substantial torsion very characteristic changes in shape occur (191) : these, too, depend on the orientation of the crystal; but they have not been investigated in detail so far.

42. *Critical Shear Stress of Metal Crystals*

Table IX gives a summary of the values so far measured of the critical shear stress of the principal glide systems of pure metals.

The figures, which are always of the order of approximately 100 g./mm.² (corresponding to a yield point of the crystal, with the glide system at 45°, of approximately 200 g./mm.²), reveal how very slight is the resistance of pure metals to deformation [cf. in this connection Section 45 and (192a)].

The table includes also, for the non-cubic metals, the elastic shear at the yield point parallel to the glide elements. In every case it is of the order of 10^{-5}. The significance of these minute values in the theoretical understanding of crystal plasticity will be discussed in Section 74. The shear is calculated with the aid of the equations

TABLE IX

Critical Shear Stress of Metal Crystals

Metal.	Purity of the initial material.	Method of production of crystal.	Glide elements.		Critical shear stress at the yield point, kg./mm.².	Shear modulus kg./mm.².	Elastic shear at the yield point.	Literature.
			T.	t.				
Copper .	>99·9	Solidified in vacuo.	(111)	[101]	0·10	—	—	(174)
Silver .	99·99				0·060			
Gold .	99·99				0·092			
Nickel .	99·8				0·58	—	—	(175)
Magnesium	99·95	Recrystallization.	(0001)	[11$\bar{2}$0]	0·083	1700	$4·9_5 . 10^{-5}$	(168)
Zinc .	99·96 *		—	—	0·094	4080	2·3	(169)
Cadmium .	99·996 *	Drawn from the melt.	—	—	0·058	1730	$3·3_5$	(170)
β-Tin .	99·99 *		(100) (110) } [001]		0·189 0·133	1790 1790	10·6 $7·4_3$	} (178)
Bismuth .	99·9 *		(111)	[10$\bar{1}$]	0·221	970	22·8	(177)

* " Kahlbaum " Brand.

$(7/2)$. These equations, which refer to the principal crystallographic system of co-ordinates, must, for the present purpose, be transformed to a system in which the z-axis is placed at right angles to the operative glide plane, while the y-axis coincides with the direction of glide. γ_{yz} is then the required elastic shear in the glide system. The six stress components occurring in the expression for γ_{yz}, if σ is the applied tensile stress and χ and λ the angles between tensile direction and the elements, are given by :

$$\sigma_x = \sigma . (\sin^2 \lambda - \sin^2 \chi) ; \quad \tau_{yz} = \sigma . \sin \chi \cos \lambda$$
$$\sigma_y = \sigma . \cos^2\lambda \qquad \qquad \tau_{zx} = \sigma . \sin \chi \ \sqrt{\sin^2 \lambda - \sin^2 \chi} \qquad (42/1)$$
$$\sigma_z = \sigma . \sin^2 \chi \qquad \qquad \tau_{xy} = \sigma . \cos \lambda \ \sqrt{\sin^2 \lambda - \sin^2 \chi}$$

The transformation of the modulus of elasticity (s_{ik}) from the principal crystallographic system of co-ordinates to the new co-ordinate system (s'_{ik}) proceeds according to the usual transformation formulæ (192).

If these calculations are carried out for the crystal types and glide systems contained in Table X [1] (180), it is found that five of the six elasticity coefficients disappear and that only s'_{44} differs from zero. This means that in all such cases the shear γ_{yz} in the glide system is proportional to the shear stress τ_{yz}. Constancy of shear stress and constancy of shear strain are therefore identical conditions. It

TABLE X

Shear Coefficient s'_{44} in the Glide System

Lattice type.	Crystal class.	Glide elements.		s'_{44}.
		T.	t.	
NaCl lattice	O_h	(101)	[10$\bar{1}$]	$2\,(s_{11} - s_{12})$
Hexagonal close-packed . .	D_{6h}	(0001)	[11$\bar{2}$0]	s_{44}
Tetragonal (β-tin) . . .	D_{4h}	(100) (110)	[001] [001]	s_{44} s_{44}
Rhombohedral . . .	D_{3d}	(111)	[101]	s_{44}

naturally follows that the elastic energy of shear at the yield point is also a constant independent of orientation; it amounts to 10^{-6} cal./g. for the pure metals listed in Table IX.

However, there is no proportionality between shear stress and shear strain for the octahedral glide of the cubic crystals. A decision between the conditions of constant shear stress or constant shear strain at the yield point can therefore be made as a result of tests with cubic face-centred metal crystals. Shear in this case is calculated by :

$$\gamma_{yz} = \left[\frac{4}{3}(s_{11} - s_{12}) + \frac{1}{3}s_{44}\right].\tau_{yz} + \frac{\sqrt{2}}{3}[s_{44} - 2(s_{11} - s_{12})].\tau_{xy} . \quad (42/2)$$

Significant deviations from the proportionality between shear stress and shear strain occur only when the second term of the expression for γ_{yz} can no longer be neglected beside the first. This is the case with crystals of a decidedly anisotropic character [isotropy is represented by $s_{44} = 2(s_{11} - s_{12})$], and with crystals having orientation areas in which λ is much greater than χ (marked deviation of the direction of glide from the projection of the tensile direction on the glide plane).

[1] In addition to the glide of metal crystals the table contains particulars of the dodecahedral glide of cubic crystals of the rock salt type.

The average errors of the mean values of shear stress and shear strain for the four examples in which the dependence of yield point on orientation has been examined, are collected in Table XI. For

TABLE XI

Mean Shear Stress and Shear Strain at the Yield Point of Cubic Face-centred Metal Crystals with Octahedral Glide

Metal.	Critical shear stress, kg./mm.^{2}.	Elastic shear strain.	Degree of anisotropy $s_{44} - 2(s_{11} - s_{12})$ cm.^{2}/Dyn.
Silver (174)	$0 \cdot 060 \pm 5 \cdot 8\%$	$2 \cdot 69 \cdot 10^{-5} \pm 8 \cdot 5\%$	$-43 \cdot 4 \cdot 10^{-13}$
Gold (174)	$0 \cdot 092 \pm 2 \cdot 4\%$	$4 \cdot 12 \cdot 10^{-5} \pm 7 \cdot 0\%$	$-43 \cdot 2 \cdot 10^{-13}$
α-Brass (173)	$1 \cdot 44 \pm 1 \cdot 9\%$	$43 \cdot 3 \cdot 10^{-5} \pm 10 \cdot 8\%$	$-41 \cdot 6 \cdot 10^{-13}$
Al–Cu alloy (age-hardened) (172)	$9 \cdot 2 \pm 2 \cdot 1\%$	$355 \cdot 10^{-5} \pm 2 \cdot 0\%$	$-6 \cdot 8 \cdot 10^{-13}$

the elastically anisotropic metals, silver, gold, α-brass, the difference between the mean errors is considerable : the mean error for the shear strain always exceeds substantially that for the shear stress. In the case of the practically isotropic aluminium there is no difference. This table would therefore appear to justify the assumption that it is constancy of shear stress and not constancy of elastic shear that characterizes the yield point for crystal glide.

The numerical value of the critical shear stress depends in a very marked degree on various circumstances. We shall deal in subsequent sections with the importance of impurities (alloying), temperature, speed of deformation and mechanical pre-working. At present we shall discuss, briefly, the influence of the method of production, the speed of growth during formation of crystals, and annealing.

Magnesium crystals that had been drawn from the melt, and which contained the same impurities as the recrystallized crystals mentioned in Table IX, had an appreciably higher critical shear stress than these, amounting to 103·3 g./mm.^{2} (193). Table XII contains the critical shear stress of *cadmium* crystals drawn from the melt at varying speeds : the slower the rate of crystal growth, the lower the shear stress.[1] Subsequent heat treatment can also greatly reduce

---

[1] Zinc crystals behave in a similar manner (194).

the yield stress of the basal plane if the crystal has been drawn quickly from the melt. Annealing at an elevated temperature (300° C.), on the other hand, causes an increase in the critical shear

TABLE XII

Influence of the Speed of Withdrawal from the Melt and of Annealing on the Critical Shear Stress of Cadmium Crystals (170)

Speed of withdrawal.	Critical shear stress.	Heat treatment of crystals drawn from the melt at the rate of 20 cm./hour.		Critical shear stress.
20 cm./hour	$58 \cdot_4$ g./mm.^{2}	0 hours	275° C.	$58 \cdot_4$ g./mm.^{2}
5–10 ,,	$39 \cdot_7$,,	16 ,,	275° C.	$43 \cdot_5$,,
1·5 ,,	$25 \cdot_5$,,	24 ,,	275° C.	$27 \cdot_4$,,

stress with increasing duration of the heat treatment, amounting to approximately 75 per cent. after 6 hours' treatment. It would appear that this phenomenon could be explained by crystal recovery (Section 49) or by the temperature dependence of the solubility of impurities (195).

The critical shear stresses contained in Table IX relate to the *principal glide system*, at room temperature, of the metal crystal

TABLE XIII

Critical Shear Stress and Closeness of Packing of the Glide Elements of Tin Crystals (178)

Glide system.		Critical shear stress, g./mm.^{2}.	Closeness of packing.		Spacing of the lattice planes from T, Å.
T.	t.		T.	t.	
(100)	[001]	189	1	1	2·91
(110)	[001]	133	0·706	1	2·06
(101)	[10$\bar{1}$]	160	0·478	0·478	2·09 and 0·70
(121)	[10$\bar{1}$]	170	0·346	0·478	1·84 ,, 0·61

concerned. The yield stresses of the next best glide systems (crystallographically non-equivalent) have so far only been determined quantitatively for the tin crystal (Table XIII). In four crystallographically different glide systems the critical shear stresses differ only slightly. The relationship between capacity for glide and closeness of packing, which was mentioned earlier and found to have been more or less confirmed, is now revealed by the example of tin to be only a rough approximation. Only the lower limits of the critical stress can be given for the further glide systems of other crystals. For instance, these can be deduced from the orientation range in which the main glide system remains operative. Thus the critical stress of the pyramidal plane type I order 1 of the cadmium crystal (which, however, has not yet been observed as a glide plane) is at least 4·7 times that of the basal plane. When we come to

discuss, in Section 48, the question of extension at elevated temperatures, we shall be in a position to assess more accurately the capacity for glide of the second-best glide system, using the aluminium crystal as an example.

A graphical representation of the dependence of the yield point of crystals on their orientation, as deduced from the Shear Stress Law, can now be given in two figures. Fig. 86 illustrates the plastic yield surface of cubic crystals with octahedral glide (or

Fig. 86.—Plastic Yield Surface of Cubic Crystals with Octahedral Glide (183).

dodecahedral glide with the body diagonal as glide direction). The radius vector from the centre of the solid is a measure of the magnitude of the yield point in the direction concerned. The model illustrates clearly the way in which resistance to plastic deformation depends upon direction. The minimum yield point values occur in directions which include the angles 20° 46′, 65° 52′ and 84° 44′ with the three axes of the cube (glide plane and glide direction are in this case at 45° to the direction of stress); the maximum yield point lies on the body diagonals. The ratio of the yield point of the strongest crystal to that of the weakest amounts to 1·84.

An example of the dependence of the yield point on the orientation of hexagonal crystals with basal glide is given in Fig. 87, which represents a section through the plastic-yield

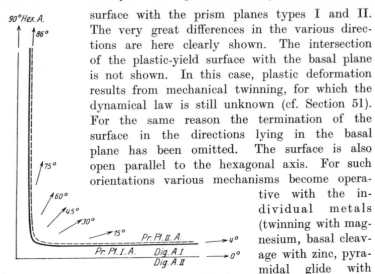

FIG. 87.—Sections through the Plastic Yield Surface of Hexagonal Crystals with Basal Glide (170).

surface with the prism planes types I and II. The very great differences in the various directions are here clearly shown. The intersection of the plastic-yield surface with the basal plane is not shown. In this case, plastic deformation results from mechanical twinning, for which the dynamical law is still unknown (cf. Section 51). For the same reason the termination of the surface in the directions lying in the basal plane has been omitted. The surface is also open parallel to the hexagonal axis. For such orientations various mechanisms become operative with the individual metals (twinning with magnesium, basal cleavage with zinc, pyramidal glide with cadmium).

43. *The Progress of Glide. The Yield–Stress [1] Curve*

Having established that the yield point of magnesium is bound up with the attainment of a definite shear stress in the glide system, the question now arises whether this shear stress is operative for the further course of extension, or whether the shear stress in the glide system changes with increasing deformation. A direct answer to this question is supplied by a comparison of stress–strain curves obtained experimentally, with curves calculated on the assumption that the shear stress in the operative glide system is constant (196). This calculation is as follows. In first place the equation (40/1) gives

$$\sigma = \frac{S_0}{\sin \chi_0 \cos \lambda_0}.$$

$\dfrac{1}{\sin \chi_0}$ represents the area of a glide which remains constant during

[1] *Footnote of Translator.* " Yield point " is according to the preceding sections the stress at which substantial plastic deformation begins *in a previously undeformed* crystal. " Yield stress " is the stress at which plastic deformation continues *in a previously deformed* crystal (or poly-crystalline specimen). The yield stress depends on the magnitude of the preceding strain; its value for the strain 0 is the yield point. In what follows " yield stress " will usually mean the resolved shear stress in the glide system at which plastic glide ontinues after a given glide strain.

extension.   As a first approximation the *stressed* portion of the glide ellipse is also given by this expression, since the sickle-shaped area which becomes exposed in the course of glide is negligible compared with the total area.   By means of the factor cos λ_0 the components in the glide direction are obtained from the tensile stress. Owing to the lattice rotation which accompanies glide, this factor does not remain constant.   Its variation is given by the extension formula (26/1), so that ultimately we obtain

$$\sigma = \frac{S_0}{\sin \chi_0} \cdot \frac{1}{\cos \lambda} = \frac{S_0}{\sin \chi_0} \cdot \frac{1}{\sqrt{1 - \frac{\sin^2 \lambda_0}{d^2}}} \quad . \quad . \quad (43/1)$$

This formula represents the equation of the stress–strain curve for

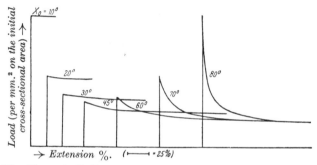

FIG. 88.—Theoretically Determined Stress–Strain Curves, Based on Constancy of the Shear Stress of the Glide System (197).   ($\lambda_0 = \chi_0$ is assumed—a permissible simplification.)

various initial orientations of the glide elements given by angles χ_0 and λ_0.   Fig. 88 contains these theoretical stress–strain curves, characterized by constancy of shear stress, for a number of orientations.   It will be observed that *in all cases* extension takes place at a continuously falling stress; the more transverse the original position of the glide elements in the crystal, the greater the drop of stress.

The stress–strain curves in Fig. 89 which were obtained experimentally on cadmium crystals, reveal once more the great importance of the initial position of the glide elements; the energy of deformation necessary to obtain a given amount of extension (area below the curve) is very dependent on the orientation.   In addition, the curves show that in most cases the stress *increases* appreciably in the course of extension.   It is true that, where the initial position of the basal plane is transverse, a drop of stress can be observed, but here, too, the stress increases again as extension proceeds. A comparison of Figs. 88 and 89, therefore, reveals that when cadmium crystals are stretched the shear stress in the operative glide system remains by no means constant, but increases substantially with increasing deformation.

These two observations (exemplified by cadmium crystals), namely, that the shape of the curve depends on orientation, and that the operative shear stress increases with increasing deformation, have been confirmed by all stress–strain curves obtained with other metals. We meet here for the first time the technologically important phenomenon of the *work-hardening* of metal crystals by plastic

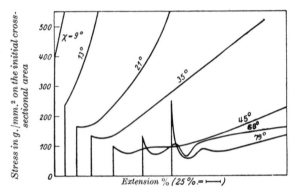

FIG. 89.—Experimental Stress–Strain Curves of Cd Crystals (170).

deformation [(198), (199)]. Moreover, it is a question of shear-hardening, as expressed by an increase, in the operative glide system, of the shear stress needed for further deformation.

Before embarking upon a detailed analysis of the stress–strain curves two observations should be made. The first refers to certain *discontinuities* in the *course of extension*. It was found that zinc crystals, especially after previous deformation by bending backwards and forwards, stretched in a series of regular jumps, so long as extension was confined to the lower range (200). The magnitude of the jumps was about 1 μ. With very pure zinc crystals (99·998 per cent.) this stepwise deformation was also observed in the normal tensile test (201). The magnitude of the jumps depends in a large measure upon the orientation of the crystal. While there are no discontinuities if the glide elements are in an oblique position, they have been observed to amount to as much as 80 μ where the glide plane was more transverse. The fact that the jumps also occur when the surface of the crystal is dissolved during extension shows that they are not due to a surface effect. The phenomenon disappears at a temperature of −185° C. Common to all cases of markedly jerky glide is the fact that they occur within a range of the stress–strain curve where the stress is falling, or at least stationary. An oblique initial position of the glide plane, reduction of the

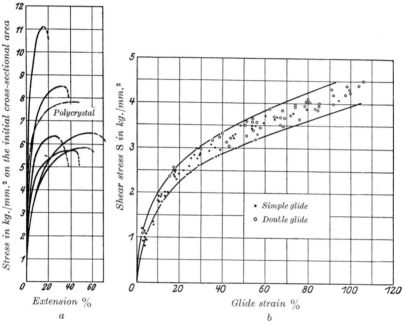

FIG. 90 (a) and (b).—(a) Extension, and (b) Yield–Stress Curves of Al Crystals (203).

temperature and substantial deformation, all of which lead to a relatively steep *increase* of the stress, prevent jerky extension. On the other hand, a decrease of stress accompanying increasing deformation is not a sufficient condition for the occurrence of jerky glide. This is shown by cadmium (or tin) crystals which, in spite of exceptionally large reductions in load, exhibit only very small jumps at the onset of stretch.

The second observation relates to *local contractions* (necking) which occur with crystals of oblique orientations at the start of extension (Figs. 79 and 121). In this case, owing to the initial decrease in stress, the extension is limited at the outset to those portions of the crystal which are first deformed; not until extension has proceeded further does the entire length of the crystal participate in the deformation. However, where there has been only slight hardening it may happen that the extension will not spread and that fracture will result from continued glide at the first contraction (202) (cf. also Fig. 123). If the glide elements are initially in an

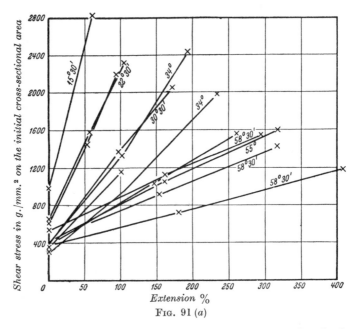

FIG. 91 (a)

oblique position, the crystal will as a rule stretch uniformly from the start.

Having ascertained that the stress–strain curve of metal crystals, like their yield point, depends on the position of the glide elements, we must now enquire whether this curve cannot be explained in terms of orientation, in the same way as the yield point was related to the critical shear stress. Obviously, it will be useful to employ co-ordinates which express the process of glide more adequately than do extension and stress. Suitable " crystallographic " co-ordinates for this purpose have been found in the glide strain and in the resolved shear stress in the glide system [(203), (204)], with which we have already dealt in Section 26. One arrives in this way at a method by which the whole set of the stress–strain curves for various orientations can be satisfactorily represented by a *single* curve, the " yield–stress " curve. The extension and the yield–

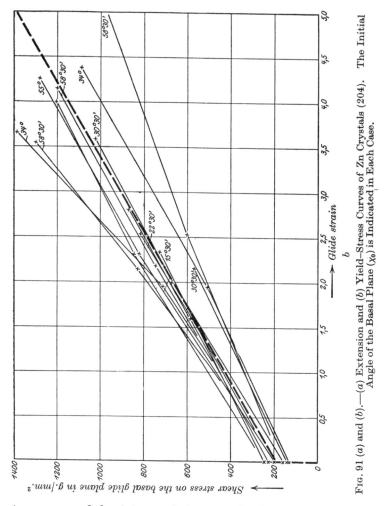

Fig. 91 (*a*) and (*b*).—(*a*) Extension and (*b*) Yield–Stress Curves of Zn Crystals (204). The Initial Angle of the Basal Plane (χ_0) is Indicated in Each Case.

stress curves of aluminium and zinc crystals of various orientations shown in Figs. 90 and 91, may serve as an example.[1] The shear

---

[1] The calculation of the shear stress from stress, extension and initial position of the glide elements is readily obtained by solving the equation (43/1) with respect to S : $S = \sigma . \sin \chi_0 \sqrt{1 - \dfrac{\sin^2 \lambda_0}{d^2}}.$

The glide strain is calculated according to the formula (26/3).

stress of the operative glide system continues therefore to be independent of the normal stress *during* glide. This is most impressively demonstrated in the case of aluminium crystals (Fig. 92), for which the yield–stress curves in compression (where the normal stress in the glide plane is compressive) coincide with those from the tensile tests (where the normal stress is *tensile*). As will be seen from Fig. 90 the yield stress continues without a break even after double glide starts.

Systematic deviations from the mean curve occur with cubic crystals only for initial orientations in which more than two glide systems are almost equally favoured, and in which, consequently,

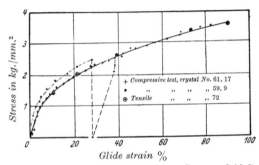

FIG. 92.—Tensile and Compressive Shear–Stress Curves of Al Crystals (205).
Orientation of the crystals : 61, 17: $\chi_0 = 44°$ $\lambda_0 = 46\cdot5°$
                                59, 9 $33°$ $35\cdot2°$
                                72           $48\cdot1°$ $52\cdot1°$

disturbances in the progress of extension may be expected.   In the case of hexagonal metal crystals the scatter of the individual curves about the mean curve is to some extent due to the formation of twins—a phenomenon to which we shall refer later.

A summary of the yield–stress curves obtained hitherto with crystals of pure metals is given in Fig. 93.   The substantially better glide capacity of the single-glide system of hexagonal crystals and of the principal glide systems of the tetragonal tin is very clear.[1]

The yield–stress curve describes the increase in the shear stress of the glide system that is operative in extensions.   Close observation of the lattice rotation in the course of stretching gives an indication of the magnitude of the shear hardening in crystallographically equivalent latent glide systems.   If the lattice rotation correspond-

[1] An attempt has been made in (214) to explain, in terms of atomic displacements, this striking difference in the behaviour of crystals of the two closest-packed systems.

ing to the new system occurs *before* the achievement of a geometrically equally favourable position, then this system has become
less hardened than the operative one; if, on the other hand, while
the first system is exclusively operated, the symmetrical position is
exceeded, then the latent system has become *more* hardened.

The results obtained with cubic metals (aluminium, nickel, copper,
silver, gold) indicate that the second glide system always begins to
operate when the symmetrical position has been either reached or

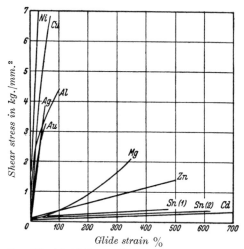

FIG. 93.—Yield–Stress Curves of Metal Crystals.

Cu (207); see also (206, 208)
Ag (207)
Au (207)
Al (209)
Ni (208)
Sn (1) : $T = (100)$; $t = [001]$
Sn (2) : $T = (110)$; $t = [001]$ (213)
Mg (210)
Zn (211)
Cd (212)

only slightly exceeded (cf. Figs. 62 and 63). This means that in
such cases the latent octahedral glide system becomes as hardened
as, or even slightly more hardened than, the operative system
[(215), (216)]. In the case of tin crystals the end position which is
sometimes observed with crystals stretched in a [101] direction
indicates that a latent glide system with $T = (101)$ $t = [10\bar{1}]$
hardens *substantially* more than the crystallographically equivalent
system. No details are yet available regarding the hardening of
crystallographically *non-equivalent* latent glide systems.

Attempts have often been made to express the yield–stress curve
by an equation. With such an equation, which represents S as a
function of a and hence of d, the dependence of the usual stress–

strain curves on orientation could be mathematically represented by substituting it for S_0 in (43/1). For the linear increase of the shear stress, which is often approximately valid for hexagonal metals:

$$(S = S_0 + ka) \quad . \quad . \quad . \quad . \quad (43/2)$$

there results (197)

$$\sigma = \frac{d}{\sin^2 \chi_0} \left(k + \frac{S_0 \sin \chi_0 - k \cos \lambda_0}{\sqrt{d^2 - \sin^2 \lambda_0}} \right) \quad . \quad . \quad (43/3)$$

k, the tangent of the angle of inclination of the yield–stress curve, becomes the coefficient of hardening.

The yield–stress curve for aluminium crystals shown in Fig. 90 is expressed approximately by the equation (217)

$$S = 4 \cdot 2_8 a^{0, \, 33} \quad . \quad . \quad . \quad . \quad . \quad (43/4)$$

Other formal representations of the yield–stress curve of aluminium crystals will be found in (218), (219) and especially in (220), which also contains a theoretical foundation (cf. Section 76).

44. *Termination of Glide*

Such simple and general principles as the Critical Shear Stress Law and the yield–stress curve, which govern the inception or progress of glide, do not hold for its conclusion, since very diverse processes may bring this about. We will discuss in the first place the relatively simple behaviour of the hexagonal metals, and afterwards that of the cubic metals, where rupture is accompanied by necking.

With magnesium crystals glide terminates with the fracture of the crystal, the fracture occurring variously according to the orientation of the basal plane to the direction of tension. Where the initial angle is greater than ~12° a shear fracture showing a stepped surface occurs in the basal slip plane; as a result of the steps the surface of fracture lies more obliquely than the basal plane in the crystal (Fig. 94). Where the basal plane of the crystal is very oblique, the fracture either passes more or less transversely through the crystal or it follows a twin plane. With zinc and cadmium crystals the basal glide is limited at room temperature solely by the operation of the second crystallographic mechanism of deformation: mechanical twinning. In both cases deformation twins are formed on a (10$\bar{1}$2) plane; a new secondary basal glide develops in the twins which, accompanied by a drop of load, leads to a marked necking of the crystal ribbon and to final rupture (cf. Fig. 73; for further details see Section 52).

With these crystals and the magnesium crystal, which fails by a shear fracture, the primary basal glide generally comes to an end when a limiting shear stress characteristic of the metal is reached in the glide system. This statement, however, applies only to such crystals where the basal plane was originally at an angle of more

Fig. 94.—Fracture of Magnesium Crystals [(221), (210)].

than 15–20° to the longitudinal direction. It is based not only on experiments at room temperature, but also on a number of experiments designed to test the relationship between crystal plasticity and temperature (Section 48).

Consequently it is not only the start (yield point) and the slope (coefficient of hardening) which are characteristic for the yield–

TABLE XIV

Conclusion of the Basal Glide of Hexagonal Metal Crystals

Metal.	Shear strength of the basal plane in g./mm.²		Upper limit of glide strain, %.	Work of deformation, cal./g.
	at the start of glide.	at the conclusion of glide.		
Magnesium (210) .	83	2100	350	4·09
Zinc (222) .	73	1220	380	0·84
Cadmium (223) .	58	420	500	0·26

stress curve of these hexagonal crystals : the conclusion of the curve is also substantially independent of crystal orientation over a wide range. Therefore the upper limit of the glide strain and the work of deformation (the surface below the yield–stress curve) are likewise constants independent of orientation. The relevant values at room temperature for magnesium, zinc and cadmium will be found in Table XIV.

On the basis of these values it is now possible, by a simple calculation, to arrive at the dependence of final orientation, maximum load

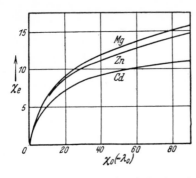

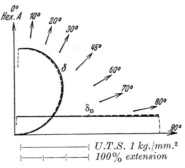

FIG. 95.—Final Angle of the Basal
Plane as a Function of its Initial
Angle in the Extension of Hexagonal
Crystals.

FIG. 96.—Sections of the Ultimate
Tensile Stress and Extension Surface
of Zn Crystals with Prism-plane
Type I (Continuous Line) and Prism
Type II (Discontinuous Line) (224).

and extension, upon the initial position of the lattice. Assuming
for the sake of simplicity that $\lambda_0 = \chi_0$ (this does not substantially
impair the generality) then formula 26/4a
gives the values shown in Fig. 95 for the
final angle of the basal plane. These
correspond substantially with experimental
results. The extension is derived from the
initial and the final orientation according
to the extension formulæ (26/1, 2), while
the maximum load per square millimetre of
the initial cross-section (the ultimate tensile
stress) is obtained by inserting the limiting
shear stress S_e in the equation (43/1). Fig.
96 gives as an example sections of the
extension and ultimate tensile-stress surface
of zinc crystals; while Fig. 97 represents a
model of the extension surface (cadmium).

FIG. 97.—Model of the
Extension Surface of
Cd (212).

The true stress corresponding to the
maximum load, unlike the ultimate tensile
stress, is substantially independent of the
initial orientation, owing to the constant final shear stress of
the basal plane and to the very similar final orientations of
stretched crystals.

We shall discuss in Sections 53 and 54 the question of the termination of primary basal glide of zinc crystals by basal cleavage, which occurs at low temperatures.

The extension of *cubic* metal crystals is in no wise limited by a condition of constant final shear stress in the glide system, independent of orientation. In this case, double glide will always occur sooner or later in two crystallographically equivalent glide systems which are geometrically equally favourable, leading to necking and so finally to fracture. With such crystals the end of the uniform extension occurs when the maximum load has been reached, since necking itself is accompanied by a reduction of load. Table XV

Table XV

Ultimate Tensile Stress and Elongation of Cubic Metal Crystals

Metal.				Ultimate tensile stress, kg./mm.².	Elongation, %.
Aluminium (226)	.	.	.	5·9–11·5	19–68
Copper (227)	.	.	.	12·9–35·0	10–55
a-Iron (228)	.	.	.	16–23	20–80
Tungsten (229)	.	.	.	105–120	—

shows the dependence of the ultimate tensile stress (as given by maximum load and initial section) and of the uniform extension on the orientation. In addition, Fig. 98 illustrates, with the aid of models, the experimental results of tensile tests on copper crystals of various orientations.

A mathematical determination of the maximum-load point can be attempted by differentiating the equation of the load–extension curves :

$$\sigma = f(d) = \frac{S(a)}{\sin \chi_0 \sqrt{1 = \frac{\sin^2 \lambda_0}{d^2}}}$$

with respect to the extension, and putting $\frac{d\sigma}{dd} = 0$ [cf. (43/1); $S(a)$ is the equation of the yield–stress curve of the operative glide system]. The following is then obtained :

$$\frac{dS}{da} = S(a) \cdot \frac{\sin \chi_0 \sin^2 \lambda_0}{(\cos \lambda_0 + a \sin \chi_0)(1 + 2a \sin \chi_0 \cos \lambda_0 + a^2 \sin^2 \chi_0)}$$

If the equation of the yield–stress curve is known, this expression connects the glide strain which corresponds to the maximum tensile stress with the initial position of the glide elements. Whether the

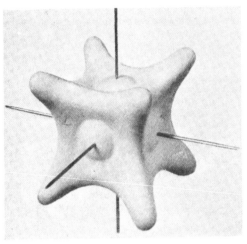

Fig. 98 a

values determined in this manner really do correspond to a *maximum* of the extension curve, is seen from the sign of $\dfrac{d^2\sigma}{dd^2}$. For the extreme case we have

$$\frac{d^2\sigma}{dd^2} = \frac{d^2S}{da^2} \cdot \frac{1 + 2a \sin \chi_0 \cos \lambda_0 + a^2 \sin^2 \chi_0}{(\cos \lambda_0 + a \sin \chi_0)^2} +$$
$$+ 3S \cdot \frac{\sin^2 \chi_0 \sin^2 \lambda_0}{(\cos \lambda_0 + a \sin \chi_0)^2 (1 + 2a \sin \chi_0 \cos \lambda_0 + a^2 \sin^2 \chi_0)}$$

This expression, which in the case of a maximum load would have to be *negative*, has *always a positive* sign whenever the resolved shear stress increases linearly or with a higher power of the glide strain. In agreement with what is known of hexagonal crystals, a maximum load for such yield–stress curves can never be reached with simple glide; the extreme value obtained from $\dfrac{d\sigma}{dd} = 0$ represents a minimum in the extension curve. A maximum load can be reached only when the increase of the shear stress with the glide strain is less than linear.

But even in the cases of the yield–stress curve given above for aluminium crystals a maximum in the extension curve will not be reached with simple glide. On the other hand, a maximum load

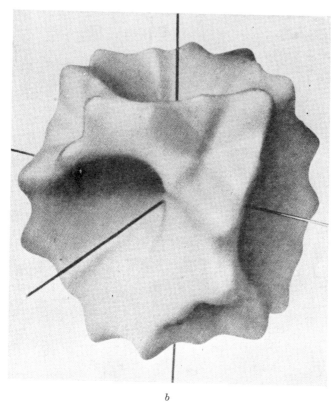

b

Fig. 98 (*a*) and (*b*).—(*a*) Ultimate Tensile Stress and (*b*) Extension Surfaces of Cu Crystals; as Determined Experimentally (225).

will be reached in the region of double glide. The result of the calculations, which cannot always be carried out exactly and which we do not propose to discuss here, is graphically represented by the ultimate tensile stress and extension surfaces of aluminium crystals shown in Fig. 99. In general, the results are satisfactorily confirmed by experience, although in the case of extension the discrepancies amount to as much as 50 per cent.

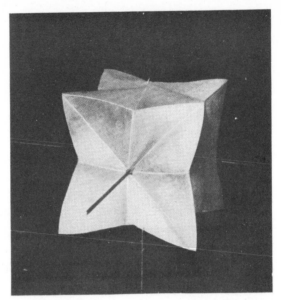

a

b

FIG. 99 (*a*) and (*b*).—(*a*) Ultimate Tensile Stress and (*b*) Extension Surfaces of Al Crystals; Calculated (217).

(b) EFFECT OF ALLOYING

Hitherto in dealing with the beginning, progress and termination of glide, we have confined our observations to the " pure " metals. In Table IX will be found the purity percentage of these metals. These materials do not afford a safe guide to the behaviour of *absolutely* pure metal crystals, since even very small impurities have an extraordinarily large effect. The influence of alloying on the plasticity of metal crystals will be described in the following two sections.

45. The Start of Glide in Alloyed Metal Crystals

The study of the initiation of glide in alloyed metal crystals is greatly facilitated by the observation, so far invariably made, that the yield point of such crystals is much more pronounced than that of similarly oriented crystals of the pure metal.

The example of the critical shear stress of *zinc* crystals alloyed with cadmium shows how strongly even small additions can influence the plastic behaviour of metal crystals (Fig. 100). A content of 0·60 atomic-per cent. cadmium (1·03 weight-per cent.) raises S_0 from the value of 94 g./mm.^{2} (for the " Kahlbaum " material with 0·03 weight-per cent. cadmium) to 1150 g./mm.^{2}—a more than twelve-fold increase.[1]   Linear extrapolation to a cadmium content of 0 per cent. gives a shear stress of 40 g./mm.^{2} for the yield point of the cadmium-free zinc crystal—which, however, still contains traces of lead and iron. Recently this value was all but achieved (49 g./mm.^{2}) with crystal material produced from " Kahlbaum " zinc refined by distillation, the cadmium content of which was less than 5×10^{-4} atomic-per cent. (201).

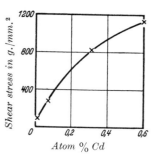

FIG. 100.—The Critical Shear Stress of Cd–Zn Solid Solutions as a Function of Concentration (230).

Alloying has a much smaller hardening effect if the added element

[1] That cadmium was present in solution in the alloy crystals drawn from the melt was established by X-ray examination (244). We have here supersaturated solid solutions, since below 100° C. cadmium is practically insoluble in zinc.

forms a second phase in the matrix. Fig. 101 shows the polished section of a zinc crystal alloyed with 2 per cent. tin, in which layers of zinc–tin eutectic have come to lie parallel to the basal plane. The initial critical shear stress of the basal plane undergoes as a result no more than a four-fold increase.

Fig. 102 relates to the binary solid solution systems of aluminium–magnesium and zinc–magnesium. It gives the hardening ($v_{all.}$) due to alloying as a function of the concentration of the added metal; the hardening is expressed by the quotient of the critical shear stress

FIG. 101.—Polished Section of a Zn Crystal in which Zn–Sn Eutectic is Present (230).

of the solid solution ($S_{0.all.}$) and that of the pure magnesium crystal (S_0): $\left(v_{all.} = \dfrac{S_{0.all.}}{S_0}\right)$. The increase of the shear stress is approximately linear with increasing content of foreign metal. The specific effect of the zinc greatly exceeds that of aluminium. The diagram also indicates the dependence upon concentration of the lattice parameters c and a, for both solid solutions. Corresponding to the smaller atomic radius of the added metals (Mg, 1·62 Å.; Al, 1·43 Å.; and Zn, 1·33 Å.) the lattice contracts in both cases. It is not yet clear whether the hardening effect of the zinc is connected with the greater reduction of c (*i.e.*, of the spacing of the glide planes) in the zinc–magnesium solid solutions.

Ternary aluminium–zinc–*magnesium* solid solutions (232) exhibited, at a concentration of $2·5_4$ atomic-per cent. Al and $0·36$ atomic-per cent. Zn, an $S_0 = 766$ g./mm.^{2}; and at a concentration of $5·0_2$ atomic-per cent. Al and $0·38$ atomic-per cent. Zn an $S_0 = 1153$ g./mm.^{2}.[1] These values were obtained by means of a good approximation from the behaviour of the two binary solid solutions on the assumption that the increases in the critical shear

---

[1] These are the technical alloys AZ.31 (with 2·8 per cent. Al and 0·9 per cent. Zn by weight), and AZM (with 5·6 per cent. Al and 1·0 per cent. Zn by weight).

stress due to the two alloying constituents were additive, and that consequently there was no reciprocal influence.

$$S_{0,\,\text{tern.}} - S_0 = (S_{0,\,\text{Al}} - S_0) + (S_{0,\,\text{Zn}} - S_0)$$

and therefore

$$v_{\text{all. tern.}} = v_{\text{afl. Al}} + v_{\text{all. Zn}} - 1.$$

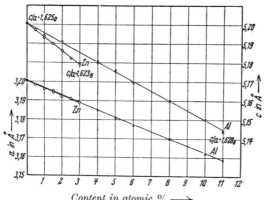

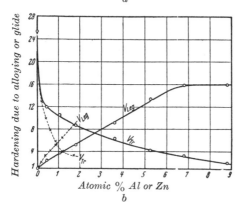

——— Al–Mg, – – – – Zn–Mg.

Fig. 102 (a) and (b).—Solid Solutions of Al–Mg and Zn–Mg.
(a) Lattice constants; (b) hardening as a function of concentration (231).
$s_0 = 82 \cdot 9$ g./mm.^{2}.   (N.B. In the text $v_{\text{Leg.}} = v_{\text{all.}}$.)

This formula leads to $v_{\text{all. tern.}} = 8 \cdot 8$ for the first of the two alloys, and to $v_{\text{all. tern.}} = 14 \cdot 5$ for the second, whereas experimentally the values 9·2 and 13·9 were obtained. Thus if the saturation limit of the ternary solid-solution region is known, it is possible to calculate, from the behaviour of the binary solid solutions, the concentration needed to give maximum shear resistance to the glide plane.

In the case of N additions, which do not substantially influence each other in their hardening effect, let it be assumed that the formula for calculating the hardening effect is

$$v_{all.} = v_{all. A} + v_{all. B} \ldots + v_{all. N} - (N - 1)$$

in which $v_{all. A} \ldots v_{all. N}$ represent the hardening of the appropriate

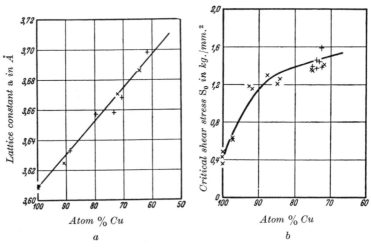

FIG. 103 (a) and (b).—α-Brass Crystals.

(a) Lattice constants [× (234), + (245), ● (246)]; (b) critical shear stress [× (234), + (233)] as a function of concentration.

binary series, which corresponds to an alloying metal content of A, . . .,N.

The hardening of cubic solid solutions is represented in Fig. 103 by the case of α-brass. Here the critical shear stress does not increase linearly with increasing concentration of the added metal; the greater the zinc content the smaller is the further increase in shear stress. The lattice constant indicated in the diagram shows an expansion due to the formation of the solid solution, and corresponding to the greater atomic radius of zinc (Cu, 1·27 Å.; Zn, 1·33 Å.). This expansion increases linearly with the atomic concentration. In the case of α-brass, therefore, hardening is bound up with an increase in the spacing of the operative glide plane.   At a zinc content of 18 atomic-per cent. the strength has increased approximately three-fold.

Fig. 104 contains the critical shear stress and the lattice constant for the complete silver–gold solid-solution series.   The pure metal values are followed by a steep increase in the critical shear stress of

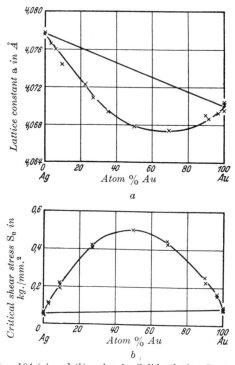

a

b

Fig. 104 (*a*) and (*b*).—Ag–Au Solid-solution Crystals.
(*a*) Lattice constants (247), (*b*) critical shear stress (235), as a function of concentration.

the octahedral glide system; the maximum is reached at approximately equal atomic concentration of both metals.   The lattice constant exhibits a minimum at an intermediate concentration.   A curve for critical shear stress similar to that shown in Fig. 104 has been obtained for the copper–nickel solid-solution series (236).

An instructive example of the significance of the *atomic arrangement* in the crystal for its plastic behaviour is found with gold–copper crystals, whose composition is given by the formula $AuCu_3$

(237).   The disordered arrangement of the atoms at high temperature becomes ordered as the temperature falls; the cubic face-centred lattice, however, is retained (248). The appearance of super-structure lines corresponding to the ordered distribution is seen clearly in Fig. 105.   The critical shear stress of these crystals falls

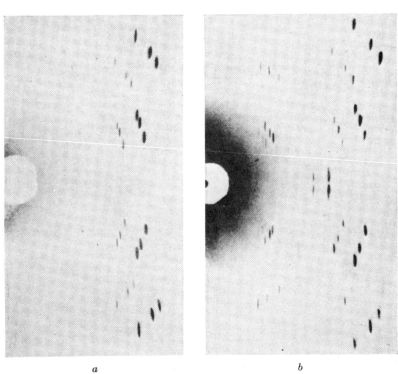

a   b

FIG. 105 (a) and (b).—Rotation Photographs of a $AuCu_3$ Crystal (237).
(a) With random atomic distribution; (b) with ordered atomic distribution.

from 4·4 kg./mm.^{2} to 2·3 kg./mm.^{2} when passing from the disordered to the ordered arrangement.

The effect of *ageing* on the critical shear stress has been demon-strated by experiments carried out on crystals of an aluminium alloy with 5 per cent. copper (238), which, in the annealed state (slow cooling from 525° to 300° C., and kept at this temperature for an hour), gave a value of $S_0 = 1·9$ kg./mm.^{2}, but after quenching from 525° C. followed by precipitation treatment for half an hour at

100° C., gave $S_0 = 9\cdot3$ kg./mm.[2].   Fig. 106, referring to a quenched Al–Mg solid solution with 10 per cent. Al, shows the change in the

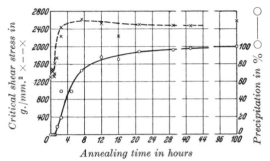

FIG. 106.—Shear Hardening and Precipitation (242).   Supersaturated Al–Mg Solid Solution (10% Al) Annealed at 218° C.   The " Precipitation " is the Amount of Al actually Precipitated (as Measured by Determination of the Lattice Constant) Expressed as a Percentage of the Amount Required to Establish Equilibrium.

critical shear stress of the basal plane as a function of the duration of heat treatment at 218° C.   It is seen from the diagram, which also includes a curve giving the amount precipitated from the solid solution, that hardening and precipitation go hand in hand.

46. *Progress of Glide and Fracture in Alloy Crystals*

Fig. 107 contains extension curves of zinc crystals, of approximately similar orientation, alloyed with varying percentages of cadmium. They reveal once more the extremely strong dependence of the critical shear stress on the Cd content; they also show that the rate of hardening due to extension is less, the higher the original critical shear stress resulting from alloying.

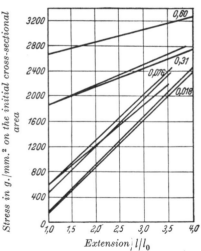

FIG. 107.—Extension Curves of Solid Solutions of Cd–Zn of Approximately Similar Orientation.   Cd-content Given in Each Case in Atomic-% (230).

This reduction of the capacity for work hardening with increased initial hardening has been confirmed by all subsequent extension tests with alloy crystals $\left[\text{cf. curves in Fig. } 102b \text{ with } v_{\text{Tr}}\left(=\frac{S_{e,\,\text{all.}}}{S_{0,\,\text{all.}}}\right)\right].$

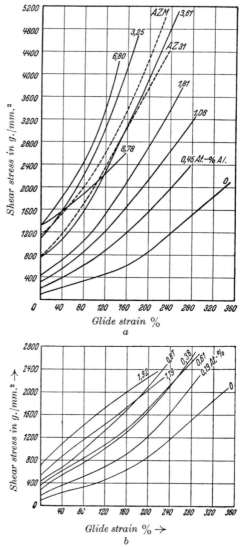

FIG. 108 (a) and (b).—Yield–Stress Curves of Mg Solid Solutions.
(a) Binary Al–Mg and ternary Al–Zn–Mg alloys (231, 232); (b) binary
Zn–Mg alloys (231).

The behaviour of *magnesium* crystals is shown in Fig. 108. The mean ultimate glide strain decreases clearly with increasing alloy content, a phenomenon due to the growing tendency to glide fracture along the basal plane. In order to assess technologically the effect of alloying, one has to bear in mind not only the increase in the critical shear stress of the pure metal ($v_{all.}$) but also the subsequent capacity for work hardening. Accordingly the total hardening which can be achieved by alloying *and* glide to the point of fracture $\left(v_{all.} \cdot v_{Tr} = \dfrac{S_{0,\,all}}{S_0} \cdot \dfrac{S_{e,\,all.}}{S_{0,\,all.}} = \dfrac{S_{e,\,all.}}{S_0}\right)$ is plotted in Fig. 109 as a function of the concentration.

In both cases it will be found that it is not the maximum but a

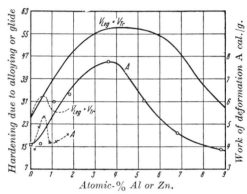

FIG. 109.—Work-hardening Capacity of Mg Solid Solutions (231).
———— Al–Mg, – – – – Zn–Mg.

medium concentration to which the maximum value of this total hardening corresponds. The cause of this must no doubt be sought in the increase of brittleness, already referred to, which accompanies an increase in alloying content. The work of deformation behaves in a similar way to that of the product $v_{all.} \cdot v_{Tr}$.

The relationship between the yield–stress curves and the concentration of cubic face-centred *zinc–aluminium* solid solutions is shown in Fig. 110. The hardening caused by zinc additions, particularly noticeable at the start of extension, is apparent from these curves. The lattice rotation which accompanies stretching of the crystal corresponds to the normal case of octahedral glide which was described in Section 36. At higher zinc contents the symmetrical position of the crystal axis, which leads to double glide, is not attained, because the crystals had fractured within the region of simple glide without necking. This fracture appears to be a sort of shear fracture on the operative glide plane, similar to that observed with magnesium solid solutions.

The yield–stress curves of α-brass crystals of various zinc concentrations are shown in Fig. 111. In spite of the initially increased yield stress, the shear stress in the operative octahedral system remains during extension below that of the pure copper crystal. The higher the zinc content the lower is this shear stress. The lattice rotation which accompanies glide again corresponds essen-

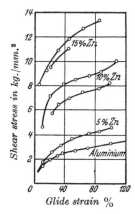

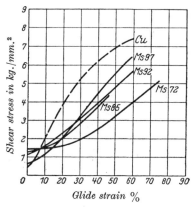

FIG. 110.—Yield–Stress Curves of Zn–Al Solid Solutions (239).

FIG. 111.—Yield–Stress Curves of α-Brass Crystals (234).

tially to the typical case for octahedral glide. The differences relate only to the interaction between the two operative glide systems. The higher the zinc content, the more will the exclusive operation of the first system extend beyond the symmetry position (Fig. 112). This means that the more zinc there is in the crystals the more will the hardening of the initially latent octahedral system exceed that of the operative system. Fig. 113 gives an example for the variation of the critical shear stress in the two glide systems. The second system becomes active when its resolved shear stress exceeds that of the first system by 30 per cent. In the course of further extension the symmetry position (equality of both shear stresses) is again exceeded, this time in the opposite direction because it is now the first system, temporarily latent, which experiences the greater hardening. Only under the influence of a shear stress that is 10 per cent. higher than that of the operative system will it again become active. The higher the shear stress at the start of glide, the less marked will be the subsequent hardening with increasing extension. The new bands which appear with each change in the glide system are shown in Fig. 114. The ultimate tensile stress of crystals with 72 per cent. copper is between 12·9 and 32·9 kg./mm.^{2}, the elongation between 67 and approx. 122 per cent.—according to their initial orientation (233).

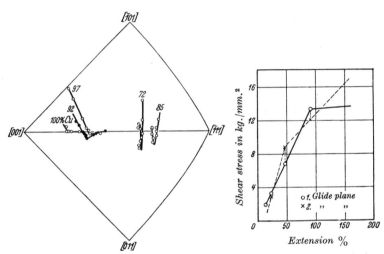

FIG. 112.—Unequal Rate of Hardening of the Operative and Latent Octahedral System in the Extension of α-Brass Crystals, Expressed in Relation to Concentration.

FIG. 113.—Variation of the Critical Shear Stress in the Double Glide of α-Brass Crystals with 70% Cu (240); cf. also (233).

Copper–aluminium solid solutions exhibit a behaviour substantially analogous to that of the α-brass crystals (241). In this case, too, if the initial shear stress is high the yield–stress curve soon intersects that of the pure copper crystal, and exhibits for the higher glide strains lower shear stresses than this. Here, too, the lattice rotation of the alloy crystals is characterized by a substantial transgression of the symmetry position, thereby revealing the greater hardening of latent glide systems.

The extension of silver–gold solid solutions differs only comparatively slightly from the behaviour of the pure metals (235). Only at small deformations does the hardening of the solid solution remain noticeably below that of the pure metals. The lattice rotation, too, is little affected by the formation of solid solutions. The hardening of the latent octahedral system in this case exceeds that of the active system by about 12 per cent. at the most (at medium concentrations). The *copper–nickel* series of solid solutions appears to behave in a similar manner, *i.e.*, there is at medium concentrations more pronounced hardening of the latent octahedral system (236).

The significance of the *atomic arrangement* for shear hardening in plastic extension is evident from Fig. 115. Whereas the cubic

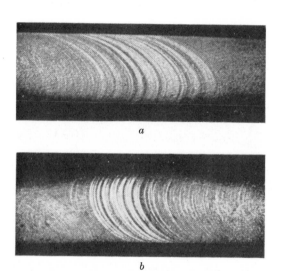

a

b

FIG. 114 (*a*) and (*b*).—Change of the Operative Glide System when Extending *a*-Brass Crystals (234). Appearance of New Bands.

(*a*) After the first change ; (*b*) after the second change.

face-centred $AuCu_3$ crystal with disordered atom arrangement (quenched from 800° C.) shows the type of yield–stress curves characteristic of solid solutions (high initial shear stress and rather slow increase of the shear stress with increasing glide), in the case of ordered atom arrangement (produced by an ageing treatment of the crystal at 325° C.) the hardening curve rises steeply with increasing glide strain from a low initial value—as with pure metals. The extent, too, to which the symmetrical position is transgressed, corresponds to this behaviour : with disordered atomic arrangement it is twice that of the ordered arrangement.

The extension of *aged* copper–*aluminium* solid solutions (with

5 per cent, Cu) can also be represented by a mean hardening curve (238). However, the fracture of these crystals does not obey a condition of maximum load, as in the case of pure aluminium. Instead, as with Zn–Al solid solutions, fracture occurs in certain orientation ranges, without previous necking, along one or more nearly plane surfaces. The mechanism of this fracture is not known. The ultimate tensile stress, according to the initial orientation, lies between 33·0 and 45·5 kg./mm.^{2}, the elongation between 7 and 26 per cent.

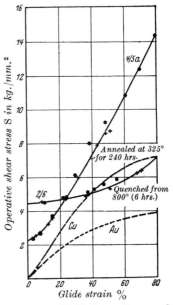

FIG. 115.—Yield–Stress Curves of AuCu$_3$ Crystals (237).

(c) INFLUENCE OF TEMPERATURE AND TIME

The description of the effect of composition and structure upon the glide of metal crystals will now be followed by a discussion of the significance of temperature and speed of deformation. The dynamic properties have hitherto referred to room temperature and to a limited range of deformation speeds. However, it has been precisely the experiments conducted over a wide range of temperature and time which have led, if not to an explanation, then at least to a physical interpretation of crystal glide.

47. The Influence of Temperature and Speed upon the Start of Glide

The influence of the test temperature on glide has hitherto been widely investigated mainly with hexagonal crystals. It can be said of them in general that the sharp bend in the stress–strain curve at the yield point becomes more accentuated with rising temperature. It was seen in Section 43, in which we discussed the dependence of the stress–strain curve upon the orientation of the crystal, that the sharpness of this bend was decisively influenced also by the position of the glide elements. Where the glide elements are more or less transverse to the direction of tension, an initial

reduction in load with increasing extension is possible in spite of
the shear *hardening* due to glide, whereas the same yield–stress curve

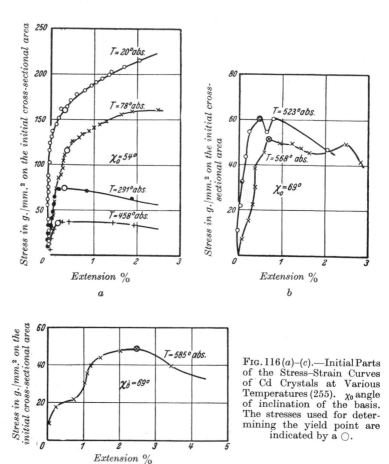

Fig. 116 (a)–(c).—Initial Parts
of the Stress–Strain Curves
of Cd Crystals at Various
Temperatures (255). χ_0 angle
of inclination of the basis.
The stresses used for determining the yield point are
indicated by a ○.

leads at once to a steep increase in load if the initial position of the
glide elements is very oblique.

A characteristic example of the initial parts of stress–strain curves
of several fractions of the same cadmium crystal, obtained at various
temperatures, is found in Fig. 116, *a*, *b* and *c*. The test temperature
ranged from 20° to 585° K., *i.e.*, to within 9° of the melting point.
A stress–strain curve obtained with special equipment at still lower

P. 147 :

Para <u>1</u> : Considerable plasticity persists even at low temperatures. In particular, the critical shear stress is of the same order of magnitude as at the other temperatures

temperatures down to 1·2° K. is shown in Fig. 117. It is seen that considerable plasticity persists even at these temperatures and that in particular the critical shear stress is of the same order of magnitude as at the other temperatures. This fact has also been confirmed for zinc crystals [(251), (253)].

Fig. 118 shows in diagrammatic form the relationship between critical shear stress and temperature. It will be observed that the shear stress of the basal glide system of hexagonal metals is only slightly affected by temperature. The increase in the shear stress

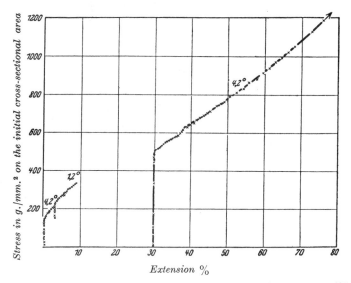

FIG. 117.—Stress–Strain Curve of a Cd Crystal at Low Temperatures (254).

of cadmium, for instance, is slightly less than four-fold, despite a reduction in temperature from just below the melting point to 20° K. A similarly slight dependence of S_0 on temperature has been observed with the rhombohedral bismuth (263) and tetragonal tin (262).

Although the experimental data available for very low temperatures are scanty (a serious deficiency, especially where cubic crystals are concerned), extrapolation down to absolute zero for zinc and cadmium can at least be attempted. It reveals that even at this temperature the low yield stress of the basal plane persists.

In the neighbourhood of the melting point there was in all cases a zone of approximately constant critical shear stress. A decrease to zero value (that of the melt) must proceed very suddenly, at any rate for cadmium and bismuth; down to temperatures of 9° or 6° below melting point there is still the same value as in the range in which there is no dependence on temperature. In view of the temperature dependence of the solubility of impurities, their role in the above tests cannot at present be assessed.

It has already been mentioned that *aluminium* crystals show no sharp yield point at room temperature (see Section 40). This has also been observed for the temperature range −185 to 600° C. (257). However, the yield–stress curves shown in Fig. 128 indicate that

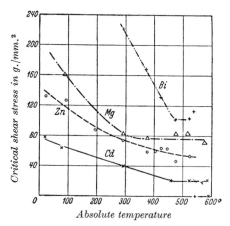

FIG. 118.—Effect of Temperature on the Critical Shear Stress.

for aluminium, too, there is only slight dependence of the start of glide on temperature. Whereas at 600° C. a shear stress of 100 g./mm.^{2} suffices for perceptible deformation, at −185° C. a stress of approximately 800 g./mm.^{2} is necessary to produce observable octahedral glide.

Having described the effect of temperature, we will now deal briefly with the effect of the speed of deformation on the critical shear stress. It will be explained later (Section 50) that, owing to their similar influence, time and temperature are probably very closely connected. It has been found that in general the yield point increases with increasing speed of testing. It is probable that the decisive factor is not the increase in the stressing speed within the range of purely elastic deformation, but the greater speed of deformation within the narrow range that lies between the elastic limit and

the yield point.   Some idea of the magnitude of the influence of
this speed can be gained from Fig. 119 in the case of cadmium
crystals.   A hundred-fold increase in the speed of testing leads,
within a wide range of temperature, to an increase of the critical
shear stress of 20–30 per cent.   Fig. 119 reproduces, also, as a
broken line, the curve which appears in Fig. 118; since it relates to
crystals which were drawn from the melt much more slowly, it is
not directly comparable with the other two curves.

It is difficult to give figures for aluminium crystals, owing to the
absence of a sharp yield point in this case.   From available tests

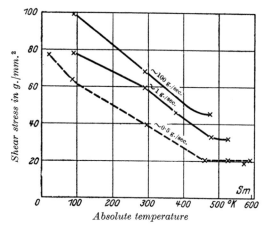

FIG. 119.—Dependence of the Critical Shear Stress of Cd Crystals upon
the Speed of Stressing (Deformation) (252).

in which the test speed varied from 1 : 23,000 it appears that, in
spite of this wide range, glide begins at approximately the same
stress (256).

48. *Progress and Termination of Glide in Relation to Temperature and Speed*

An example of the energy needed to continue extension at various
temperatures is shown in Fig. 120.   It represents the extension of a
cadmium crystal in stages at various temperatures.   The change in
temperature was always effected after removal of the load from the
crystal.   It will be noticed that the relatively slight dependence of
the critical shear stress upon temperature in the initial unworked state
persists at larger glide strains.   On the other hand, the work needed

to continue the extension is strongly dependent upon temperature. Whereas at room temperature and even more so at elevated tempera-

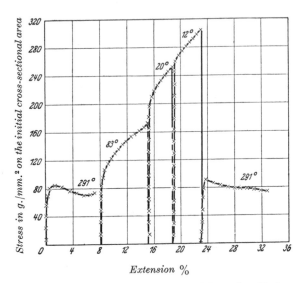

FIG. 120.—Stress–Strain Curves of a Cd Crystal at Various (Absolute) Temperatures (254).

tures stretching can take place initially with a falling load (however, here too the shear stress in the operative basal glide system increases) a very considerable increase in load is necessary at low

FIG. 121.—Local Glide of a Cd Crystal Extended at 250° C. (255).

temperatures. Consequently at low temperatures stretching usually proceeds much more uniformly along the whole crystal than at temperatures at which, owing to the reduction in stress, it is confined initially to the weakest portions of the crystal which experience the first extension (Fig. 121).

As a rule the glide bands become coarser as the temperature rises (Fig. 122 compared with Fig. 35a). Quantitative statements regarding the mean thickness of the glide packets are not available. Difficulty arises from the circumstance that, with increasing magnification, new glide lines becomes visible.   In the case of regular zinc

FIG. 122.—Zn Crystal Extended at 300° C. (252).

polyhedrons which were grown from vapour, values were obtained which fluctuated around 0·8 μ., or a multiple thereof (267).   Tin crystals usually show glide bands; sometimes, however, they appear perfectly smooth in spite of substantial deformation (extension up to ten times the initial length); the thickness of the glide packets must then be below microscopic resolution (261).   Reduction of the

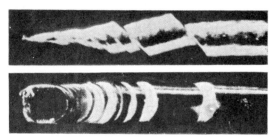

FIG. 123.—Formation of Coarse Glide Packets and of a Shear Fracture in the Course of Very Slow Extension of a Sn Crystal (see 260).

speed of straining, like an increase in temperature, produces coarse glide bands.   Fig. 123 contains two illustrations of zinc crystals which have been stretched extremely slowly.

The yield–stress curves obtained for *hexagonal* metal crystals at different temperatures are shown in Fig. 124.   They reveal a very pronounced influence of temperature on the shear hardening of the basal plane.   The hardening coefficient derived from the slope of the yield–stress curve changes 400-fold within the temperature range investigated. This dependence is most marked at intermediate temperatures; it decreases rapidly at both high and low tempera-

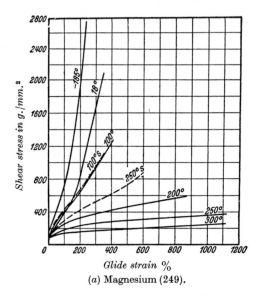

(a) Magnesium (249).

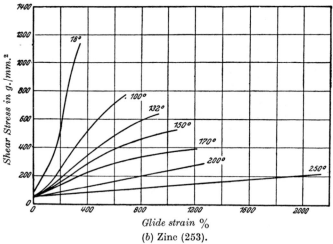

(b) Zinc (253).

tures, and becomes insignificant where these are extreme. For very low temperatures this is apparent from Figs. 117 and 120 (cadmium crystals).

The amount of basal glide that can occur at different temperatures, which in the case of magnesium is limited by shear fracture along the basal plane, or by the start of pyramidal glide, and in the case of zinc and cadmium by secondary glide in twin lamellæ (see Section

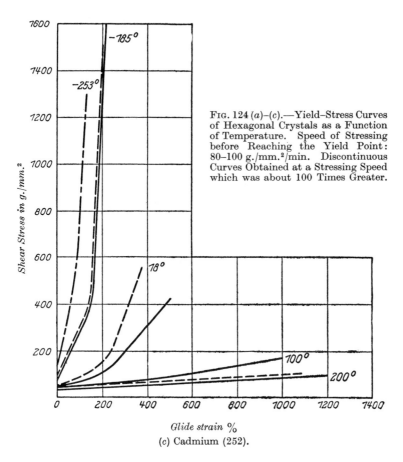

Fig. 124 (a)–(c).—Yield–Stress Curves of Hexagonal Crystals as a Function of Temperature. Speed of Stressing before Reaching the Yield Point: 80–100 g./mm.²/min. Discontinuous Curves Obtained at a Stressing Speed which was about 100 Times Greater.

Glide strain %
(c) Cadmium (252).

44), is again well expressed by a limiting shear stress which is independent of orientation. Thus within a wide range of temperature the mean yield–stress curves for basal glide are characterized not only by their starting point and slope, but also by the point at which they end. The relationship between temperature and the limiting values for shear stress and glide strain is shown in Fig. 125. In the temperature range investigated extension by basal glide increases by an order of magnitude, while at the same time hardening decreases equally rapidly. This entails even with the flattest yield–stress

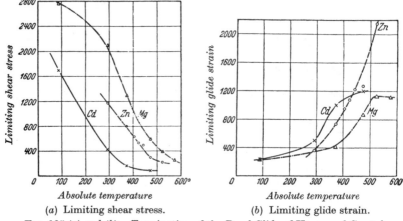

(a) Limiting shear stress.   (b) Limiting glide strain.

FIG. 125 (a) and (b).—Termination of the Basal Glide of Hexagonal Crystals as a Function of Temperature.

curve an increase in the shear stress during extension of three times the original value.

While the slope of the yield–stress curve and of the limiting glide

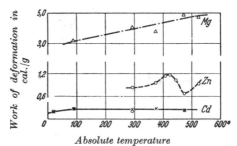

FIG. 126.—Work of Deformation Required for Basal Glide as a Function of Temperature.

(For Cd : × = slow fracture, in about 20 min.; ○ = rapid fracture in about 10 sec.)

strain and shear stress markedly depend upon temperature, the *work of shear* for the three hexagonal metals for basal glide is found to be *approximately* independent of temperature (Fig. 126). Zinc crystals, however, show a flat maximum at 150° C. It is not known for certain whether this is bound up with a possible deviation from pure basal glide which may occur in this range of temperature. The magnitude of the mean shear energy amounts to 4·4 cal./g. for magnesium (that is 18 times the specific heat, or 1/10th of the latent heat of fusion); for zinc it is 0·97 cal./g. (11 times the specific heat, or 1/24th of the heat of fusion), and 0·24 cal./g. for cadmium (4 times the specific heat, or 1/44th of the latent heat of fusion). This purely empirical finding can hardly represent a limiting condition, because in the course of deformation the greater part of the added energy is transformed into heat and is therefore no longer present in the crystal at the conclusion of glide. It would be more reasonable to suppose a *saturation* with internal energy, which in the cases described here, of course, would have to amount always to the same fraction of the total work of deformation. (For further details see Section 60.)

In Fig. 124, *a* and *c*, the dependence of the yield–stress curve on time at different temperatures is also shown by curves obtained at a one hundred-fold speed of deformation. It is found that within the range in which the yield–stress curve is very dependent upon temperature there is also a marked dependence upon speed; in the case of magnesium, for instance, the shear stress after 600 per cent. glide at 250° C., in the crystal that has been stretched rapidly, is three times greater than in that which has been stretched slowly.[1] Within the range in which the yield–stress curve is less dependent upon temperature the speed of deformation is also without considerable influence. The work of shear, which is independent of temperature, is also largely independent of the speed of deformation, as is seen in the case of cadmium in Fig. 126.

In the extension of zinc and cadmium crystals represented in Fig. 124, *b* and *c*, the primary basal glide is restricted by the start of a new basal glide within the twin lamellæ formed towards the end of primary glide. If the temperature exceeds the maximum values given in the diagram, then the normal course of basal glide is disturbed by recrystallization during the extension. At temperatures above 300° C., cadmium crystals break transversely without considerable elongation. With zinc crystals, decrease of temperature leads at −80° C. to pronounced reduction of the ductility and to the appearance of cleavage fracture along the basal plane—a phenomenon

[1] Great importance therefore attaches to the constancy of test speed in certain temperature ranges if the tests are to be comparable. In accordance with the crystallographic process of glide, the value to be maintained constant is the rate of glide strain which is independent of the initial position and of the change in orientation in the course of stretching (256).

which is observed at all lower temperatures. (For further details
see Section 53.)

Only approximate details are available for *tin* crystals. While the
slope of the yield–stress curve remains more or less constant, the
limiting glide strain falls appreciably with rising temperature (up
to 200° C.). In this case the range within which the hardening

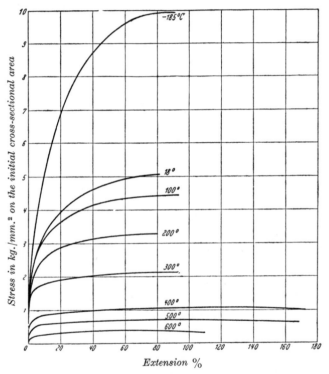

Fig. 127.—Stress–Strain Curves of Al Crystals in Relation to
Temperature (257).

coefficient is only slightly dependent on temperature is reached
already between 20° and 200° C. (262).

The relationship between temperatures and the stress–strain
curves of *aluminium* crystals is given in Fig. 127. From these it
will be seen that hardening is greatly reduced as temperature rises,
but that the total extension between −185° and 300° C. remains
noticeably constant. This clearly distinguishes aluminium crystals
from the hexagonal crystals, which, in general, experience sub-

stantial extension at higher temperatures. The increase in the extension of aluminium crystals at 400° and 500° C. is followed at 600° C. by a substantial reduction, due to the premature elimination of primary octahedral glide by the emergence of a new (cubic) glide. Recrystallization was observed in this case only at the highest temperatures, and even then only at the apex of the extension cone in the neck.[1]

The relationship between temperature and the yield–stress curve based on primary octahedral glide (glide along the system which first becomes operative) is shown in Fig. 128. While at the start of glide the influence of temperature is only small, here, too, the effect of temperature on the increase of the shear stress becomes very marked as the strain increases.

The influence of the *test speed* on shear hardening has been studied at room temperature only (256). The yield–stress curve obtained in dynamic stressing at a 23,000-fold speed (450 per cent. elongation per second) exhibits shear-stress values

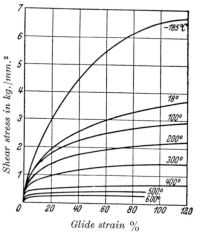

FIG. 128.—Yield–Stress Curve of Al Crystals in Relation to Temperature (257).

which are approximately 16 per cent. higher than for the static curve. A test which was carried out at an exceptionally reduced speed of loading (test period : 234 days) is described in (258).

Particulars of the qualitative influence of time on glide are available also for α-iron (259), tin (261) and bismuth (250). In all cases an increase of the speed of testing leads to more pronounced hardening and to a reduction of plasticity. This is particularly noticeable

[1] The lattice position in which cubic glide undoubtedly starts enables us to estimate its yield stress relatively to that of an octahedral system, as was mentioned in Section 42. It is true that this will apply to extended crystals only; the corresponding ratio for the intact initial crystal need not be the same. This estimate is based on the fact that, in the end position [112] towards which octahedral glide tends, cubic glide occurs, and leads to a new end position [111]. Consequently, the shear stress in the cube plane in the direction of a face diagonal exceeds that of an octahedral glide system by only about 15 per cent.

in the case of the tin crystal, owing to the simultaneous operation
of several glide systems in rapid stretching. The same effect is
produced by a reduction of the test temperature to $-185°$ C.

49. Crystal Recovery

In the previous section we dealt with the significance of temperature and time in the plasticity of crystals, and we gave particulars
of the yield–stress curves in relation to temperature and speed of
deformation. In so far as an influence exists, it takes the form of a
reduction of the hardening with rising temperature or diminishing
speed of deformation, for a given deformation. The experiments

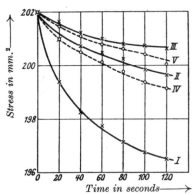

FIG. 129.—Flow Curves of a Sn Crystal
after Hardening and Recovery (261).
Between Tests III and IV the Unloaded
Crystal was Heated for 1 Minute at
60° C.

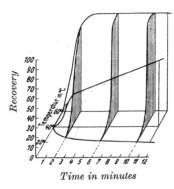

FIG. 130.—Recovery Diagram of a
Sn Crystal (250).

which we are about to describe, and which have led to the conception of " crystal recovery ", illustrate even more directly the
softening effect of temperature and time [(264), (266), (265)]. In
addition, they clearly reveal the close connection between these two
factors.

Fig. 129 contains flow curves of a tin crystal after repeated stretching at the same stress, and after a rest of 1 minute at 60° C. The
curves were recorded in a Polanyi filament-extension apparatus
(see Fig. 75). All extensions took place below the yield point of the
crystal. The initial reduction in the mean speed of flow reveals the
hardening which accompanies the deformation ; while the weakening
effect of intermediate annealing is shown by the resulting increase in

flow. This weakening, which is not a result of recrystallization or a change in orientation of the crystal, can be modified within a wide range by changing the conditions of annealing. The measure of the softening (e) is obtained from $e(\%) = \dfrac{v'_0 - v_e}{v_0 - v_e} \cdot 100$, where v_0 represents the original flow speed, v_e the flow speed of the hardened crystal, and v'_0 that of the recovered crystal. In the limiting case of $v'_0 = v_e$ there is no recovery; if, on the other hand, the original flow speed is again reached ($v'_0 = v_0$) recovery has been 100 per cent. A three-dimensional diagram, based on such tests, which shows the recovery of zinc crystals as a function of time and temperature, is given in Fig. 130. Recovery can be considerable even at room temperature over sufficiently long periods; it becomes perceptible after 10 minutes and amounts to about 50 per cent. after 20 hours. Analogous flow tests proved that *bismuth* crystals also possess a marked capacity for recovery (250).

An even more striking demonstration of the softening effect of time and temperature is shown in Fig. 131, which shows the exten-

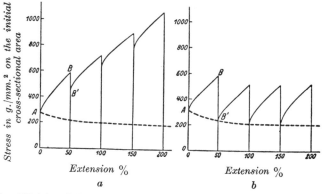

FIG. 131 (*a*) and (*b*).—Extension of a Zn Crystal; Tests Interrupted for 30–40 sec. in (*a*), and for 24 hours in (*b*), to Permit of Recovery.

sion curves of a *zinc* crystal. After each extension of 50 per cent. the load was removed from the crystal for 30–40 seconds in the one case, and for 24 hours in the other. The softening which occurred during the interval is expressed in the reduction of the stress necessary for further extension (from B to B'). The broken curves in the diagrams represent the extension curves calculated theoretically on the assumption of constant shear stress of the glide system (formula

43/1). It will be seen that the *short* pause resulted in only a slight softening, whereas in the 24-hour interval the strongly increased yield stress of the basis reverted to its original value. Therefore the crystal recovered completely during this period. The measure of recovery in these tests is given by the expression

$$e(\%) = \frac{S_1 - S'_1}{S_1 - S_0} \cdot 100,$$

S_0 being the initial yield stress of the unstressed crystal (Point A), S_1 the yield stress at the close of the first stage of extension at 50 per cent. (Point B), and, finally, S'_1 the yield stress of the pre-worked and recovered crystal (Point B').

The occurrence of recovery in *aluminium* crystals is clearly apparent in creep tests (flow under constant load) at elevated temperature (250° C.). In this case the extension of the crystals proceeds in three different stages. Upon application of the load the crystal at first flows rapidly, and in so doing forms numerous glide bands; then, as hardening begins, there follows a stage of small deformation in which only a few new glide bands occur, while finally, as a result of crystal recovery, rapid flow occurs which leads to fracture (258).

Tungsten crystals, too, exhibit very distinct recovery after suitable heat treatment (at least 10 minutes at 2100° C.). This was revealed by the considerable reduction in the ultimate tensile stress of crystals that had hardened by drawing through diamond dies. The original softness of the crystals was restored by this heat treatment, indicating that here, too, the shear stress in the glide systems had been reduced (266).

Although the direct experimental investigation of crystal recovery has been limited to a few metals only, there can be no doubt that we are faced here with a very general phenomenon which is technically of great importance. In summing up it may be said that the essential feature of crystal recovery is that, while preserving the lattice, it *continuously* reduces, during heat treatment, the mechanical properties of the work-hardened crystals approximately to the initial values. Consequently, crystal recovery contrasts sharply with recrystallization which involves the formation of new grains and reduces the properties of the hardened metals down to their original values *in stages* (cf. Section 65).

Hitherto the weakening effect of recovery has been discussed in relation to the resistance of the crystals to plastic deformation; it will be shown later that recovery influences also their tensile strength.

50. *Glide Regarded as the Superimposition of Recovery upon an Athermal Fundamental Process*

In the previous sections the influence of temperature and time on the progress of glide have been discussed. The main facts which have been established, chiefly with hexagonal crystals, can be summed up as follows : the critical shear stress which characterizes the start of glide is only slightly dependent upon temperature (and the speed of loading). There is no reason why it should not be assumed that the shear stress in the glide system at absolute zero is of the same order of magnitude as at elevated temperatures.

In marked contrast to this limited dependence on temperature of the initial yield stress, the increase in the shear stress in the course of glide (hardening) depends very strongly on the temperature, at any rate within certain ranges. At very low temperatures both the start and the slope of the yield–stress curve are practically independent of temperature. On raising the temperature a range is reached in which the hardening is appreciably reduced as the temperature increases; and finally the hardening curve, which now rises only very slightly, again becomes almost independent of temperature.

The speed of deformation has a very similar influence on the yield–stress curve : there is an insignificant effect at very low temperatures and near the melting point, but a very strong influence exists in the intermediate range.

The following physical description of crystal glide is suggested by these facts (251). The basic process is athermal, *i.e.*, independent of time and temperature. It operates undisturbed at very low temperatures, and finds expression in a characteristic function of the material, the yield–stress curve, which indicates that the low yield stress of the glide system of the unworked crystal greatly increases with increasing deformation. Upon this basic process, which is independent of temperature, there is imposed, as the temperature rises, a thermal-softening process : *crystal recovery.* In general, therefore, two fundamentally distinct processes are simultaneously involved in the deformation of a crystal by glide : the one concerns the actual crystal and is structurally determined ; the other is based on the thermal vibrations of the lattice. As temperature rises, the thermal recovery increasingly cancels out the hardening caused by glide, leading ultimately in extreme cases to a deformation which is free from hardening and which takes place at a constant shear stress of the glide system.

From the above it will be immediately apparent that an increase of time, *i.e.*, a reduction of the speed of testing, has the same effect upon the softening process as an increase of temperature. This is a plausible explanation of the influence which speed exerts on the yield–stress curve.

C. THE DYNAMICS OF TWINNING

51. *The Condition of Twinning*

Whereas glide could be characterized dynamically by the critical shear stress and the yield–stress curve, similar quantitative principles governing mechanical twinning are not yet available. As already explained, the formation of a deformation twin represents, contrary to glide, a process that is also macroscopically discontinuous. Fig. 132*a* illustrates this by the extension curve of a cadmium crystal (Polanyi filament-stretching apparatus), which clearly reveals the formation of twin lamellæ (accompanied by audible crackling) resulting from discontinuous changes of load. The twins themselves are usually narrow lamellæ which often do not traverse the entire crystal; and from these lamellæ initially only occasional small bands develop, which sometimes widen as the stress increases, or are replaced by the formation of new twins. During the formation of a twin lamella a further increase of load is unnecessary. It will suffice, therefore, for a dynamic characterization of twinning if an account is given of the conditions which govern its *initiation*. The reasons why we are unable at present to make any statement regarding this " critical " state are to be sought on the one hand in the very pronounced sensitivity of twinning to inhomogeneities within the crystal, as a result of which the deformation twins always develop in a more or less extended stress range. On the other hand, the attempt to formulate quantitatively the conditions which govern twinning is greatly hampered by the circumstance that, as a rule, the crystals also exhibit glide, in which event, of course, twinning cannot be studied in the intact initial state but only in an already hardened state. An example of this will be found in the extension curves of variously oriented cadmium crystals shown in Fig. 132, *b* and *c*.

The few indications at present available for the initial law originate from alternating-torsion tests. These tests, which were carried out on zinc and bismuth crystals, would appear to indicate that a single initial law based on a limiting stress, such as governs glide, does not exist for mechanical twinning [(268), (269)]. It is thought that twinning obeys an energy condition for which the path

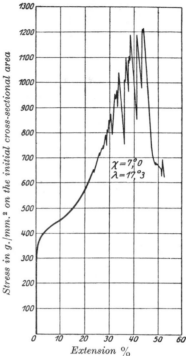

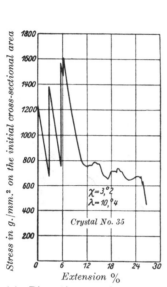

(a) Discontinuous change of load in the formation of deformation twins.

(b) Twinning after substantial glide.

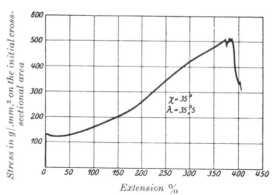

(c) Twinning after substantial glide.

FIG. 132 (a)–(c).—Stress–Strain Curves of Cd Crystals (270).

of the atoms from their starting points to their final position is also of importance. One fact does emerge, however, from the scanty material so far available regarding mechanical twinning of metal crystals, namely, that as with glide (see Section 42), the shear of the two neighbouring lattice planes is not essentially elastic. The elastic shear value which is reached at the onset of twinning is many orders of magnitude below that of the shear resulting from twinning. Let the cadmium crystal serve as an example : the reciprocal shear modulus in the twinning system is certainly not greater than 60×10^{-13} cm.^{2}/dyne, the maximum value for $1/G$. The shear stress needed to produce twin lamellæ in the twinning system is also substantially smaller than 500 g./mm.$^2 \cong 5 \times 10^7$ dyne/cm.^{2} (270), so that the elastic shear attained at the moment of twinning is

FIG. 133.—Fe Crystal Extended at $-185°$ C., Showing Deformation Twins. $K_1 = (112)$ (269a).

certainly smaller than 3×10^{-4} against $s_{Cd} = 0.175$. Differences of the same order occur with other metals which form deformation twins.

Temperature appears to influence twinning even less than it affects the start of glide, for twinning is superseded by glide at high temperatures (Zn, Cd), but occurs preferentially at low temperatures (α-Fe; Fig. 133).

52. *Mutual Influence of Twinning and Glide*

The width of the stress range in which mechanical twinning can occur in hexagonal crystals under static tensile stress depends to a very marked degree upon the orientation of the crystals, or rather upon the shape of the extension curve as determined by the orientation of the glide elements. This can be seen from Fig. 132, *b* and *c*, which shows that twinning starts early where the initial position of the basal plane is very oblique, and the extension curve therefore steep, but that it is restricted to a very narrow range just below the

maximum load if the position of the basal plane is rather transverse
and the extension curve consequently flat.

This phenomenon can perhaps be explained as follows : if, taking
as a basis the yield–stress curve which is valid for basal glide, we
calculate the extension curves in the true tensile stress–strain
diagram, the result shown for cadmium in Fig. 134 is obtained.[1]
If allowance is made for the change in orientation which accom-

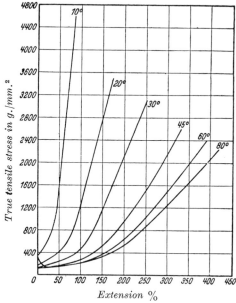

FIG. 134.—Calculated True Tensile Stress–Strain
Curves for the Basal Glide of Cd Crystals of
Various Orientations.

panies extension, it is also possible to deduce from these curves the
behaviour of the shear and normal stress in the preferred twinning
system. The curves obtained for these stresses exhibit in their
general course no substantial difference from the true stress curves.
In all three cases the high values are obtained with much smaller
extensions if the initial position of the basal plane is oblique than if

---

[1] The true stress $\sigma_{tr.}$ (referred to the current cross-section) is obtained from
the " conventional " stress σ referred to the initial section by multiplying by
$d = \dfrac{q_0}{q}$.   In order to obtain the equation of the true stress curves, the equation
(43/3) must be multiplied by d.

it is transverse. Consequently, each of the three most obvious assumptions : *i.e.*, that either the true tensile stress, or the shear stress, or the normal stress in the twinning system, is responsible for the lower limit of the stress range in question, is compatible qualitatively with the experimental data relative to the width of the stress range that leads to twinning. The experiments show further, however, that the range within which twinning occurs decreases with increasing χ_0 angle much more rapidly than would follow from either of the three assumptions. If the effect is attributed to an increase of the lower stress limit, it is then possible to say that *twinning is made more difficult* by previous *glide*.

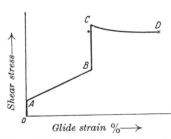

FIG. 135.—Sketch showing the Behaviour of the Resolved Shear Stress on the Basal Plane in the Course of Primary Glide (AB), Mechanical Twinning (BC) and Secondary Glide (CD) of Zn Crystals (272).

Only a few details are as yet available regarding the converse case of the influence of twinning on glide. An examination of zinc and cadmium crystals suggests that the shear stress of the basal plane experiences a considerable discontinuous increase as a result of twinning [(271), (272), (270), cf. also (273)]. This shear hardening by twinning amounts at room temperature to about 100 per cent. for zinc crystals, and about 200 per cent. for cadmium crystals. It is true that the resultant twin bands, especially in the case of cadmium, are usually very thin, so that it would appear possible that the mechanical properties are influenced by the size of the crystal investigated. Subsequent glide in the twin is accompanied by only a slight shear hardening on the basal plane, frequently even by a reduction of the shear stress (a softening during secondary basal glide occurred in nine out of twelve cases with zinc crystals, and in four out of twenty-three cases with cadmium crystals). A diagram showing the behaviour of the resolved shear stress of the basal plane of a zinc crystal is given in Fig. 135.

The fact that secondary glide on the basal plane is accompanied by only slight hardening (or even by some softening) furnishes a further example of the limitation of the hardening capacity of crystals. Analogous examples have already been noted in the glide of solid solutions (Section 46). Whereas in the case of solid solutions the small extent of the further increase of the yield stress was due to primary hardening by alloying, or to a substantial hardening of a latent glide system, in the present case it is the pronounced hardening of the basal plane due to twinning that is responsible for the reduction of its capacity for further hardening.

Scarcely any information is yet available on the question of *the influence of mechanical twinning on further twinning*—that is, whether a deformation twin or the initial crystal is more resistant to mechanical twinning. The answer depends on whether the plasticity which can be realized through repeated twinning is parallel or opposed to the direction of stress. It is also necessary to distinguish between the untwinning of a deformation twin and the development of a secondary twin which is distinct from the original crystal. Tests carried out on zinc sheets (274) show that it is possible, by bending backwards, to cause previously produced deformation twins to disappear; in this case, therefore, untwinning proceeds more easily than further twinning in the parent crystal. Similar observations have been made with antimony crystals (275). A more quantitative investigation of these relationships confirmed this result (276). In the course of an alternating bend test carried out on coarse-grained zinc sheets it was found possible to produce, and untwin, the same twin lamellæ as many as twenty times. Besides sudden broadening (or narrowing), in the course of these experiments a phenomenon was observed which, under the microscope, appeared to be a continuous change of the width of twin lamellæ with increasing (or again decreasing) bending moment. The moment required for the complete untwinning of a lamella is greater than that needed for its production. If the same lamella is repeatedly produced, weakening will be observed in so far as the required moment will diminish steadily. Consequently, if the test is repeated with equal bending moments, broader lamellæ will be produced. On the other hand, the bending moments required for untwinning appear to increase as the test is repeated.

This chapter shows that we have still a very great deal to learn about the dynamics of mechanical twinning.

D. FRACTURE ALONG CRYSTALLOGRAPHIC PLANES

Hitherto we have been concerned with crystal *deformation* processes, a consideration of which included glide and shear fractures. In the present section we shall discuss that type of rupture which consists of a fracture of the crystal along plane surfaces which coincide invariably with one of the cleavage planes listed in Table VI

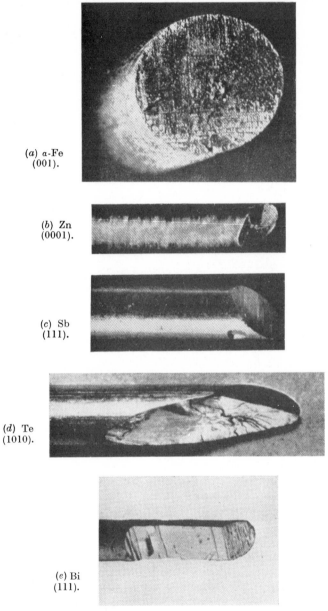

(a) α-Fe
(001).

(b) Zn
(0001).

(c) Sb
(111).

(d) Te
(1010).

(e) Bi
(111).

FIG. 136 (a)–(e).—Fracture (Cleavage) Planes of Metal Crystals.

(see Fig. 136). This fracture either occurs without any previous perceptible deformation (brittle crystals), or it terminates a more or less large deformation the magnitude of which substantially affects the value of the tensile strength.

Nothing is yet known regarding the period of time covered by the process of rupture along cleavage planes. Although as a rule fracture occurs rapidly, it is, nevertheless, certain that the speed of the operation is finite (see Section 69). Markings on the cleavage planes and the occasional presence of flaws in the crystal fragments indicate that the process of fracture often consists in the gradual deepening of cracks.

53. Conditions Covering Brittle Fracture. Sohncke's Normal Stress Law

Tensile tests carried out on rock-salt crystals of various orientation led Sohncke to formulate the Normal Stress Law (277). According to this law these crystals break when a certain critical normal stress has been reached in the cubic-cleavage plane which is invariably the surface of fracture. The fracture strength therefore depends strongly upon the angle (χ) between cleavage plane and the direction of the tension. If N is the critical normal stress, equation (40/2) gives for the fracture strength σ the expression $\sigma = \dfrac{N}{\sin^2 \chi}$ (53/1). (For further particulars of experiments with rock salt see Chapter VII.)

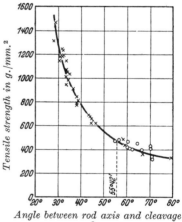

Angle between rod axis and cleavage plane

FIG. 137.—Tensile Strength of Bi Crystals as a Function of Orientation (284).

$\times$ = at − 80° C., $\bigcirc$ = at 20° C.

Systematic experiments with metal crystals have shown that here, too, the occurrence of brittle fracture along smooth cleavage planes can readily be brought within the scope of the normal stress law. For example, *bismuth* crystals fracture in a brittle fashion at low temperatures and also at room temperature within a certain orientation range. Fig. 137 shows the extent to which the fracture strength of bismuth crystals depends upon the position of the (111) plane which functions as cleavage plane (basal plane in the hexagonal representation). The curve calculated according to equation (53/1) on the assumption of a constant normal stress on the cleavage plane, agrees with the observations. Since the (111) plane is also the best glide plane of the bismuth crystal, the lattice positions which at 20° C. separate the area of brittle fracture from that of incipient

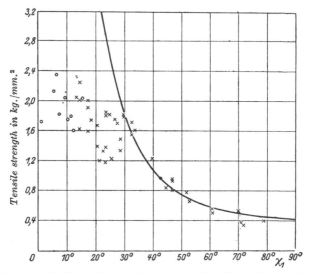

FIG. 138:—Tensile Strength of Te Crystals as a Function of Orientation (283). χ_1 angle between tensile direction and the most transverse prism plane type I.

glide can be calculated from the critical shear and normal stress of this plane. From the condition

$$\sigma = \frac{S}{\sin \chi \cdot \cos \lambda} = \frac{N}{\sin^2 \chi} \quad \text{the relation} \quad \frac{\sin \chi}{\cos \lambda} = \frac{N}{S}$$

is obtained for the boundary of the areas of glide and of fracture. With the aid of the critical stress values for bismuth we arrive at a limiting angle of 55° 40′, for $\chi = \lambda$. Where the initial position of the (111) plane is more transverse, it is the critical normal stress that will be reached first, while with more obliquely oriented crystals it will be the critical shear stress.

Results similar to those with bismuth were obtained with tellurium crystals, which exhibited tensile-strength differences of 1 : 4. The cleavage plane here is one of the (10$\bar{1}$0) planes (prism planes type I).

The tensile strength increases, therefore, as the direction of tension approaches the hexagonal axis. Down to an angle of inclination of 29° for the most transverse ($10\bar{1}0$) plane the normal stress law continues to apply (Fig. 138). If the position of the cleavage plane becomes still more inclined, the strength of the material [1] is exceeded before the critical normal stress is reached. In this range of orienta-

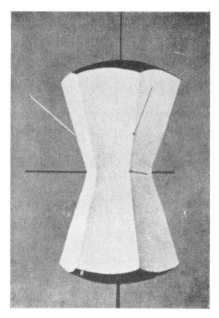

Fig. 139.—Tensile Strength Surface of the
Te Crystal (283).

tion the tensile strength remains noticeably constant at approx. 1·8 kg./mm.² A spatial representation of the dependence of the tensile strength of tellurium crystals upon the orientation is given by the surface illustrated in Fig. 139.

Zinc crystals, which at 20° C. and at higher temperatures do not as a rule fracture along smooth crystal surfaces, exhibit at −80° C. (and at lower temperatures) smooth cleavage fractures along the basal plane. This fracture is mostly preceded by glide along the basal plane, which at −185° C. can amount to a glide of approx. 200 per cent. Therefore the basal normal stress at which fracture

[1] *Translator's footnote.* With respect to fracture processes other than cleavage along the ($10\bar{1}0$) plane.

172   Plasticity and Strength of Metal Crystals

occurs (obtained from experiments) does not usually apply for the undeformed initial state. It may depend very largely on the extent of previous deformation, as will be explained in greater detail in Section 54. At −185° C., however, this dependence was found to be only slight, so that the tests at this temperature can be discussed

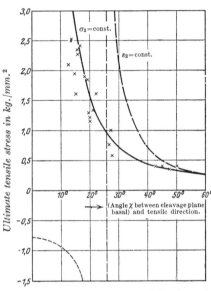

here (278). Fig. 140 represents the tensile strength as a function of the angle of the cleavage plane. The observations again agree satisfactorily with the curve calculated on the assumption of the normal stress law. In particular, these tests with zinc reveal substantial disagreement between the results of the tests, and a fracture condition of constant *normal dilatation* perpendicular to the cleavage plane. A discontinuous line in Fig. 140 indicates the course of the tensile strength resulting from this assumption. Agreement between the two statements is assumed for the transverse position of the basal cleavage plane. It will be seen that constancy of the normal

FIG. 140.—Tensile Strength of Zn Crystals at −185° C. as a Function of Orientation [(278), (282)].

————: constant normal stresses : $\sigma = \dfrac{\epsilon_2}{\sin^2 \chi}$;

– – –: constant normal dilatation :

$$\sigma = \frac{\epsilon_2}{s_{11} \cos^2 \chi + s_{33} \sin^2 \chi}$$

dilatation would demand a very much more pronounced dependence of the tensile strength upon orientation. For a position angle of $\sim 26° \left(tg\chi = \sqrt{-\dfrac{S_{13}}{S_{33}}} \right)$ an infinitely great tensile strength is obtained (dilatation resulting from longitudinal extension, and reduction of area resulting from transverse contraction of the test bar, are here equal to each other) and for a still more oblique position of the cleavage plane normal dilatation becomes possible only under compressive stress. Whereas in the case of the initial condition for glide, constancy of the shear stress can at present be described only as much more probable than constancy of strain, in the case of brittle fracture of metal crystals the balance is clearly in favour of the constant-stress law.

The zinc crystal affords also an opportunity for determining the critical normal stress of the second-best cleavage plane, the prismatic-plane type I, which appears as cleavage plane when the basal positions are very oblique. At $-185°$ C. á value is obtained for this which is about ten times greater than for the basal plane; it is measured, of course, after slight basal glide. There is an order-of-magnitude preference for the basal plane, not only for glide but also for cleavage fracture.

Taking into account the critical normal stresses of basal and prismatic-plane type I it is now possible to calculate generally the

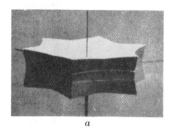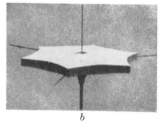

a   b

Fig. 141 (a) and (b).—(a) Tensile-strength Surface, and (b) Fracture Surface, of the Zn Crystal at $-185°$ C. (278).

dependence of the tensile strength upon orientation. The result is represented by the surface illustrated in Fig. 141a, which gives the tensile strength corresponding to various orientations and clearly reveals the strong influence of crystal orientation. A surface representing the tendency to fracture in a wider sense is shown in Fig. 141b. It connects the *initial* orientation of the crystal with its tensile strength with respect to brittle cleavage along the basal or prism planes, and (in far the greater part of the field of orientation) with its yield point with reference to extension by basal glide. This is given by the constant critical value of the shear stress in the glide system (Section 40).

For bismuth, too, the second-best cleavage plane—the rhombohedral plane $(11\bar{1})$—could be characterized quantitatively by means of its critical normal stress (285). It is true that, with the orientations investigated, mechanical twinning always started before fracture occurred, as could be detected from narrow striations on the cleavage plane. The value obtained for the critical normal stress of the rhombohedral plane, which is more than double that of the (111) plane, can therefore be taken only as an approximation.

Only scanty experimental information is so far available regarding the dependence of the critical normal stress on temperature, and in any case we have seen that such dependence can be only slight. Thus the tensile strengths for bismuth at 20° and −80° C. can be represented by the same value of the critical normal stress of the basal plane (Fig. 137). And even for the zinc crystal there appear to be no noticeable differences in the basal normal stress within the range −80° to −253° C. (Section 54).

TABLE XVI

Critical Normal Stress of Metal Crystals

Metal.	Method of producing the crystal.	Cleavage plane.	Temperature, ° C.	Critical normal stress at fracture, kg./mm.²	Literature.
Zinc (0·03% Cd) .	Drawn from the melt.	(0001)⎫ (10Ī0)⎭	−185	0·18–0·20 1·80	(278) (281)
Zinc + 0·13% Cd		(0001)	−185	0·30	⎫ (280)
Zinc + 0·53% Cd		(0001)	−185	1·20	⎭
Bismuth . .	Drawn from the melt and solidified *in vacuo*	(111)	+ 20 − 80	⎫ 0·324 ⎭	(284)
Antimony . .		(111)⎫ (11Ī)⎭ (11Ī)	+ 20 + 20	0·29 0·69 0·66	⎫ (285) ⎭
Tellurium . .	Solidified slowly	(10Ī0)	+ 20	0·43	(283)

Scarcely anything is known of the influence of metals in *solid solution* on the critical normal stress. In the case of zinc crystals alloyed with cadmium there appears to be quite a substantial increase in the fracture strength (280).

A summary of the critical normal stresses so far measured is given in Table XVI. In addition to the examples already mentioned, zinc (fracture along basal and prismatic planes, alloying effect), bismuth (fracture along basal and rhombohedral planes, influence of temperature), and tellurium, the table contains an approximate value for the critical normal stress of the rhombohedral plane of antimony. It should be particularly noted that the values for the normal stresses are less reliable than those for the critical shear stresses. This is partly due to the difficulty of ensuring correct gripping of a cylindrical specimen when testing brittle cleavable crystals. The magnitudes of the normal strengths of the cleavage

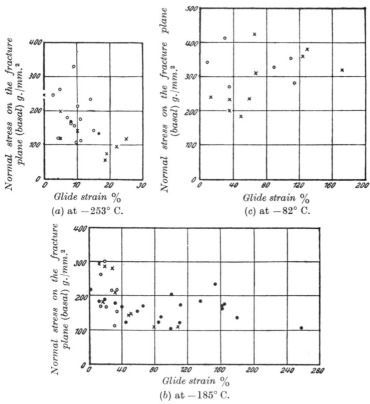

FIG. 142 (a)–(c).—Influence of Previous Glide on the Critical Normal Stress on the Basal Plane of Zn Crystals at Various Temperatures (281).

planes are of the same order as those of the shear strengths of the principal glide systems (Table IX).

Further tests are necessary if the significance of the critical normal stress in a quantitative description of *cleavability* is to be elucidated.

54. *Influence of Previous Deformation upon the Critical Normal Stress.* " *Strain Strengthening* "

It was pointed out in the previous section that the fracture of metal crystals on cleavage planes is often preceded by plastic deformation. This is specially obvious in the case of zinc crystals.

The effect of this deformation on the critical normal stress of the basal plane is seen from Fig. 142. Although the results of the individual tests are scattered, the following facts do at least emerge. As was pointed out in Section 53, at −185° C. the influence of preceding glide is only slight; at −82° C. there is a marked increase, at −253° C. a pronounced decrease in the critical normal stress with increasing glide, which in this case just exceeds 20 per cent. This shows that, according to the temperature, the normal strength of the glide plane can be increased or reduced by preceding glide. The influence of strain hardening, due to deformation, on the critical normal stress of cleavage planes (" strain strengthening ")

TABLE XVII

Strain Strengthening and Recovery of Zinc Crystals (279)

Extension, %.	Fracture strength at −185° C.		Recovery, %.
	Immediately after extension, kg./mm.².	After recovery for 2 minutes at 80° C., kg./mm.².	
50	2·6	—	—
100	3·2	—	—
160	5·5	3·4	38
300	7·3	—	—
350	8·2	—	—
380	8·4	6·2	26
400	9·6	5·4	44
500	9·8	—	—

can therefore be negative, in which case it represents a weakening. This contrasts remarkably with the effect of strain on the yield stress of glide systems. In the latter case there is nearly always a hardening (shown by the yield–stress curve) which increases with decreasing temperature and finally reaches a limiting value (see Section 48). If, at the yield point of zinc crystals, normal stresses of about 500 g./mm.² on the basal plane at 20° C. do not lead to fracture, this may perhaps be attributed to a strain strengthening which is caused at this temperature by the slight extension which occurs before the yield point is reached.

With the prismatic cleavage plane of zinc crystals as well as the basal plane there appears at −185° C. to be no marked influence of the basal glide which precedes fracture. For extensions between 11 and 48 per cent. the critical normal stresses scatter unsystematically between 1330 and 2130 g./mm.² (278). On the other hand, considerable basal glide at room temperature has a very pronounced influence on the normal strength of the prismatic plane (determined at −185° C. immediately after stretching). The varying amounts of extension were achieved through the use of crystals of different initial orientations. Table XVII gives the fracture-strength values obtained, to which the critical normal strengths of the prism planes are proportional owing to the very similar lattice position in stretched crystals. The increase of the strain strengthening of the prism plane with increasing extension is very clear.

From a knowledge of the behaviour of the critical normal stresses for the basal and prismatic planes it is possible to understand tensile

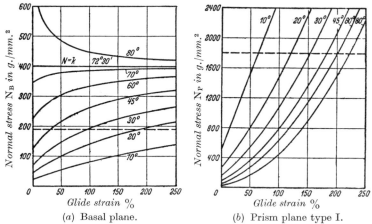

(a) Basal plane.   (b) Prism plane type I.

FIG. 143 (a) and (b).—Theoretical Course of the Normal Stress when Extending Zn Crystals of Various Orientations at −185° C. The Initial Angle between Basal Plane and Longitudinal Direction is Indicated for the Respective Curves.

tests with zinc crystals, and, in particular, the deformation which precedes fracture at those temperatures where fracture occurs in one of the cleavage planes. In Fig. 143 is plotted the increase, during the tensile test, of the normal stress on the basal plane and on the prismatic-plane type I which lies most transversely, deduced from the yield–stress curve for crystals of various initial orientations at −185° C.[1] If the critical normal stresses (N_B and N_P for the two

---

[1] N_B was calculated with the aid of the relationship

$$N_B = \frac{S_0 + ka}{a + \cot g \, \chi_0} \quad (S_0 = 126 \text{ g./mm.}^2, \, k = 400 \text{ g./mm.}^2),$$

which is easily deduced from the formulæ (40/2) and (43/3) : while N_P was calculated from the equation $N_P = \cos^2 30 \, (S_0 + ka) \cdot (a + \cot g \, \chi_0)$.

cleavage planes are independent of the glide strain, as is more or less the case at this temperature, then the limiting glide strain which is achievable for crystals of various initial orientations is obtained by intersection of the series of curves with the horizontal straight lines at distance N_B or N_P from the abscissa axis. It will be seen that where the initial positions of the *basal plane* are oblique, its critical normal stress (180 g./mm.^{2}) is obtained only after considerable glide. Fracture would have to be preceded by a very substantial deformation. It has been found, however, that even after relatively small deformations the critical normal stress (1800 g./mm.^{2}) for the *prismatic plane* is reached. These crystals fracture therefore by cleavage along the prism plane.

Beyond an χ_0 angle of approximately 25° this prismatic cleavage is then replaced by fracture along the basal plane. Whereas formerly the glide strain increased with increasing χ_0, it now begins to diminish. For a χ_0 angle corresponding to the equation $tg\chi_0 = \dfrac{N_0}{S_0}$ the critical normal stress is already reached at the yield point, so that for all χ_0 angles $>$ arc $tg\dfrac{N_0}{S_0}$ ($\sim$58°) brittle fracture of the crystals occurs.[1] The critical normal stress was here assumed constant; however, even if it depended on the preceding deformation the examination would follow a similar course.

Although the interpretation of the fracture process at low temperatures as one of shear fracture is contradicted by the appearance of the fracture surfaces of zinc crystals, it should be pointed out that this view is also untenable for another reason. If the onset of fracture resulted from the limiting shear stress, then this would always have to operate (owing to the yield–stress curve which is valid at low temperatures) upon reaching a certain point on this curve, that is, at a *constant limiting glide which is independent of orientation*. This does not agree with experimental results.

Apart from this case of the zinc crystal, which made possible an investigation of strain strengthening of an operative glide plane (the basal plane) and of a plane which crossed this (prismatic-plane type I), there are no other examples of the strain strengthening of metal crystals by glide. The question of the influence of mechanical twinning upon the normal strength of cleavage planes is still quite obscure.

[1] It should be noted that all curves in Fig. 143a have the straight line $N = k$ as asymptote. Whereas sooner or later fracture occurs in all orientations for values of the critical normal stress $N < k$, for $N > k$ a fracture about the plane under consideration is possible within only a restricted range of orientation, and then only if there has been no previous glide.

M. Polanyi (287) created the conception of strain strengthening and contrasted it with shear hardening. He observed strain strengthening on deformed tungsten and rock salt crystals (286). Only with rock salt is the surface of fracture a smooth plane. The strengthening of tungsten is assessed on the basis of the strength related to the final section.

Finally, it may be mentioned that in deformed crystals strain strengthening, like shear hardening, can be gradually removed by *crystal recovery*. Figures relating to this phenomenon, for the fracture of zinc crystals on the prismatic-plane type I, are contained in Table XVII. After extension at room temperature, and before fracturing at $-185°$ C., the crystals were heated for 2 minutes at $80°$ C. This produced a tensile weakening of 30–40 per cent. without recrystallization of the crystal. Just as deformation produces both strain hardening and strain strengthening, recovery leads to both a softening and a weakening of the crystal (decrease of critical shear and normal stress).

E. AFTER-EFFECT PHENOMENA AND CYCLIC STRESSING

Hitherto we have examined the behaviour of metal crystals under substantial plastic deformation. We will now describe those closely related phenomena which occur within a range of small deformation upon removal of the load or change of the direction of stress. In the first place it is a question of *after-effect* and *hysteresis* which, although usually termed elastic processes, must no doubt be regarded essentially as plastic. The Bauschinger effect will also be discussed in this context since it is probably due to the same causes. The available material relating to these phenomena in metal crystals is at present confined to a few experiments. A close enquiry into this behaviour will, however, certainly contribute greatly to our knowledge of crystal deformation and hardening.

Subsequently we shall describe investigations carried out on metal crystals in regard to the technically very important property of fatigue. In discussing these experiments we shall deal first with the external crystallographic features of the crystals resulting from cyclic stressing (deformation phenomena, crack formation), and then with the changes which take place in the mechanical properties as the stress increases.

55. *After-effect and Hysteresis.   Bauschinger Effect*

The elastic *after-effect*, which consists of an " after-shortening ", *i.e.*, establishment of equilibrium length as a function of time after removal of the stress, has been investigated on zinc and tungsten crystals (288). The crystal wires were twisted by certain angles (5–10° in the case of 10-cm.-long zinc crystals, 180° for the 50-cm.-long tungsten crystals) and then released. Whereas the tungsten crystals immediately regained their initial position, the zinc crystals retained a substantial permanent torsion. In both cases, however, contrary to comparative tests carried out with polycrystals, the final position was already reached 1 minute after removal of the load. No further changes occurred during the investigation, which extended over 36 hours, whereas in the case of polycrystalline wires over the same period there was a subsequent reverse deformation amounting to 40 divisions on the scale (corresponding to an angle of torsion ∼2°). These experiments show that the torsional after-effect of metal single crystals, if it exists, is very much smaller than with polycrystalline material.

Elastic *hysteresis* means that the shape of a solid is dependent not only upon the instantaneous value of the stress but also upon the stress–strain history of the material. So far investigation of the effect in single crystals has been limited to aluminium (291). Tensile loads were applied to the crystals slowly and then slowly removed, and the resulting changes in length were observed by means of the Martens mirror apparatus. The stresses were kept within the range 0·028–1·43 kg./mm.². The total extension after a large number of load cycles amounted ultimately to 2·2 per cent. The results of the tests can be summarized as follows : plastic extension already occurred at the first application of load, even within the minimum stress range of 0·028–0·144 kg./mm.²; therefore a range of a purely elastic deformation was not observed. Repeated loading and unloading, after the initial occurrence of permanent deformation, led ultimately to closed hysteresis loops similar to those observed with polycrystals. The existence of these loops is not compatible with a homogeneous elastic deformation of the lattice.

The Bauschinger effect consists in the observed fact that, after a test bar has been deformed, its resistance to deformation (elastic limit, yield point) is greater for further deformation in the same direction than for deformation in the reverse direction (289). In certain circumstances, even a weakening can be observed in the reverse direction. For single crystals the Bauschinger effect was studied on α-brass (72 per cent. copper). Cylindrical bars 9·2 mm. in diameter were subjected to tension–compression tests, the changes in the 10-mm. gauge length being observed by means of a Martens mirror apparatus. The results obtained with the single crystal substantially resembled those observed on the polycrystal. Fig. 144 shows the effect of previous compression on the limits of plasticity determined in the tensile test. The re-duction of these limits well below their initial values by increasing compression appears very clearly.

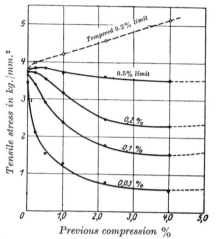

FIG. 144.—Effect of Previous Compression on the Limits of Plasticity of an α-Brass Crystal Subjected to Tensile Testing (290).

The importance of in-creasing the available ex-perimental data on this subject must again be emphasized. It is particu-larly important for our conception of the funda-mental processes involved to decide whether the freedom from after-effect and the occurrence of hysteresis in one and the same crystal material are really compatible with each other. An interpretation of the data presented here will be attempted in Section 76, while a discussion of these properties in relation to the technical polycrystal will be found in Section 82.

56. *Crystallographic Deformation and Cleavage Processes under Cyclic Stressing*

Experiments with cyclic stressing have been carried out on a large number of metal crystals. Usually the experiments have consisted of alternating-*torsion* tests on cylindrical specimens in which either the torque amplitude or the angle of torsion was maintained constant. Aluminium crystals were examined under tension–compression stress also. In the extensive series of tests by Gough the crystals were stressed in technical fatigue-testing machines. The specimens were here of the standard shape (cylindrical bar approx. 8 mm. in diameter, with thickened ends for clamping). The fatigue testing of thin crystal wires was carried out in suitably simplified machines, the angle of torsion being kept constant during the test. The number of stress reversals per minute in the various series of tests was 400–2300.

It should again be stated that an important result of these tests was the establishment of the fact that the same glide and twinning elements operate under alternating as under static stress. *Aluminium* clearly shows glide bands which, in full agreement with the effects

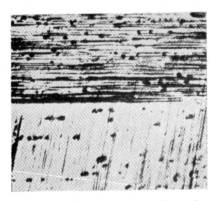

observed in tensile and compression tests, correspond to glide in the octahedral glide system as determined by the locally prevailing maximum shear stress [(292), (291)]. Therefore in alternating torsion the same glide system does not operate at all points on the circumference of the crystal. The transitions of the operative systems are often very clearly marked (see Fig. 145), and their positions always confirm the theoretical expectation.[1] The crystals

FIG. 145.—Transition of the Operative Glide Systems : Al Crystal Subjected to Alternating Torsion (292).

fracture by forming cracks which in the main run along the operative glide planes (see Fig. 146). This is shown clearly by an alternating-torsion test carried out under tap water. The areas of maximum deformation are indicated by the greatly increased corrosion (cf. Fig. 147). In this test the effects of deformation and corrosion reinforce one another, and the fracture, which, in this case, occurs much earlier, is found to have occurred almost exclusively along the operative glide planes. This observation is important, because

it shows that, contrary to the usual opinion, the cause of reduction of the fatigue strength under corrosive attack is not to be sought in the presence of stress concentrations at notches and holes (at any rate in so far as single crystals are concerned), but in the attendant crystallographic processes.

FIG. 146.—Fracture in an Al Crystal Subjected to Alternating Stress (291).

With a *silver* crystal, as in the case of aluminium, octahedral glide was observed after it had been subjected to torsional cycles. In this

[1] See Section 41 for a calculation of the shear stresses in the individual glide system.

case also the choice of the glide elements operative at various parts of the exterior of the crystal is determined by the condition of maximum shear stress; the resultant fracture again follows partially the operative glide planes. As with aluminium, mechanical twinning was not observed (295).

Magnesium crystals which had been subjected to alternating torsion exhibited both basal glide and the formation of deformation twins. The fractures are usually jagged and composed of several planes. Basal plane, prismatic plane $(10\bar{1}0)$, pyramidal planes $(10\bar{1}1)$ and $(10\bar{1}2)$, etc., have been observed as cleavage planes (296).

With *zinc* crystals, too, the same mechanisms of deformation

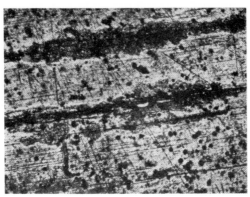

Fig. 147.—Fracture Produced in an Al Crystal by
Corrosion Fatigue (294).

usually appear in alternating torsion as in the static tensile test, namely, glide in the basal plane and mechanical twinning with $K_1 = (10\bar{1}2)$. Dependent on the crystal orientation, several types of cracks accompany the traces of these crystallographic deformations. They result from cleavage along the basal plane, the secondary basal plane in twin lamellæ and also along the twin plane. Part of such a crack, composed of consecutive primary and secondary basal planes, is illustrated in Fig. 148. With this are associated cracks which follow the prism planes. These, however, are not so much fissures as flat troughs enclosed within swellings, although it is true that, after appreciable deformation, cracking makes its appearance at the bottom of the troughs (see Fig. 149). The crystals then fracture either by smooth cleavage along a primary or secondary.

basal plane, or stepped fractures occur, in which, however, the basal plane is always involved [(298), (297), (299)].   Both the striations

Fig. 148.—Zn Crystal Subjected to Alternating Torsion :
Fracture along the Primary Basal Plane, and along a
Secondary Basal Plane in Twin Lamellæ (298).

observed on the surface of *cadmium* crystals subjected to alternating torsion, and the change of the shape of the sections of these crystals

Fig. 149.—Fracture along Prism-plane Type II
in a Zinc Crystal Subjected to Alternating
Torsion (298).

(rib formation), were found by analysing the state of stress to be basal glide (300).

An *antimony* crystal exhibited no glide bands in alternating torsion, but deformation twins appeared on all three (011) planes. Secondary twinning was observed within the twin lamellæ. Crack formation and fracture followed definite crystal planes, as was indeed to be expected with a crystal having such good cleavage as antimony. The first cracks ran parallel to the (111) plane in its original or twinned position. Fracture occurred finally by cleavage about a twin plane (301).

Glide could no more be detected in *bismuth* crystals subjected to alternating torsion than in antimony. Numerous deformation twins on the (011) planes made their appearance in the first stages of the test. The cracks were oriented mainly parallel to the twinned (111) plane. Fracture occurred by cleavage along the basal plane (111) or along a plane which cannot be expressed by simple indices (302).

In conclusion, attention must also be drawn to the important fact that cracking under cyclic stressing is not confined to the crystallographic cleavage planes, but can also occur on a large scale along operative glide planes. This phenomenon points clearly to a close connection between the fatigue fracture of the single crystals and the deformation by which it was preceded.

57. *Phenomena of Hardening under Cyclic Stressing*

The crystallographic processes which attend the cyclic stressing of metal crystals were described in the previous section. Attention has already been drawn to the dynamics of these tests when discussing the initial conditions of twinning (Section 51). The present section will deal with the changes in the mechanical properties of crystals brought about by cyclic stressing. It will be seen that these changes are of a very marked and special nature, and they reveal the profound influence exerted by this type of stressing.

It can already be concluded from the development in time of glide bands in *aluminium* crystals under alternating stress that a substantial *shear hardening* must take place (291). This hardening seems to be less pronounced in the principally affected octahedral systems, which are characterized by maximum shear stress, than in the others, since the glide which takes place initially on several systems is soon restricted to the one which is most favourably placed. The non-equality of hardening, therefore, is of the same kind as in static tensile tests (Section 43).

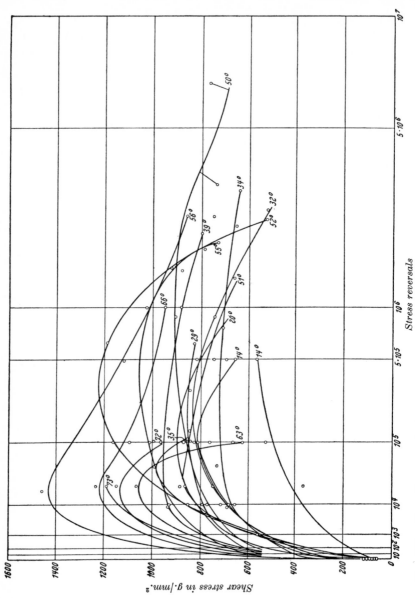

Fig. 150.—Shear Stress on the Basal Plane of Zn Crystals Subjected to Alternating Torsion (299).　Angle of the

Hardness measurements at various points on the cross-section of an *aluminium* crystal, which had been subjected to alternating torsion to the point of fracture, reveal clearly a hardening in those areas which had experienced strong deformation. The hardness values obtained by ball impressions exceeded by 30–40 per cent. those obtained in the central undeformed zone (292).

The changes in the mechanical properties of *zinc* crystals were investigated directly [(297), (299)]. Thin crystal wires (approximately 1 mm. in diameter) were subjected to alternating torsion with a constant angular amplitude over varying numbers of stress reversals, followed by static tensile testing. Up to seven test specimens could be taken from one original crystal. Fig. 150 shows the shear stress of the basal glide system after alternating torsion by 4° in either direction (gauge length 40 mm.) as a function of the number of stress reversals. From this it will be seen that in every case the yield stress increases after a few stress reversals to many times its initial value (of approx. 90 g./mm.^{2}) and that it falls again *after reaching a maximum value*. Relatively to each other the curves scatter very considerably, but it will be noticed that those of the transversely oriented crystals are in the upper part of the scatter zone, while those of longitudinal orientations appear in the lower part. Similar results were obtained with magnesium crystals (296). Thus in contrast to the independence of the critical shear stress upon orientation in the undeformed state, after identical alternating stressing the shear stress seems to increase with increasing angle between the basal plane and wire axis.

The dependence of the critical normal stress of the basal cleavage plane upon the number of stress reversals was determined by fracturing the stressed crystal pieces at −185° C. in order to avoid the extensions which usually occur in tension at room temperature. The normal stress which is shown in Fig. 151 as a function of the number of reversals reveals, in general, a behaviour similar to that of the shear stress; it increases steeply to a maximum, but falls away again as the number of reversals increases. The dependence of the strain strengthening upon orientation also appears to resemble that of the shear hardening; as the angle of the basal plane increases so, too, does the maximum normal stress. At identical orientations the maximum values for shear and normal strength occur after the same number of reversals.

Whereas, therefore, cyclic stressing initially causes a substantial *hardening* and strengthening, in the further course of stressing a new phenomenon which leads to the *softening* and weakening of the crystal becomes increasingly predominant. That the process which causes this weakening should be sought preferably in the lattice and not merely in the appearance of microscopic cracks, may be inferred from the circumstance that the maximum strength of the crystal is attained long before the appearance of visible cracks. Moreover, the fact mentioned above that other metals form cracks along

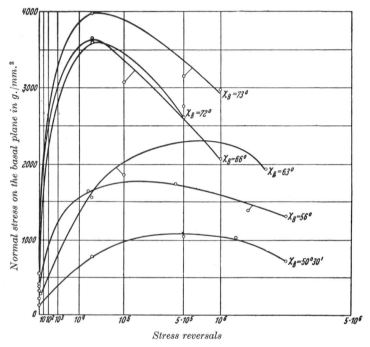

FIG. 151.—Normal Stress on the Basal Plane of Zn Crystals Subjected to Alternating Torsion (299).

operative glide planes which exhibit no cleavage in the undeformed crystal, points to a weakening of the glide planes (cf. also 293).

It follows from this that the fatigue strength of crystals may be defined physically as the stress at which, after an infinite number of reversals, the maximum of the mechanical properties is not exceeded. Statistical details of the fatigue strength of crystals are not yet available. In particular, the question of the relationship between orientation and fatigue strength is still open (cf. Section 82).

So far we have discussed the start of plastic extension and the

brittle fracture of cyclically stressed zinc crystals at low temperatures. Investigations of the *capacity for plastic deformation* at room
temperature revealed a state of affairs which, in its essential
characteristics, is illustrated in Fig. 152. First of all it will be
noticed that with an angle of torsion approximately twice that
adopted in the tests already described, the dependence of the yield
stress on the number of stress reversals (to which reference has

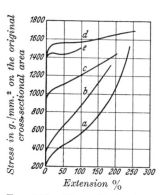

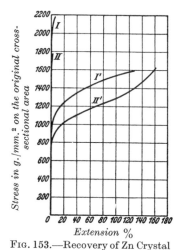

FIG. 152.—Stress–Strain Curve
of a Zn Crystal after Varying
Numbers of Stress Reversals
(Torsion) (297). (*a*) After 0, (*b*)
after 100, (*c*) after 1000, (*d*) after
10,000, and (*e*) after 50,000 twists
(within an angle of 15°).

FIG. 153.—Recovery of Zn Crystal
after Alternating Torsion (297).

I stress–strain curve after about
80,000 twists (within an angle of
20°) ; I′ after subsequent recovery
over a period of 4 days (20° C.);
II after about 140,000 twists; II′
after subsequent recovery over a
period of 2 days (20° C.).

already been made) is again substantially the same. Up to 10,000
reversals, the capacity of the crystal for extension is by no means
reduced ; only after 50,000 twists does it fall to about one half.
Where the angles of torsion are smaller, and the basal plane is not
too oblique, no decrease in extension is observed even after many
millions of reversals. At this number of reversals the shear-
strength maximum has long since been exceeded. The further
hardening by plastic extension, which is expressed in the inclination
of the curves, diminishes noticeably as initial cyclic stressing
increases. Here again we confront the fact, to which attention
has been repeatedly drawn, that with increasing primary hardening

[formation of solid solution, appreciable shear hardening of an initially latent glide system (Section 46) mechanical twinning (Section 52)] further capacity for hardening by deformation is progressively reduced.

If the cyclic stressing of the crystals is carried very far, circumstances may arise in which brittleness will occur in the tensile test even at room temperature, depending on the orientation of the crystal. Crystal recovery causes this brittleness to disappear and a crystal subjected to alternating torsion attains very nearly its original plasticity. Fig. 153 illustrates this by two examples. The capacity for recovery, together with the already mentioned retention of ductility in certain cases, conflicts with the view that the formation of cracks is entirely responsible for the observed weakening in the later stages of the fatigue test.

Two remarks remain to be made. First, in the alternating-torsion test it is not the whole section of the crystal which is deformed but only a relatively thin surface layer. Therefore, since tubular test specimens were not employed, the shear and tensile stresses obtained in the static tensile test with cyclically stressed crystals represent only the average values for the more or less deformed portions of the crystals. Consequently the changes in the mechanical properties which actually occur in the deformed zone are greater than those shown here. Secondly, when the same zinc crystal is subjected to torsion and subsequent extension it is not usually the same digonal-axis type I which becomes effective as a glide direction. Owing to cyclic deformation, glide on the basal plane, which is always the operative glide plane, is also greatly hampered in initially latent directions.

F. CHANGE IN PHYSICAL AND CHEMICAL PROPERTIES BY COLD WORKING

In the preceding sections we dealt with the crystallographic laws governing the plastic deformation of crystals, and the dynamics of the processes themselves. The appreciable changes produced in the mechanical properties by plastic deformation (hardening) have already been discussed. But deformation also modifies to a greater or lesser extent most of the other physical and chemical properties of the crystals.

In regard to the directional properties, an important distinction must here be made between the changes resulting from cold working as such, and those due to the re-orientation of the crystal in the course of deformation. If it is desired to ascertain solely the effect of cold working, allowance must always be made for the changes resulting from lattice rotation in interpreting the observed effects.

TABLE XVIII

Elastic Coefficients of Metal Crystals (10^{-13} $cm.^2/Dyn.$)

Metal.	$s_{11}.$	$s_{12}.$	$s_{44}.$	$s_{13}.$	$s_{33}.$		Litera-ture.
Aluminium . .	15·9	— 5·80	35·1₆	—	—	—	(6)
Al + 5% Cu .	15	— 6·9	37	—	—	—	(303)
Copper . .	15·0	— 6·3	13·3	—	—	—	(7)
Silver . .	23·2	— 9·93	22·9	—	—	—	(8)
Gold . . {	24·5	— 11·3	25	—	—	—	(6)
	22·7	— 10·35	22·9	—	—	—	(8)
Ag + 25 At.-% Au	20·7	— 8·91	20·52	—	—	—	
Ag + 50 At.-% Au	19·7	— 8·52	19·66	—	—	—	(8)
Ag + 75 At.-% Au	20·5	— 9·09	20·63	—	—	—	
α-Brass (72% Cu) .	19·4	— 8·35	13·9	—	—	—	(9)
α-Iron . .	7·57	— 2·82	8·62	—	—	—	(10)
Tungsten . {	2·534	— 0·726	6·55	—	—	—	(11)
	2·573	— 0·729	6·604	—	—	—	(12)
Magnesium . {	22·3	— 7·7	59·5	— 4·5	19·8	—	(13)
	20·4	— 5·2	87·8	— 5·2	20·4	—	(14)
Zinc . . {	8·4	1·1	26·4	— 7·75	28·7	—	(15)
	8·23	0·34	25·0	— 6·64	26·38	—	(11)
Cadmium . {	12·3	— 1·5	54·0	— 9·3	35·5	—	(15)
	12·9	— 1·5	64·0	— 9·3	36·9	—	(11)
Bismuth . .	26·9	— 14·0	104·8	— 6·2	28·7	$s_{14} = $ 16·0	(11)
Antimony . .	17·7	— 3·8	41·0	— 8·5	33·8	$s_{14} = -$ 8·0	
Mercury * . .	154	— 119	151	— 21	45	$s_{14} = -100$	(304)
Tellurium . .	48·7	— 6·9	58·1	— 13·8	23·4	$s_{14} = ?$	(11)
β-Tin . . .	18·5	— 9·9	57·0	— 2·5	11·8	$s_{66} = $ 135	

* At −190° C.

TABLE XIX

Maximum and Minimum Values for Young's Modulus and the Shear Modulus of Metal Crystals

Metal.	Emax.		Emin.		Gmax.		Gmin.	
	kg./ mm.².	Direc-tion.	kg./ mm.².	Direc-tion.	kg./ mm.².	Direc-tion.	kg./ mm.².	Direc-tion.
Aluminium .	7,700		6,400		2,900		2,500	
Copper . .	19,400		6,800		7,700		3,100	
Silver . .	11,700	[111]	4,400	[100]	4,450	[100]	1,970	[111]
Gold . .	11,400		4,200		4,100		1,800	
α-Iron . .	29,000		13,500		11,800		6,100	
Tungsten . .	40,000		40,000		15,500		15,500	
Magnesium .	5,140	0° *	4,370	53·3°	1,840	44·5°	1,710	90°
Zinc . .	12,630	70·2°	3,560	0°	4,970	90°	2,780	41·8°
Cadmium .	8,300	90°	2,880	0°	2,510	90°	1,840	30°
β-Tin . .	8,640	[001]	2,680	[110]	1,820	45·7° †	1,060	[100]

* Angle with the hexagonal axis.
† Angle with the tetragonal axis in prism plane type II.

In the first place, therefore, we will consider the possible extent of the effects of orientation.

58. *Anisotropy of the Physical Properties of Metal Crystals*

The dependence upon orientation of Young's modulus and of the shear modulus has already been expressed, in Section 10, by the equations (10/1, 2) and (10/4, 5) for the case of cubic and hexagonal crystals. With regard to the specific electrical resistance, the cubic crystals are isotropic. Hexagonal, trigonal and tetragonal crystals exhibit axially symmetrical behaviour with respect to the principal axis. If ϕ is the angle formed by the investigated direction with the c axis, then the specific resistance ρ is expressed by—

$$\rho = \rho_\perp = (\rho_\| - \rho_\perp) \cdot \cos^2 \phi \quad . \quad . \quad . \quad (58/1)$$

$\rho_\perp$ and $\rho_\|$ signify the specific resistance perpendicular and parallel to the principal axis. The same dependence upon orientation applies to the specific resistance ω for the thermal conduction,[1] to the coefficient of thermal expansion α, to the thermoelectric power e,[2] to the constant σ of the Thomson-effect,[3] and to the magnetic susceptibility κ.[4] In respect of all these properties cubic crystals are isotropic; in crystals with a principal axis (hexagonal, trigonal, tetragonal) the value of the pro-

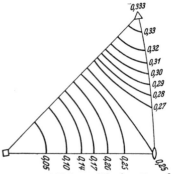

perty in question, in any direction whatsoever, is represented by two constants (the values parallel and perpendicular to the principal axis) and by the angle formed by the direction investigated with the c-axis, in accordance with the above formula.

Tables XVIII–XX contain a summary of the values so far obtained for metal crystals. A graphical representation of the anisotropy of Young's modulus and the shear modulus is found in Fig. 154, b and c.

Fig. 154 (*a*) Values of the function of orientation $(\gamma_1^2\gamma_2^2 + \gamma_2^2\gamma_3^2 + \gamma_3^2\gamma_1^2)$ in the equations (10/1, 2) for cubic crystals.

---

[1] $\omega = \dfrac{1}{\lambda}$; λ = thermal conductivity, *i.e.*, the quantity of heat which flows per second through a unit section of the conductor with a temperature gradient of 1° C. per cm.

[2] Thermoelectric power against a standard metal (*e.g.*, copper) when the difference in temperature at the two junctions is 1°.

[3] σ is the heat developed in 1 cm. of conductor by the passage of 1 unit of electricity at a temperature gradient of 1° per cm.

[4] $\kappa = \dfrac{M}{H}$, when M represents the magnetic moment produced in the unit volume through a magnetic field of intensity H.

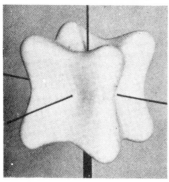

Young's modulus, Au

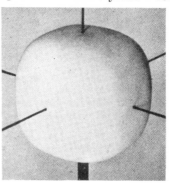

Young's modulus, Al

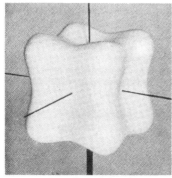

Young's modulus, Fe

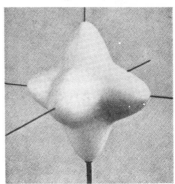

Shear modulus, Fe

(*b*) Young's modulus and shear modulus surfaces of cubic crystals (10).

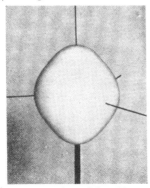

Young's modulus, Mg

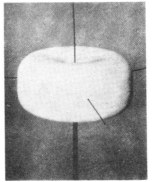

Young's modulus, Zn

(*c*) Young's modulus surfaces of hexagonal crystals (305).

FIG. 154 (*a*)–(*c*).—Elastic Anisotropy of Metal Crystals.

Anisotropy of the Physical Pro

Metal.	Specific electrical resistance, $10^{-6}\ \Omega/\text{cm}$.			Specific thermal conductivity, watt/cm., ° C.		
	‖.	⊥.	Literature.	‖.	⊥.	Literature.
Magnesium {	3·80	4·58	(306)	—	—	—
{	3·85	4·55	(307)	—	—	—
Zinc . . {	6·06	5·83	(309)	1·24	1·24 *	(309)
{	6·13	5·91	(311)	—	—	—
Cadmium . {	8·36	6·87	(309)	0·83	1·04	(309)
{	8·30	6·80	(311)	—	—	—
Mercury † . .	0·0557	0·0737	(315) (304)	0·399	0·290	(316)
Bismuth . .	138	109	⎫	—	—	—
Antimony . .	35·6	42·6	⎬ (311)			
Tellurium . .	56,000	154,000	⎪			
β-Tin . . .	14·3	9·9	⎭			

* At −252° C.; 7·09 or 5·65.
† All measurements at −188° C.

A further property, not referred to in the tables, is the rate of solution. This, too, depends substantially on the crystallographic

Fig. 155.—Anisotropy of the Rate of Solution of the Copper Crystal in Acetic Acid (317).

direction. With cubic metals it may vary very considerably, as is seen from Fig. 155. The directions of maximum rate of solution,

perties of Metal Crystals (at 20° C.)

Temp. range, ° C.	Thermal expansion, 10⁻⁶.			Thermoelectric force compared with copper, 10⁻⁶ V/° C.			Thomson effect 10⁻⁶, cal./coul. ° C.		
	‖.	⊥.	Litera-ture.	‖.	⊥.	Litera-ture.	‖.	⊥.	Litera-ture.
20–100	26·4	25·6	(306)	1·87	1·66	(308)	—	—	—
about 20	27·1,	24·3	(308)	—	—	—	—	—	—
20–100	63·9	14·1	(310)	1·32	− 0·50 ‡	} (312)	0·34	0·86	(313)
about 20	57·4	12·6	(311)	—	—	—	—	—	—
20–100	52·6	21·4	(310)	1·60	− 1·74		1·64 §	1·96	(314)
−188 to −79	47·0	37·5	(304)	−17·9¶	−15·1	(316)	—	—	—
about 20 {	14·0	10·4	} (311)	—	—	—	—	—	—
	15·6	8·0							
	−1·6	27·2							
	30·5	15·5							

‡ The minus sign shows that the current at the cold junction is flowing towards the copper.
§ Determined at 100° C.
¶ Measured against constants.

however, are by no means always the same; the nature of the solvent is of decisive importance [cf. (318), (319)]. Among hexagonal metals the anisotropy of the rate of solution has so far been investigated in the case of magnesium only (320).

If it is desired to examine the effect of cold work upon the physical properties of a polycrystalline material, then only non-directional properties should be measured directly. Otherwise attention must be paid to the development of deformation textures, that is, to the preferred orientations which result from deformation (see Section 78). This orientation effect is superimposed upon the actual change in the properties which has been produced by cold working. Several investigations into the effect of cold deformation on polycrystals have to-day lost much of their value through neglect of this fact, the significance of which has been recognized only in the past fifteen years.

59. Cold Work and the Crystal Lattice

The mere fact that, even after very appreciable distortion, the deformation of crystals is strictly determined by crystallographic factors, is a clear indication that the lattice structure is essentially retained during cold working. This has been proved quantitatively by measurements of the symmetry and parameters of the lattice. Fig. 156 shows by way of example Debye–Scherrer diagrams of copper in the form of fine-grained casting, and of wire subjected to maximum cold working; in both cases the position of the lines is identical. Exact determinations of the lattice constants show that their value varies less than 1 part per thousand even after very substantial deformation (Al, Zn, Cu).

It is true that the diagrams of both single and polycrystals in the cold deformed state reveal changes which indicate that the lattice structure is to some extent influenced by deformation. In the diagrams obtained with monochromatic X-rays these differences

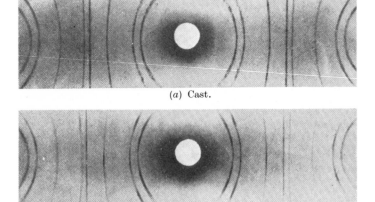

(a) Cast.

(b) Wire drawn cold to nearly 2000 times the original length.

FIG. 156 (a) and (b).—Retention of the Lattice Structure During Cold Working. Debye–Scherrer Patterns of Copper (see 321).

relate to both the length of the X-ray reflexions and to some extent to their width and intensity also. In the Laue diagrams of deformed crystals obtained with " white " X-rays, long distorted spots appear (asterism) (see Fig. 157).

The *lengthening* of the reflexions in the rotation photographs of deformed single crystals indicates the presence of positions which differ from the lattice positions as determined theoretically from the initial position and the degree of deformation. Investigations on aluminium crystals have shown that these deviations from the theoretical position are due to a rotation about an axis which lies in an operative glide plane perpendicular to the glide direction (323). This may be a consequence of a bending of the crystal or of a rotation of separate crystal fragments. In the Laue diagram a *streak-shaped distortion* of the reflexions along curves of the fourth order is to be expected from a lattice curvature (Fig. 158); in fact, the observed asterism conforms entirely to this expectation [(324), (325)].[1] The

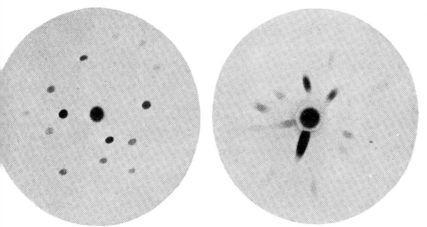

(a) Unstressed Al crystal.   (b) The same crystal after extension (20%) and rolling down by about 30%.

FIG. 157 (a) and (b).—Asterism in a Laue Photograph of a Deformed Crystal (322).

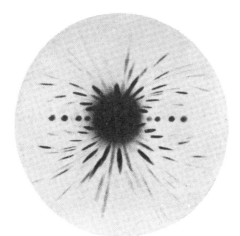

FIG. 158.—Laue Photograph of a Bent Crystal of Gypsum (326).

frequency of the positions diminishes rapidly as the distance from the theoretical orientation increases; the maximum of the deviation

---

[1] For special positions of the axis of bending in relation to the incident ray, cf. (327) and (328).

increases with the extent of the glide. The curvature does not greatly exceed 20°. If multiple glide occurs in the further course of the deformation, several axes of rotation become available for the displacement of the lattice components (329). If deformation is

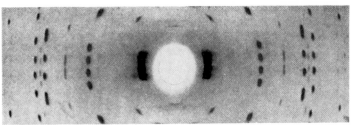

(a) Initial state, high angle lines θ.

(b) Fractured specimen, low angle lines θ.

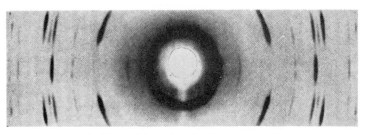

(c) Fractured specimen, high angle lines θ.

FIG. 159 (a)–(c).—Divergences from the Theoretical Position of the Lattice in an Extended Al Crystal. Rotation photographs, Cu-radiation.

followed by reverse deformation the asterism is again reduced in spite of increased hardening (330). The divergencies from the theoretical position of the lattices are particularly noticeable for reflexions with large angles of deviation.[1] This is illustrated in

---

[1] It follows from the differentiation of formula (19/1) that for a given angular range of the normal of the reflecting plane, the length of the reflected spot increases with the angle of reflexion.

Fig. 159 by rotation photographs of an extended aluminium crystal, and in Fig. 160 by Laue photographs of the same crystal; in the back-reflexion photographs the effect of the curvature is stronger.

The *broadening* of X-ray diffractions has been studied mainly with polycrystalline material [(331), (332), (333)], but this effect makes its appearance also in single crystals. It is found only in high angle lines where it is observed as a diffusion of the K_α doublet (see Fig. 161). This broadening of the lines increases mainly at the start of deformation with increasing degree of working [(334), (335)]; however, it by no means attains the same magnitude with

(a) Transmission photograph.   (b) Back-reflexion photograph.

Fig. 160 (a) and (b).—Laue Photographs of an Extended Al Crystal.

all metals. Whereas with iron, copper, silver, tungsten and platinum, considerable diffusion is observed, with aluminium and zinc the doublet lines remain clearly separated even after strong deformation. Reduction of the temperature of deformation, however, results in distinct broadening of the lines in the case of aluminium (336). Alloys of aluminium in the quenched and heat-treated state exhibit very substantial broadening of the lines and reduction of their intensity even after small deformations (337). Just as the orientation of the grains differs in the different layers of cold-worked components (cf. Section 78) so too does the broadening of the X-ray-diffraction lines (338).

There are three possible explanations of the broadening of the lines as a result of cold working.

1. *Elastic distortion* caused by internal stresses (Heyn's residual stresses) which, *over sufficiently large volumes*, produce approximately

constant lattice spacings, although of slightly different value from the original. These divergencies occur in both directions; the volumes within which the stress is approximately constant must be sufficiently large to produce visible interferences (approx. 0·5 μ) (cf. also Section 82).

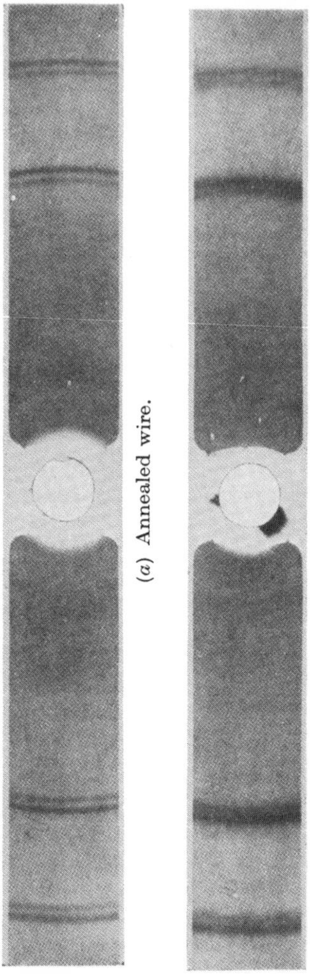

2. *Fragmentation.* A broadening of the lines becomes perceptible at grain sizes below approximately 10^{-5} cm.

3. *Inhomogeneous distortions of the lattice* (warpings, strong distortions within the intact lattice). The calculation of models representing this type of lattice disturbances revealed that their effect on the width of the line depends largely on the type of disturbance assumed; lattice bending leads to a broadening; if, on the other hand, the distortions are partial and periodic, so that less than half of the lattice points are shifted from their normal position, then as a rule there will be no change in the definition of the lines. Broadening on this assumption differs fundamentally from that arising out of assumption 1. Whereas in that case broadening is due to normal diffraction in large volumes, with which only a slightly changed lattice constant is associated, in the present case it is explained by the super-imposing of "ghosts" upon the normal diagram. The intensity of these new lines, caused by the periodicity of the disturbances, is particularly great in the neighbourhood of the original Debye–Scherrer lines into which they fuse (333).

(a) Annealed wire.

(b) Cold-drawn wire.

Fig. 161 (a) and (b).—Broadening of the X-ray Diffraction Rings by Cold Working. Exemplified by Copper (see 321).

A final decision in favour of one of these assumptions, or of a combination of them, cannot yet be made. Difficulties arise with the first two assumptions from the occasional absence of broadening [cf. also (339)]. Assumption 3, in so far as it admits of various types and distributors of the lattice disturbances, can give an account of this diversified behaviour of the metals.

On the basis of assumption 1 the magnitude of the elastic stress can be estimated from the broadening of the lines (change in the lattice constant), a procedure which would be inadmissible if assumption 3 applied. The values obtained in this way for the internal stress of copper after approximately 100 per cent. extension were about 6 kg./mm.^{2} for the single crystal, and about 13 kg./mm.^{2} for the polycrystal (334).

Since a displacement of a part of the lattice points from their normal position weakens the high-angle interferences more than the low-angle ones,[1] changes in the relative intensities of lines must also be expected as cold working increases. Actually it has been found that the intensity of a reflexion of higher order, compared with that of a lower order from the same plane, is smaller after cold working than in the unworked (annealed) state of the metal (Ta, Mo, W) (340). Systematic study of the changes in intensity which accompany deformation should afford a suitable method for a more detailed investigation of the accompanying disturbances.

60. *Cold Work and Physical and Chemical Properties*

Attention was drawn in Section 58 to the need for taking into account the changes in orientation when studying the physical and chemical properties of crystalline material after cold work. In this connection mention should be made of the internal cavities which occur in deformed metals, for instance in overdrawn wires, and which, of course, affect numerous properties (density, specific resistance, etc.). It has already been pointed out in Section 38 that purely crystallographic processes may be the cause of these internal cavities (Rose's channels, which accompany multiple twinning).

A summary of the changes produced by cold deformation in the physical and chemical properties of metals, omitting the obviously trivial effects, is contained in Table XXI. In conformity with the observed invariance of the lattice constants under cold working, which has already been described, *the changes in density* are only

---

[1] A previous deviation from the normal position involves a relatively greater disturbance for high-indexed planes (with small lattice spacings) than for low-indexed planes which succeed each other at wider intervals.

TABLE XXI

Change Produced by Cold Work in the Physical and Chemical Properties of Metals

Property.	Metal.	Deformation, %.	Change.	Literature.
Density	Al single crystal	alternate stressing	0	(341)
	„ „	27% stretched	0	(342)
	„ „	~40% „	within a margin of error of ($\pm 0.02\%$)	(343)
	Polycrystal	~40% „	-0.3%	(343)
	Fe ingot iron Carbon steels	} >90%	-0.4%	(344)
	Armco-iron Carbon steels	} compression	at rising pressure initially -0.1% followed by a further slight increase	(345)
	Armco-iron	77% rolled	-0.12%	(346)
	„	4% stretched	-0.36%	(347)
	Steel	5% „	-0.51% }	(347)
	Cu	60% hammered	-0.2%	(348)
		drawn	0	(349)
		60% drawn	-0.3%	(350)
		4% stretched	-0.13%	(347)
	W	substantially swaged and drawn	initially $+0.5\%$ then -2%	(351)
		„	with a margin of error of ($\pm 0.3\%$)	(352)
	Bi	extruded wire	slight reduction	(353)
	α-Brass 91% Cu single crystal polycrystal 85% and	}70% compressed{	$+0.13\%$ no definite change }	(342)
	72% Cu single crystals	simple glide multiple glide	$0(\pm 0.002\%)$ -0.03% }	(354)
	63% Cu	4% deformed	-0.16%	(347)
Moduli of elasticity and rigidity	Fe single crystal	7.5% stretched	-3%	(355)
	Carbon steel	90% drawn	~0	(344)
	Substantial increases or decreases in several tests on polycrystalline material. Development of deformation textures disregarded.			(356) (357)

TABLE XXI (*continued*)

Property.	Metal.	Deformation, %.	Change.	Literature.
Thermal expansion	Fe ingot iron	50%	$\pm$ 0·1%	(358)
	W	drawn	+11% (?)	(359)
	Bronze	50%	$\pm$ 0·1%	(358)
Specific heat	Zn	?	$\pm$ 0	(360)
	Fe ingot iron (up to 0·5% C)	up to 90%	$\sim$ 0	(361)
	Armco iron and steel (up to 0·6% C)	forged	+ 3%	(362)
	Ni	99·5% drawn	within the limits of experimental error of ($\pm$0·5%)	(363)
	W	94% drawn		
	Bronze (5–6% Sn)	up to 90%	$\sim$ 0	(361)
Internal energy	Al single crystal Polycrystal	$\sim$55% stretched $\sim$20% ,, $\sim$70% drawn	$\sim$0·11 cal./g. $\sim$0·10 ,, $\sim$0·11 cal./g.	(365) (366)
	Cu polycrystal	$\sim$20% stretched $\sim$70% drawn	$\sim$0·07 cal./g. $\sim$0·23 ,,	(365) (366)
	Fe	stretched until necking occurred	0·02 up to 0·09 cal./g.	(364)
Heat of solution	W	substantial	within the limits of experimental error	(369)
Heat of combustion	W	99% swaged and drawn	within the limits of experimental error of ($\pm$1%)	(370)
Electrode potential	Cu Zn Mo Ag Cd Pb	abraded sheets	16 (1) * 2 (4) 1 (0) 20 (9) 9 (1) 2 (0)	(371)

* The figures 16 (1) indicate that out of 17 tests the surface of the abraded sheet was less noble in 16 cases, and that of the untreated sheet less noble in only one case.

TABLE XXI (*continued*)

Property.	Metal.	Deformation, %.	Change.	Literature.
Rate of diffusion	Increases, sometimes considerable, have usually been obtained in numerous tests with polycrystalline materials; but decreases have also been observed. Effect of the solvent. Development of deformation textures hitherto disregarded. Cold working modifies considerably the capacity of crystals for etching.			(372) (344) (373) (374) (375)
Thermo-electric force (compared with annealed metal)	Al Ni Cu Ag Au	drawn and rolled	$+20$ to $+60 . 10^{-8}$ V/° C.	(376); also for copper (377)
	Fe		$-35 . 10^{-8}$V/° C.	
Specific electrical resistance	Fe Steels	$>90\%$ drawn —	$+2\%$ has not been satisfactorily determined	(344)
	Armco Smaller increase	4% stretched in resistance with increasing C content	$+0.96\%$	(347)
	1·3% C	4.5% stretched	$+ 0.29\%$	
	Ni	$>99\%$ drawn	$+ 8\%$	(378)
	Cu	82% drawn 40–80% drawn 4% stretched	$+ 2\%$ $+ 2\%$ $+ 1.5\%$	(379) (380) (347)
	Mo	99% drawn	$+18\%$	(378)
	Ag	60% drawn	$+ 3\%$	(380)
	W	99% drawn	$+50\%$	(378)
	Pt	99% drawn	$+ 6\%$	(378)
	a-Brass (63% Cu) $(a + \beta)$-Brass (57% Cu)	4.3% stretched 4.0% ,,	$+ 1.6\%$ $+ 1.0\%$	(347)
Temperature coefficient of electrical resistance	Ni Mo W Pt	$>99\%$ drawn	$- 5\%$ -16% -35% $- 7\%$	(378)
Thermal conductivity	Cu crystal	swaged	-73%	(381)

small. If the observed decrease of the density exceeds the extent of the change in the lattice constant, the explanation must be sought either in unobserved cavities, or in precipitations (from supersaturated solid solutions) initiated by cold working.

Owing to the generally substantial elastic anisotropy of the metal crystals (cf. Table XIX and Fig. 154) the observed changes caused by cold working in *Young's modulus* and the *shear modulus* can be evaluated only if the determinations are carried out on polycrystalline material. When performing these tests hitherto little account has been taken of the formation of deformation textures. In any case the available results show that the changes brought about in the elastic properties by cold deformation are only small.

A few important remarks have yet to be made regarding the change of the *energy content* of cold-worked metals. Sections 44 and 48 contain some figures for the mechanical energy of deformation taken up by various hexagonal crystals to the point of fracture. They were of the order of four to eighteen times the specific heat of the metal. Most of the energy of deformation is converted into heat; only a fraction (1–15 per cent.) remains as internal energy in the material [(364), (365), (366)]. Whether saturation with internal energy represents the physical condition of fracture for metallic materials, as has been surmised for steel on the basis of cyclic bending tests (367), and for copper and steel on the basis of static torsion and compression tests (368), is one of the basic problems of the theory of mechanical properties. In any case experiments show that the increase of internal energy diminishes as cold working progresses and a state of saturation is approached. The quantity of total energy of deformation, which leads to the saturation of copper with internal energy at 15° C. in the torsional test, amounts to slightly more than 14 cal./g. In the compression test, after approximately the same amount of *energy of deformation* has been taken up, no further increase of the compressive yield stress occurs in the course of further deformation. A first attempt to estimate what proportion of this internal energy is stored in the form of elastic stress energy appears to indicate that this is very small, being about 1 per cent. of the increase in the internal energy (382).

Exploratory experiments carried out on brass crystals give some idea of the connection between heat development and the mechanism of deformation (383). It was found that at each change in the mechanism of glide much more heat was liberated than during glide in one system (cf. the behaviour of density in Table XXI). The changes which occur in the specific resistance of the α-brass crystals when worked in several stages are entirely analogous; they are closely related to the mechanism of deformation. In the region of simple undisturbed glide the changes in resistance are slight, despite appreciable hardening; the increase of the resistance becomes substantial (about 2 per cent.) as soon as disturbances of the glide process occur (384).

As will be seen from the changes given in the table for the *specific electrical resistance* (ρ) and its *temperature coefficient* (α), the product $\rho . \alpha$ remains on the whole constant. The Matthiessen law, originally stated for the formation of solid solutions ($\rho . \alpha$ const.), is thus confirmed for cold working also [(385), (378)]. This makes it possible to infer the change in the specific resistance from the temperature coefficient which can be reliably determined independently of the existence of internal cavities.

The *colour* and the *limits of resistance* of alloys to chemical attack can also be altered by cold working. For instance, Au–Ag–Cu alloys become distinctly yellower during cold working [(386), (371)]. The alloy-composition limits within which chemical attack becomes effective are modified by cold working, and their sharpness is reduced (387).

Changes in the *magnetic properties* have also been omitted from the table as they have not yet been investigated with sufficient care. The saturation of magnetization remains substantially unaltered by cold deformation, coercive force and hysteresis increase (3–4-fold), while the maximum permeability is greatly reduced (to approx. 1/3) [(344), (388)]. As regards magnetic susceptibility, metals and alloys can be divided into two groups; in the first group (Cu, Ag, Bi, Pb, the brasses) cold working increases the magnetic susceptibility of paramagnetic and decreases that of diamagnetic metals. Copper and the brasses even pass from the diamagnetic into the paramagnetic condition. With the metals of the second group (Al, Au, Zn, W, Mo, nickel-silver) there is practically no change in susceptibility [(389), (390) and especially (391)]. The most probable explanation of this behaviour, and one which is confirmed by ample experimental evidence, is that ferro-magnetic phases are precipitated by cold working—even in the case of the so-called pure metals—a phenomenon which is not observed in the unworked state at room temperature owing to the exceedingly small rate of nucleation. The metals of the first group possess low solubility of the ferro-magnetic phase at room temperature, the solubility increasing with temperature (391). Consequently the change in susceptibility is not a true change of a physical property explicable in terms of electron theory, but a secondary effect of cold working which is mainly a consequence of atomic rearrangement processes.

Cold working is also very important for many other processes, the mechanism of which consists of atomic rearrangements in the crystal lattice. Among these are *phase transformations, diffusion* and *recrystallization.* Thus the transformation of β- into α-tin, and of β- into α-cobalt is greatly accelerated [(392), (393)]. The speed of self-diffusion in worked lead is greater by many orders of magnitude than in the undeformed single crystal (394). This accelerated diffusion is also confirmed by the numerous experiments on the ageing of metal alloys. The precipitations responsible for this very important technical process are greatly accelerated by cold working of the super-saturated solid solutions [cf. for example (395), (396)]. We might also include in this group of phenomena the change in the axial ratio of the tetragonal Au–Cu single crystals which occurs as a result of hammering (397), and the influence of pulverization on the atomic distribution and lattice constants of a Fe–Al alloy (398). The capacity for forming new grains (recrystallization), which increases with the degree of working, and which distinguishes the worked from unworked metal, will be dealt with separately in the next section.

In conclusion, attention is drawn to a classification, often attempted in recent years, of the properties according to how they are influenced by cold working (399). Properties which are substantially altered by cold deformation (and impurities) are termed *structurally sensitive* (represented mainly by the mechanical properties, specific resistance, atomic rearrangement properties), in contrast to the structurally insensitive properties (lattice dimensions, thermodynamic properties). Although this distinction is in many ways supported by the observed facts, the material in Table XXI shows the difficulty of establishing clear and unmistakable criteria.

G. RECRYSTALLIZATION

The changes brought about in the physical, chemical and technological properties of metal crystals and crystal aggregates by cold working are not usually permanent : *the strain-hardened condition of the crystal does not represent a state of equilibrium.* We have already considered recovery (Sections 49, 50, 54), a phenomenon as a result of which the altered mechanical properties gradually approach again their original values without the formation of new grains, and while still maintaining the orientation of the lattice. Indeed, in certain cases the reversion is complete (100 per cent. recovery). Besides this gradual recovery, the hardened crystals can also return to the stable condition by a more radical method; as a result of the thermal agitation, new unhardened crystal nuclei develop individually, and, by absorbing the hardened crystal mass, produce an entirely new grain texture : *recrystallization.* This phenomenon is distinguished from crystal recovery by the discontinuity of the process in the individual crystal. The main physical principles underlying this " work recrystallization " will be discussed in the

FIG. 162.—Commencement of Recrystallization in the Glide
Planes of Low-carbon Steel (402).

following section.[1] This will be followed by a brief reference to the " grain growth " referred to in Section 12, which also leads to a re-formation of the texture. The causes of this type of recrystallization differ fundamentally from those which govern work recrystallization. Grain growth occurs as a result of the surface energy striving towards a minimum.

61. *The Nature of the Recrystallization Nuclei*

In the same way as the cast texture which develops when the melt solidifies is determined by the number of the nuclei and the speed of their growth [according to Tammann (401)], so, too, the structure which results from work recrystallization depends upon the number of nuclei formed per second and the speed at which they continue

[1] See (400) for the historical development of research on recrystallization.

to develop.   In the present section we will deal with the recrystallization nuclei.

Since the number of nuclei increases with the amount of cold work, other things being equal, it follows that they are points of very high energy content and not simply particles of crystals which have remained undeformed.   This appears clearly from Figs. 162 and 163, which show the start of recrystallization along operative glide

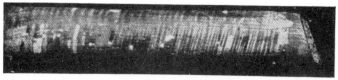

(*a*) Aluminium (cf. also 403).

(*b*) Cadmium (417).

(*c*) Cadmium : section from (*b*).
Fig. 163 (*a*)–(*c*).—Recrystallization in the Twin Lamellæ.

planes and in deformation twins.   The tendency for recrystallization nuclei to occur in deformation twins is very noticeable.   For instance, recrystallization always occurred in the twin lamellæ of extended cadmium crystals after annealing for 1 minute at temperatures of 145° C. and upwards; on the other hand, in the area free from twins new grain was observed to form only at temperatures above 240° C.

The magnitude of deformation is of secondary importance only; thus with tin crystals a compression by a few per cent. produces recrystallization more readily than an extension by many hundreds per cent. (404). It is plausible to connect the capacity for recrystallization with the hardening resulting from deformation. Hardening, too, is much more pronounced in a compressed crystal interspersed with numerous twin lamellæ, than in a crystal in which stretching has allowed the mechanism of glide to operate freely.

An exact investigation of the connection between hardening and the recrystallization texture was carried out on polycrystalline aluminium (405) and tin (406). The simple fact was established that, after stretching and recrystallization, the same grain size in the various samples always accompanied the same hardening. Fig. 164 shows this for aluminium; samples of strips having a substantially different initial grain size were subjected to the same operative stress before annealing. The number of nuclei increases with the degree of hardening. It is true that for single crystalline material the effective stress attained cannot be adduced as a measure of hardening, owing to the marked dependence of the yield point and the stress–strain curve on orientation. In this case it would have been necessary to deform until the operative glide elements had attained an equal degree of shear hardening, i.e., up to the

FIG. 164.—Uniform Hardening Produces Uniform Grain Size of the Recrystallized Texture. Al Polycrystal (407).

same point on the yield–stress curve. Experiments of this nature with the aluminium crystals have revealed the interesting fact that it is not unimportant whether the same hardening results from glide on only one or on two equivalent systems. In the second case the capacity for recrystallization is greatly reduced (see Fig. 165). With the single crystal, therefore, it is possible in the normal tensile test for more pronounced hardening to be accompanied by a reduced capacity for recrystallization. Consequently in this case hardening is not a valid measure for the capacity for recrystallization. The same result emerged from tests carried out on tin crystals. If

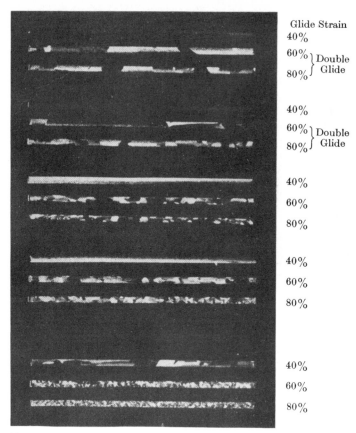

Glide Strain
40%
60%⎱ Double
80%⎰ Glide

40%
60%⎱ Double
80%⎰ Glide

40%
60%
80%

40%
60%
80%

40%
60%
80%

FIG. 165.—Recrystallized Al Crystals. Diverse Grain
Size after Identical Hardening (408).

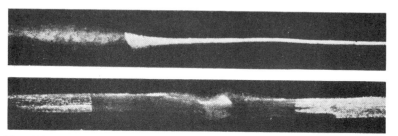

FIG. 166.—Recrystallized Sn Crystals. Formation of Nuclei in the Transition
Zones between Stretched and Unstretched Parts (404).

crystals taken from substantially stretched and from unstretched parts are annealed at low tempreature (150° C.), it is observed that nuclei always form in the transition zones (Fig. 166), while maximum hardening is certainly exhibited by those parts which have been substantially stretched.[1] Aluminium crystals subjected to plastic torsion followed by reverse torsion revealed that the capacity for recrystallization, like the X-ray asterism, was greatly reduced after the deformation had been reversed, although hardening remained unaffected (403; cf. also 410). Special attention was given to the case of bending and back-bending. Here, too, a marked reduction in the capacity for recrystallization after reverse deformation was observed, notwithstanding this had further increased hardening (measured by the hardness of the aluminium crystals) (411).

Although the results obtained with single crystals show that there is no direct connection between hardening and the capacity for recrystallization, they do not weaken the assumption of a more fundamental relationship between the tendency towards recrystallization and the increased internal energy which results from deformation. One may, indeed, conjecture that this internal energy is reduced by reverse deformation, since by this operation the elastically distorted glide layers are substantially re-straightened, and so regain their elastic stress-energy. Since it was also observed that reverse deformation did not proceed along the same glide planes as the original deformation, but along planes which lie between them (412), it is not necessary to assume the presence of strong local accumulations of energy on those glide planes which were originally operative. Moreover, after reverse deformation, the crystal may revert to a more stable condition by recovery. But previous recovery reduces the capacity for recrystallization, and may even eliminate it. In transition zones and at the junctions of irregularly extended crystals where macroscopically different lattice positions adjoin each other, such complete recovery can never occur. The reduced capacity for recrystallization after several glide systems have become operative is also understandable in the light of the above; an increase in the number of disturbed areas in the lattice,

---

[1] Recrystallization is much more difficult to produce in extended tin crystals without visible glide bands than in those with glide bands. They recrystallize only just below melting point, small irregularities in the crystal serving as nuclei (404). The experiments which failed to confirm the formation of nuclei in the transition zone described above were obviously carried out on such crystals without glide bands (409). The causes of this different behaviour of the two types of tin crystals, a difference which is also reflected in the magnitude of the extension which can be achieved, have not yet been elucidated (it is possibly due to small variations in the degree of purity).

but a decrease in the extent of the disturbance after double glide, leads to the same hardening as simple glide, but it reduces the number of nuclei (408).

The fact that in many cases the hardening which occurs in the polycrystal can serve as a quantitative measure of the capacity for recrystallization,[1] does not conflict with the results obtained with single crystals. Annealing of the deformed polycrystal reveals merely an average capacity for recrystallization, from which the differences between the variously oriented grains cannot be recognized—all the more so since the effect of the grain boundaries tends to eliminate the individual differences. As a result of the differing orientations of adjacent grains very strong distortions (hardening) occur in the neighbourhood of the grain boundaries, leading in this case to a marked increase in the number of nuclei (Fig. 167). Consequently the smaller the initial grain size, the

Fig. 167.—Increase of Nuclei in the Vicinity of the Grain Boundaries. Al (405).

smaller are the grains after recrystallization for the same degree of deformation [(405), (414)].[2] With decreasing temperature of working and constant degree of working (increased hardening), the recrystallized grain size becomes finer; the number of nuclei increases as the annealing temperature is raised (415). This behaviour is disturbed if a subsequent grain growth cannot be separated from the work recrystallization [example, aluminium of high purity (416)].

All the experiments, therefore, point to the recrystallization nuclei as being regions of maximum energy accumulation. In many cases where deformation is uniform there will be a parallel increase of internal energy and hardening. In that case the resultant hardening is a measure of the capacity for recrystallization.

---

[1] Nevertheless, even with extended polycrystalline aluminium specimens a reduction of the capacity for recrystallization was observed after reverse straining (compression) (increase in the temperature of recrystallization) (413).

[2] This is paralleled by the fact that the extension of single crystals occurs as undisturbed glide even at temperatures at which the extension of the polycrystal is attended by constant recrystallization (see Section 48).

62. *Velocity of Growth of the Newly Formed Grains*

We have stated the grounds for supposing that the nuclei for the formation of new grain in work recrystallization coincide with the areas of maximum energy accumulation. In this section we will

TABLE XXII

Recrystallization of Extended Metal Crystals. Speed of Growth of Newly Formed Grains

Metal	Anneal-ing temper-ature, ° C.	Mean speed of growth (mm./sec.) in successive intervals of				
		5–10.	5–10.	5–20.	5–30.	60.
		Seconds.				
Tin, extended by about 400% (404)	160	0·46 2·7	0·32 1·7	0·15 1·1	0·1 0·13	0·1 * —
	220	3·2 2·1 1·7	— — 0·76	— — 0·71	— — —	— — —

Metal	Anneal-ing temper-ature, ° C.	Intervals of				
		30.	120.	300.	600.	4800.
		Seconds.				
Aluminium ex-tended under a stress of up to 4 times that of the yield point (0·2% perma-nent set) (417)	520	0·022 0·045 No visible grains	0·028 0·028 >0·028 >0·031 >0·022 >0·022 >0·019 >0·011 >0·014 >0·011	— — 0·017 0·019 0·009 0·002 0·009 0·012 0·011 0·007	— — 0·0007 — — — — 0·006 — —	— — — — — — — — — —
		0·078 †	0·039	0·018	0·0006	—

* A dash indicates that no further growth of the crystal was observed during the particular interval.
† This crystal was stretched at a stress corresponding to only 3 times that of the yield point.

discuss the *speed of the growth* of the recrystallization grains which have developed from the recrystallization nuclei. That the newly formed crystals are mainly in the unhardened, stable condition is apparent from an investigation of their plastic properties, which correspond to those of undeformed crystals, and from the X-ray diagrams which show no signs of lattice disturbance.

It is often possible (if the formation of the nucleus is localized) to recrystallize a deformed crystal very largely as a single crystal. In the process the lattice of the newly formed crystal, regardless of the lattice orientations in the deformed crystal, grows parallel to itself, as has been found in tests carried out on extended and subsequently twisted tin crystals (404). The figures contained in Table XXII convey some idea of the magnitude of the speed of growth in the recrystallization of single crystals; they relate to tin crystals which had been stretched to flat ribbons, and to sheet-

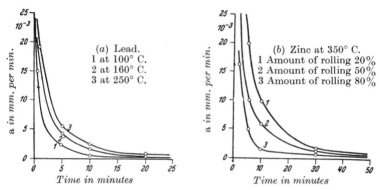

FIG. 168 (a) and (b).—(a) Speed of Recrystallization (a in mm./min.) as a Function of Temperature, and (b) Degree of Deformation (418).

shaped aluminium crystals, each of which had been deformed up to definite stresses. As will be seen from the figures in the table, the velocity of growth falls away in each case with increasing duration of the annealing period; often the growth of the new crystal comes to a complete standstill after a short time.

Similar results were obtained also with polycrystalline specimens of lead, silver and zinc. The speed of grain-boundary displacement as a function of the annealing period is given for various annealing temperatures and degrees of deformation in Fig. 168. The decrease in rate of grain growth was thought to be caused by the precipitation of impurities at the grain boundaries.[1]

We think that attention should also be directed to the recovery which precedes recrystallization and which leads, without the

---

[1] When an unstable phase type is transformed into a stable one the velocity of the grain-boundary displacement is independent of the time.

formation of new grain, to a smoothing of the lattice disturbances
(a very substantial effect in the case of a uniformly oriented
deformed single crystal), and so to a reduction of the capacity for
recrystallization.

In contrast to the above cases, measurements on polycrystalline
aluminium revealed that the newly formed grains grew at a uniform
speed (420).

The maximum velocities of growth at the start of the annealing
of a deformed single crystal (see Table XXII) exceed many times

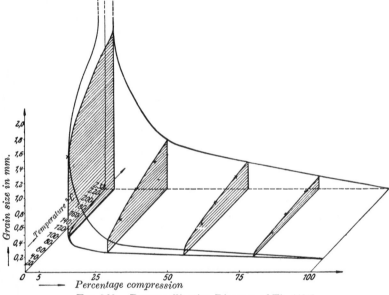

FIG. 169.—Recrystallization Diagram of Tin (415).

the speeds observed for the recrystallization of polycrystals. Thus
for polycrystalline tin (annealing temperature 100° C.) a linear speed
of growth of about 0·17 mm./sec. is obtained (419), while for
aluminium (deformation 10 per cent., annealing temperature 370° C.)
it is about 0·0007 mm./sec. (420). An explanation of all these very
striking discrepancies between the behaviour of single crystal and
polycrystal cannot be given at present; all the more so as the test
conditions (degree of purity of the metal, plastic strain, annealing
temperature) were by no means identical in the various tests. Nor
can an answer yet be given to the question whether the great speeds
observed in the single crystal are in any way connected with the

relations between the orientations of the consuming and the consumed crystal (cf. also Section 63).

The results of comprehensive research into the dependence of the velocity of growth on the plastic strain and the temperature of annealing are available for polycrystals only. They show that the velocity decreases with both decreasing strain and annealing temperature. This decrease leads even at finite deformations to a zero value, which means that there is a limit of deformation below

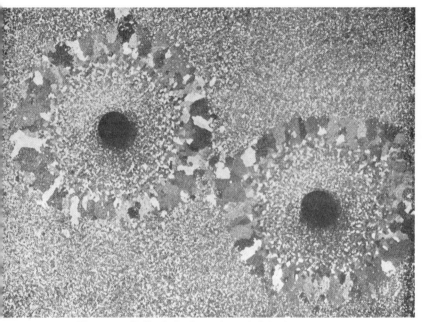

FIG. 170.—Bullet-holes in an Annealed Sn Sheet (415).

which no appreciable work recrystallization occurs even at the maximum temperature attainable [(415), (405), (406)].

As a result of the dependence of the number of nuclei and the velocity of growth on the plastic strain and annealing temperature there is naturally also a connection between these two variables and the grain size of the texture which arises after recrystallization. This relationship, which can be determined empirically, is represented by the *Recrystallization diagram* [Czochralski (419)]. As an example, the recrystallization diagram of tin is shown in Fig. 169. The general principles underlying such diagrams are, that at a given annealing temperature the grain size diminishes with increasing plastic strain, and increases for a given plastic strain with increasing temperature of annealing. Fig. 170 illustrates the first principle very clearly by a tin sheet which had been shot through and subsequently recrystallized (fine grain in the immediate vicinity of the holes, increase in grain size with distance from the holes, no recrystallization beyond the zone of coarse grains).

In this context it is important to observe that the new grains formed by recrystallization lose their capacity for further growth into the deformed material immediately they experience the slightest permanent deformation [(406), (405)]. Fig. 171 demonstrates this statement very clearly. The initial aluminium sheet had two lateral incisions in order to localize the recrystallization produced by slight stretching. After several large crystals had formed round the base of the notches, four more incisions were made (electrolytically, in order to avoid deformation), as is seen in Fig. 171a). These enabled a part of the sheet to be stretched while other parts remained undeformed. After annealing it was found (Fig. 171b) that those parts of the crystals which had remained undeformed continued to grow, while the recrystallization in the deformed parts had stopped. Slight deformation, even if it remains below the limiting value for recrystallization, deprives the newly formed crystal of the capacity to grow at the expense of the deformed environment.

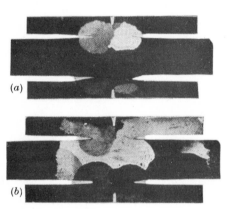

FIG. 171 (a) and (b).—Loss of Capacity for Growth of a Newly Formed Grain as a Result of Slight Deformation. Al (421).

(a) At this stage the centre portion is slightly stretched; (b) After heating the large grains grow only in the undeformed lateral strips.

63. Orientation of the Newly Formed Grains

The orientation of the newly formed grains, as well as the number of nuclei and the velocity of growth, are significant in the study of work recrystallization. It has already been remarked in the case of extended *tin* crystals which, through recrystallization, had again become single crystals, that certain relationships existed between the orientations before and after annealing [(422), (404)]. This was shown by the fact that, if the crystal was extended afresh, the original plane of the ribbon was usually retained, and the new glide striations adopted a position related to that resulting from the first extension. In nineteen out of thirty-five cases investigated the subsequent extension of the crystal was found to retain the initial ribbon plane.

The orientation relationships have been very thoroughly investigated for *aluminium*. If the plastic strain of the crystals was so small that only one or very few new crystals could form in the course of recrystallization, no relationship could be found between the initial orientation and the orientation of the recrystallized grains [(423), (424)]. Where the number of grains produced from the stretched single crystals is somewhat greater (10–30), a preference, though only a slight one, for the orientation of the mother crystal was found. There were also indications of an influence of the direction of deformation, despite the very considerable scatter of the positions of the crystallites (425). Heavy deformation of the crystals (rolling down to one-quarter of the thickness) followed by annealing at 600° C. (for a few seconds only, in order to avoid grain growth) leads to a recrystallization texture that is related to the deformation texture (after rolling) in an entirely reproducible manner. In some cases there is only a slight scatter between recrystallization texture and rolling texture, while in others the two textures diverge appreciably. By altering the direction of rolling in the original crystal, not only the rolling texture but also the arrangement of the new crystals after work recrystallization is changed (426). As in the rolling of aluminium crystals, so in their plastic compression down to approximately one-third of their thickness, regular relationship is observed between the compression texture and the recrystallization texture arising after short annealing at 600° C. [(427), (428)]. The newly formed crystals show preferred orientation which arises from the principal orientation of the deformation texture through the rotations (20–60°) connected with the process of glide, and described in Section 59. Where double octahedral glide becomes operative the orientations of the recrystallization texture are doubled compared with the deformation texture after compression and rolling. It would therefore appear as though those parts of the lattice which had rotated out of the main position, and so become especially distorted, represented favourable nuclei for recrystallization.

Investigations into the recrystallization of other metal crystals are hardly available.   Yet it is precisely such research which would be of the greatest importance for our knowledge of the recrystallization textures of deformed technical components after annealing (see Section 80).

64. Grain Growth

The three preceding sections contain the limited information available to-day about the recrystallization of single crystals after cold deformation.   Although it has been possible to establish certain fundamental laws of work recrystallization, numerous ques-

Fig. 172—Dependence of the Grain Size after Grain Growth upon the Initial Grain Size.   Al (416).

tions have still to be answered, in particular those connected with orientation relationships.   No less obscure are the phenomena connected with grain growth (secondary recrystallization, boundary migration).   This consists in a subsequent, irregular and very much slower process which, after appreciable cold deformation, and if heat treatment is continued, follows the work recrystallization and may result in extremely different grain sizes and shapes.

Aluminium, which was used in the investigation of work recrystallization, was also mainly employed for the study of grain growth. In the first place it was discovered that, provided cold deformation is not carried too far, a clear connection exists between initial grain size and the grain size after completed grain growth.   This will be seen from Fig. 172, which shows strip specimens of varying initial grain size which were subjected to prolonged heat-treatment at 600° C. after rolling down to about one-quarter of the original thickness.   Despite appreciable cold working, the grain size after this treatment is approximately equal to that which was present before deformation.   It is thought that this may be due to the fact that grain growth is greatly assisted by the similar orientation of the coagulating grains.   In fact, a similar lattice orientation exists, after initial work recrystallization, in those parts of the specimen which have developed from a single grain.

With increasing plastic strain the nature of the original grain becomes less significant for the structure which develops after prolonged heat treatment.   On the one hand, the deviation of originally parallel directions in an individual deformed grain becomes greater, whilst on the other, the differences between the orientations of neighbouring grains become smaller.   Ultimately the grain-size relationship disappears entirely, leaving the size of the crystals which form after grain growth to depend on the plastic strain.

If, therefore, a rolled aluminium crystal is subjected to complete recrystallization, according to the plastic strain the following cases will arise (416).   If deformation is slight, the number of crystals increases considerably with the strain; the grain being relatively coarse, grain growth cannot yet take place.   At a certain plastic strain a single crystal appears suddenly—as a result of grain growth —and the number of new grains increases with the deformation.

No reliable picture can yet be given of the mechanism of grain growth [see (429) for numerous relevant observations].   All that can be said is that apparently it is not a question of growth from a nucleus but of the coagulation of similarly oriented grains (426). The relationship between the orientation of the final texture and that of the texture which preceded it is still obscure.

The presence of impurities greatly influences the progress of grain growth.   It proceeds a thousand times more quickly with very pure aluminium than with technical aluminium [(416), (429)].   This means that, with very pure metal, it is still more difficult to distinguish between work recrystallization and grain growth, and that consequently the progress of recrystallization will be still more obscure.   With increasing impurity content, especially if the impurities are precipitated as a new phase, the tendency for aluminium to form large crystals by grain growth is greatly reduced (430).

65. *Reversion to the Original Properties by Heat Treatment*

After the description of the textural change due to recrystallization we will now discuss briefly the changes in the lattice structure and in the physical properties of cold-worked metal crystals produced by heat treatment. It might be expected that the changes in the properties would take place in three stages, corresponding to the three distinct processes of crystal recovery, work recrystallization and grain growth. It is impossible, however, to separate all three phases, since the corresponding processes overlap, work recrystallization being invariably attended by recovery, while grain growth can take place partly during work recrystallization.

Investigations into the *elimination* by heat treatment, of the *lattice disturbances* caused by cold working, have been carried out mainly on polycrystals. It was found with tungsten wires that recovery (revealed by a gradual reduction in strength) causes a definite resolution of the K_α doublet initially rendered diffuse by cold working. The temperature of the start of renewed resolution, determined by X-rays, is always somewhat higher than the temperature of incipient decrease in the strength of the wires even for solid solutions of tungsten with molybdenum and tantalum (431). The smoothing of the lattice becomes evident in the X-ray photograph after mechanical softening has set in; a similar observation was made when annealing extended aluminium crystals. Heating of a 15 per cent. extended crystal for half an hour at 400° C. resulted in a substantial decrease in the yield stress while the asterism in the Laue picture remained unchanged (432).

A quantitative study of the resolution of the K_α doublet during the recovery of tungsten wires revealed that the definition of the doublet tends towards a constant final value at each temperature with increasing length of the heating period; this value increases with increasing temperature of heat treatment (433). The definition was measured by the quotient of the maximum of intensity of the α_1 line and the minimum between the two lines. There is therefore a residue of lattice disturbance (stresses) in the polycrystalline tungsten wires investigated, at each temperature of recovery. These measurements cannot be extended into the region of incipient recrystallization owing to the spottiness of the Debye–Scherrer circles, which greatly interferes with the photometry. It is to be surmised, however, that upon conclusion of recrystallization the definition of the doublet will revert to that of the unworked material.

Reference has already been made in Sections 49 and 54 to the *changes* produced by recovery in the *mechanical properties* of deformed crystals. These changes are regular and gradual, and by suitable heat treatment they reproduce the initial values. In the case of recrystallization, on the other hand, the hardening of the crystals consequent upon plastic deformation is discontinuously and completely eliminated.   Table XXIII contains some figures for the yield point of tin crystals which, after the original extension, had again become single crystals by recrystallization.   After heating, the yield-stress values come within the same order of magnitude as for unstressed original crystals; the appreciable plasticity (approximately 4–6-fold), which disappeared completely after deformation, has now returned.   If work recrystallization leads to the formation of a polycrystal, the mechanical properties will depend largely upon the grain size of the resultant texture.   The same applies

TABLE XXIII

Softening of Extended Tin Crystals as a Result of Recrystallization (404)

Yield point in g./mm.^{2}.	
before	after
annealing	
($\sim$10 min., at 140° C.).	
1700	156
1480	98
1630	443
2050	137
3740	260

naturally to the grain growth which follows work recrystallization, and which in certain circumstances may even lead back to the single crystal.

Owing to the technical importance of the subject a very great deal of research has been carried out on the changes produced by annealing in the mechanical and physical properties of cold-worked poly-crystals.   Attention has already been drawn to the reduction in the strength of tungsten wire within the range of recovery.   It is not intended to examine in detail the numerous investigations relating to this matter; but we would draw attention to the comprehensive accounts in (415) and especially in (434) and (401).   It can be said in general that the modification of a property by annealing proceeds approximately as shown in Fig. 173 for the tensile strength and elongation of a hard-drawn mild-steel wire.   Following upon a range of temperature in which only slight changes occur, and in which the fibrous deformation texture is retained, at a certain temperature a sudden change takes place in the properties. An increase in the temperature of annealing is not attended by any further appreciable change, provided that the temperature is neither too high nor the duration too long. The relationship with the three processes—recovery, work recrystallization and grain growth—is quite obvious. The region of minor changes in properties within the low-temperature range corresponds to recovery; the sudden reversion to the initial values of fine-grained unworked material,

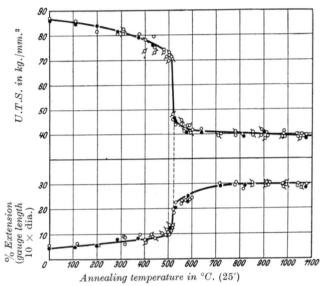

FIG. 173.—Change in the Tensile Strength and Elongation of Hard-drawn Ingot Iron (0·08% C.) with Increasing Temperature of Heat Treatment; Annealing Period 25 min. (435).

coinciding with the transformation of the texture, corresponds to work recrystallization; and the further irregular changes which occur if heating at high temperature is prolonged, and which usually have an adverse effect on the mechanical properties, are to be ascribed to the formation of a coarse grain size by grain growth.

A note must be added regarding recovery. It has been found with fine-grained materials that, as in the case of the lattice disturbances described above, a complete elimination of the changes in the properties due to cold working cannot be effected by recovery. Fig. 174 exemplifies this statement by means of the tensile strength and elongation of the mild-steel wire already referred to. It will be seen that to each annealing temperature there corresponds a degree of softening which is not exceeded even if the heating period is greatly prolonged. The capacity for recovery of the deformed polycrystal therefore differs substantially from that of the freely extended crystal (cf. also Section 62). This difference becomes intelligible if it is remembered that the grains of the deformed polycrystal, which owing to mutual interference have become warped, cannot revert to a completely unobstructed state without the formation of a new texture.[1]

Although the behaviour shown by the mechanical properties of steel wire is observed in general with other properties also, marked

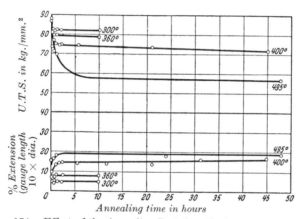

FIG. 174.—Effect of the Annealing Period at Various Temperatures (below 500° C.) on the Tensile Strength and Elongation of Drawn Ingot Iron (435).

differences arise for individual metals and alloys. It is particularly noticeable that the most rapid changes in various properties do not always occur within the same range of temperature. This happens with copper, silver and gold, but not with nickel and metals of the platinum group. With these metals the reversion to the original values takes place for different properties in different ranges of temperature [(436) to (439)].

In conclusion, reference must be made to the release of gas during the heat treatment leading to the recovery and recrystallization of worked metals (440). Recently alloys have been prepared with small

[1] Although according to this statement recovery does not restore the original properties of the technical polycrystal, this limited capacity for recovery nevertheless deserves closer attention than it has so far received [cf. also (414)].

amounts of radio-active elements; the progress of the heat treat-ment could be followed very accurately by measuring the emanations liberated (441).

As has already been mentioned in the case of cold working, when investigating the changes in properties caused by heat treatment special attention must be paid on the one hand to trivial causes, and on the other to the formation of preferred orientations. Among the former must be included the formation of cavities which are observed in grain growth at the point of contact of several grains [cf. for instance (442)]. In sections 78 and 82 particulars will be found of the regular texture which frequently results from work recrystallization, and which, owing to the marked anisotropy of many properties of metal crystals, may be attended by very pro-nounced changes in properties as a result of the preferred orientation of the crystals.

CHAPTER VII

PLASTICITY AND STRENGTH OF IONIC CRYSTALS

Hitherto we have examined the phenomena of the plasticity and strength of crystals with reference to experimental data obtained from metal crystals. It has already been emphasized more than once that such crystals yield very satisfactory experimental material, owing to the ease with which they can be produced and deformed. But it has also been mentioned that the processes of gliding and mechanical twinning which underlie the plastic deformation of crystals were originally discovered, and closely studied, long ago in ionic crystals. The mechanisms of deformation appear to be substantially independent of the type of force by which the lattice is held together, which in the case of ionic crystals are mainly electrostatic forces between heteropolar ions, while in the case of metals they arise from positively charged atomic cores in a degenerate electronic gas.

The mechanical investigations into the structure of ionic crystals which form the subject of this section were restricted initially, in the main, to a determination of the crystallographic indices of their cleavage and fracture planes, and their glide and twinning elements. Numerical data about mechanical properties were obtained for the "hardiness" of various crystal faces, determined by numerous different methods. Over a long period systematic investigations into the forces necessary to initiate plastic deformation, cleavage and fracture were carried out only occasionally, and it was not until recent years that a substantial body of experimental data referring mainly to sodium-chloride crystals became available.

66. *Gliding and Twinning Elements*

In what follows no complete account can be given of the great number of observations on the mechanism of deformation in ionic crystals; for fuller details the reader is referred to (443). Table XXIV contains particulars of the cleavage or fracture planes, and of the glide and twin elements of only the more important lattice types. In the case of ionic crystals, too, planes and directions of low indices are the most important. The significance of the lattice structure in determining the deformation elements is apparent from the table. As with metal crystals, it is still not clear by what factors the glide elements are determined. In the case of ionic crystals the sign of the ionic charges is certainly a contributory factor. In connection with the gliding of sodium chloride it has been

TABLE XXIV

The Cleavage Planes, Glide and Twin Elements of Ionic Crystals [from (443)]

Material.	Lattice-type crystal class.	Cleavage or rupture planes.	Glide elements.		Twin elements.		
			$T.$	$t.$	$K_1.$	$K_2.$	$s.$
Cubic							
Sodium fluoride, NaF							
Rock salt, NaCl							
Sylvite, KCl							
Potassium bromide, KBr	Rock salt, O_h	(001)	(110)	$[1\bar{1}0]$	—	—	—
Potassium iodide, KI							
Rubidium chloride, RbCl							
Ammonium iodide, NH$_4$I		(001), (111)?					
Periclase, MgO							
Lead sulphide, PbS		(001)	(001)	$[110]$, $[100]$			
Ammonium chloride, NH$_4$Cl	Cæsium chloride, O_h	(001)	(110)	$[001]$	—	—	—
Ammonium bromide, NH$_4$Br							
Fluorspar, CaF$_2$	Fluorspar, O_h	(111)	(001)	$[110]$			
Nantokite, CuCl	Zinc blende, T_d	(110)	—	—	—	—	—
Marshite, CuI							
Zinc blende, α-ZnS		(111)	(111)	$[112]$?			
Tetragonal							
Rutile, TiO$_2$	Rutile, D_{4h}	(110), (100)	—	—	$\begin{cases}(101)\\(101)\end{cases}$	$(\bar{1}01)$ $(\bar{3}01)$	0·908 0·190
Hexagonal							
Beryllia, BeO		(10$\bar{1}$0), (0001)?	—	—	—	—	—
Red zinc ore, ZnO	Wurtzite, C_{6v}	(0001), (10$\bar{1}$0)	—	—	—	—	—
Wurtzite, β-ZnS		(10$\bar{1}$0), (0001)	—	—	—	—	—
Greenockite, α-CdS							
Jodargyrite, α-AgI		(0001)	—	—	—	—	—
Millerite, NiS	Nickel-arsenide, D_{6h}	(10$\bar{1}$1), (10$\bar{1}$2)	—	—	$(\bar{1}012)$	$10\bar{1}0$)	0·380
Rhombo-hedral							
Brucite, Mg(OH)$_2$	Brucite, D_{3d}	(111)	(111)?	—	—	—	—
Corundum, Al$_2$O$_3$	Corundum, D_{3d}	—	—	—	$\begin{cases}(111)\\(100)?\end{cases}$	$(11\bar{1})$ (011)	0·635 0·202
Magnesite, MgCO$_3$			(111)	$[01\bar{1}]$	(011)?	(100)	0·799
Calc-spar, CaCO$_3$			(111)?	—	(011)	(100)	0·694
Manganese spar, MnCO$_3$	Calc-spar, D_3	(100)	(111)	$[01\bar{1}]$	(011)?	(100)	0·781
Chalybite, FeCO$_3$							
Zinc spar, ZnCO$_3$			—	—	(011)	(100)	0·753
Sodium nitrate, NaNO$_3$							

pointed out (444) that ions of the same sign never approach each other during glide, since this would lead to repulsion forces perpendicular to the glide plane. It is true, of course, that this selective principle, which apparently does not conflict with results obtained with other ionic crystals, is in itself inadequate. An infinite number of pairs of T and t satisfy this requirement (445).

As with the metal crystals so, too, in the case of ionic crystals an increase in temperature or pressure does not cause the disappearance of glide elements which operate under normal conditions of temperature and pressure; at elevated temperatures, on the other hand, new glide planes have been observed. For instance, in sodium chloride at elevated temperatures, in addition to the dodecahedral system which is normally operative, glide occurs also in the (111) plane (446); an increase in the number of glide planes with rising temperature also takes place with rhombic crystals of baryte ($BaSO_4$), celestine ($SrSO_4$) and anglesite ($PbSO_4$) (447). Glide in cube planes was observed with sodium chloride [(448), (449); cf. also (450)], particularly in the case of wetted crystals with certain special orientations relative to the direction of tension (451); here, too, the glide direction is the face diagonal.

Owing to the importance of *dodecahedral glide* in crystals of the sodium-chloride type, some particulars should be given of the *lattice rotation* by which it is accompanied in tension (452). First it is observed that for all orientations the same shear stress occurs in at least two glide systems, since t and the normals to T are interchangeable. Geometrically considered, therefore, single glide cannot be expected for any initial orientation of the crystal. The cube edges are found to be stable final positions; the face diagonals represent unstable positions : tensile or compressive stress parallel to the body diagonals cannot lead to dodecahedral glide, since a force acting in this direction cannot produce a non-vanishing resolved shear stress in any such glide system [three (110) planes lie parallel to the direction of force; in the remaining three the glide directions are transverse to it].

It should be emphasized in regard to *mechanical twinning* (which we do not propose to discuss further), that in many cases the movement of all lattice points is not a simple shear. Polyatomic compounds rarely exhibit a straight line movement for all atoms. It is observed, however, that centres of gravity of polyatomic ions or molecules frequently move according to a simple shear (445). As regards the dynamics of twinning, an observation concerning the magnitude of the elastic shear strain parallel to the twin elements at the moment of twinning may be mentioned. Measurements carried out on rectangular parallelepipeds of calcite revealed, for two different directions of the compressive force, elastic shear strains of $2\cdot4 \cdot 10^{-3}$ and $3\cdot5 \cdot 10^{-3}$ (453). Thus, as with metal crystals (see Section 51), the elastic shear at the start of twinning is only a small fraction of the strain due to twinning. In this case, too, the exact conditions of twinning could not be determined.

Although the *cleavability* of crystals has long been known, the principle by which the cleavage planes in the lattice are determined still remains obscure. Here again the geometrical properties of the lattice are certainly not the only relevant factors; the distribution of the electric charge is also important. A view which was originally

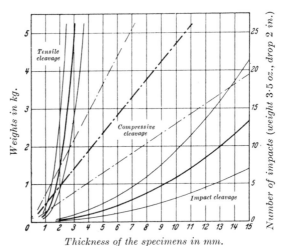

FIG. 175.—Cleavage of NaCl Crystals under Tensile, Compressive and Impact Stress (456).

expressed with reference to the cubic cleavage of sodium chloride (444) now appears to apply to many other cases also; as a result of the shear which precedes cleavage, ions of the same sign are moved so as to face each other; this produces forces of repulsion which lead to disintegration. However, like the analogous discussion in the case of glide, this condition is insufficient to explain cleavage.

A quantitative measure of cleavability with respect to a given plane is not available; qualitatively it is described as being very perfect, perfect, distinct and imperfect. In any case, cohesion perpendicular to the cleavage planes is very slight, and it is this circumstance which determines the appearance of these planes as planes of fracture. The absence of a quantitative measure of cleavability is due to the complex stress systems which arise in the usual methods of cleaving. In recent years attempts have been made to facilitate the quantitative determination of cleavability by standardizing the methods of testing (454). According to the type of test employed a distinction is made between tensile cleavage, in which the compressed portion of the crystal is hollow so that the crack appears on the underside subjected to maximum extension, compression cleavage and impact cleavage [cf. also (455)]; the relations between the necessary forces or energies and the dimensions of the crystal are then examined. A graph showing the results obtained with the three types of cleavage, using the NaCl crystal, is given in Fig. 175, which shows the influence of the thickness of the specimen and also indicates the scatter of the measurements. Similar relationships between the work of cleavage and the thickness of the specimen were also found for lead sulphide (457) and anhydrite (458).

The cleavage of NaCl crystals along a (110) plane indicates that this is only an apparent plane built up stepwise from cubic or (110) vicinal planes [(455), (459)]. It is surprising that impact cleavage along this plane should require less energy than along a cubic plane (460). This illustrates well the complexity of cleavage phenomena.

Unlike the cleavage test, the tensile and compression tests are characterized by very simple states of stress. In this type of test quantitative relationships corresponding to those described for metallic crystals are found to apply, at least for glide.

67. *Initiation of Glide.   Shear Stress Law*

The behaviour of ionic crystals under tension at room temperature differs substantially from that of the metallic crystals in the amount of extension that can be achieved. With numerous metallic crystals extensions of many hundreds per cent. are possible, as against not more than a few per cent. for ionic crystals. Systematic investigation of the latter therefore demands much more sensitive instruments. Owing to the brittleness of the material great care must be taken, in the tensile tests, to avoid bending due to eccentric application of the load. Even a slight eccentricity may reduce the strength by 50 per cent. and more (461). Pivoting of the grips into which the crystal specimen is usually cemented in universal joints or on needle points is therefore recommended. In order to ensure a uniform increase of the load, free from vibration, the tensile force is applied usually by floaters, often with additional lever transmission; the buoyancy of the floater is reduced by lowering the water level. The extension is measured, with the upper grips at rest, either directly by microscopic observation of the displacement of the lower grips, or by means of a mirror apparatus which, at very high magnifications, can at the same time be used for photographic recording [for particulars of the technique of tension and compression tests see (462), (463), (452), (464)]. The methods developed in recent years for growing large crystals have greatly facilitated the systematic testing of ionic crystals (cf. Sections 13 and 14).

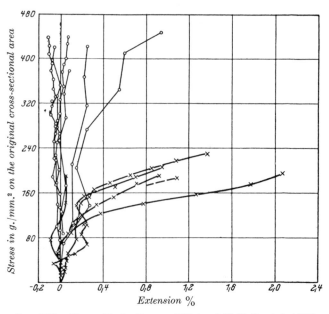

FIG. 176.—Stress–Strain Curves of Natural NaCl Crystals (452).
X = annealed for 3 days at about 600° C.; O = unannealed.

A few stress–strain curves obtained on small cylindrical bars of rock salt oriented parallel to the cube edge are shown in Fig. 176. It is seen that this material, which in the initial state breaks in a brittle manner, exhibits marked elongations after annealing at 600° C. Although in this instance the curves are not very accurate, the tests nevertheless reveal the existence of a definite *yield point*. In other ways, too, a stress value is observed beyond which the deformation increases rapidly. From this point onwards, for instance, glide bands appear on the surface of the crystals. The extension can be followed very closely in transmitted polarized light, owing to the double refraction which results from the residual

2   1

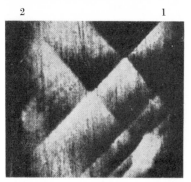

(a) Determination of the start of glide
(463).

(b) Bent crystal (465).

(c) Compressed crystal (465).

Fig. 177 (a)-(c).—Glide Bands in Deformed NaCl
Crystals, Rendered Visible by Stress Birefringence.

stresses. Fig. 177a shows the first glides in a tempered crystal
of sodium chloride. No change was observed up to a stress of

78 g./mm.². After this stress had been maintained for 20 seconds, the bands shown as 1 and 2 appeared suddenly. The other bands also appeared suddenly after varying intervals, without the load being increased (" jerky deformation by glide "). Crystals with glide planes packed very closely together are shown in Fig. 177, *b* and *c*.

Un-stressed

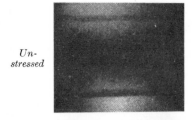

Investigation of potassium halide crystals by means of polarized light clearly revealed various stages of deformation in crystals (466). With increasing tensile stress local glide processes occur in the interior of crystals of KCl, KBr and KI (at a stress of about 50 g./mm.²), followed by the first passage of a glide plane through the entire cross-section. As the stress increases so the glide planes multiply, until finally there is a more or less sudden increase in the intensity of all the glide zones present (yielding). Continued increase of the stress gives rise from time to time to thicker packets of glide zones. By way of example, Fig. 178 illustrates the double refraction along the glide planes of a KCl crystal grown from the melt. The stage of deformation which corresponds to yielding coincides with the appearance of visible glide bands.

95·5 *g./mm.² : start of flow*

154·5 *g./mm.²*

206 *g./mm.²*

An X-ray determination of the onset of plasticity is based on the appearance of distorted Laue interferences [asterism (462)]. However, the stress limit which is thus defined does not coincide with that revealed by the stress–strain curve; it corresponds with permanent deformations which are already clearly per-

Fractured at 254 g./mm.²

Fig. 178.—Stress Birefringence in a KCl Crystal which has been Extended by Various Amounts (466).

ceptible (467). The start of plastic deformation can also be determined photo-chemically. Unlike the X-ray method, the change in colour gives very low values, which probably coincide approximately with the stresses corresponding to the first deformations revealed by polarized light (468).

The *shear stress law* which was established for metallic crystals applies also to the yield point of alkali halide crystals, as revealed by the appearance of visible glide bands [for NaCl (469), for KCl, KBr and KI (470); for wetted NaCl crystals see Section 72]. The dependence of the tensile yield point on the orientation according to the shear stress law would be represented by the *yield surface* shown

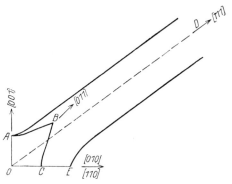

FIG. 179.—Section through the Yield Surface of Alkali Halide Crystals Assuming Dodecahedral Glide (452). $OABCO$ Intersects with (100), $O\bar{A}DEO$ Intersects with (110).

in section in Fig. 179 if dodecahedral glide alone is assumed. Crystals exhibit minimum resistance to flow parallel to the cube edges; the yield stress of rods parallel to the body diagonal is infinitely high, and directions near to this are distinguished by extremely high values.

Figures for the yield point of ionic crystals can be given only with some reserve. The critical shear stress of glide systems is, indeed, a typical example of the "structure-sensitive" properties of crystals (cf. Section 60). The tensile yield point values for natural rock salt crystals therefore vary within wide limits according to their place of occurrence and previous history [70–500 g./mm.^{2} (464)]. In order to avoid the differences existing between natural crystals, recent plasticity investigations have been carried out with synthetically produced crystals of uniform quality. From the above values, *elastic shear* strains of magnitude 10^{-4} are obtained

(471).   Therefore the same statement is true for the limit of stability of the dodecahedral plane in the [101] direction as for the glide planes of metallic crystals and twin planes : namely, that the elastic shear strain at the start of plastic yielding is only a small fraction of the displacement which occurs in the ensuing glide.

For thin crystals, a definite dependence of yield point upon the *diameter* was observed (472).   The various sections were produced

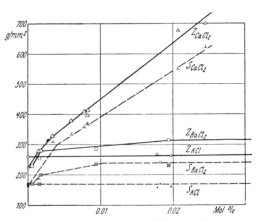

Fig. 180.—Effect of Admixtures on the Yield Point (S) and Tensile Strength (Z) of NaCl Crystals (475).

by dissolving rods of sodium chloride in their own melt.   The yield point (σ_s) increases as the initial cross-section is reduced, the observed results being well expressed by the curve

$$\sigma_s(\text{kg./mm.}^2) = \frac{0\cdot 330}{d_0} + 0\cdot 020 (d_0 = \text{initial diameter in mm.})$$

The effect of *admixtures* was exhaustively investigated on sodium-chloride crystals grown from the melt.   Shear hardening is very pronounced so long as the formation of a solid solution can be assumed; in this respect the addition of bivalent cations is much more effective than that of monovalent cations (473).   Where precipitation is ultra-microscopically observable, the effect is either nil or only very slight.   Very definite hardening results from the addition of $PbCl_2$, which in a concentration of about 10^{-3} mol.-per cent. (one molecule of $PbCl_2$ to about 100,000 molecules of NaCl) increases the yield point three-fold (474).   The effect of additions of $CaCl_2$, $BaCl_2$ and KCl is seen in Fig. 180; $SrCl_2$ behaves in the same way as $CaCl_2$; $MgCl_2$ in amounts up to $0\cdot 1$ mol.-per cent. has no effect. If two compounds which cannot be detected ultra-micro-scopically are added simultaneously, their hardening effect is additive, as with metals (cf. Section 45); this corresponds to the independent insertion of the impurities into the lattice (476). Unlike the result when the yield point is determined from the appearance of glide bands, the photochemically determined onset of plasticity of NaCl crystals is not affected by additions of $SrCl_2$ (476a).

Attention has been drawn repeatedly to the importance of a previous *heat treatment*. Fig. 176 shows the very great influence on plasticity of annealing for several days. The yield point, which had in most cases previously exceeded the strength, was reduced to about 100 g./mm.^{2} Systematic investigation of the effect of annealing at various temperatures showed that the individual differences between sodium chloride crystals of different origin are prone to disappear as the temperature of annealing rises, and tend to a common minimum value after treatment at 600° C. Still further increase of the temperature of annealing causes the yield point to rise again (464). Annealing was also found to reduce the yield point of natural sylvine. Pure crystals of KCl and KBr, grown from the melt and from solution, showed no effect, as was to be expected from their method of production (470).

The softening which results from annealing must be regarded as a recovery effect. This is seen mainly from the fact that, in the polar-ization microscope, heat treatment is revealed by the elimination of the stresses which are always present in natural crystals as a result of plastic deformation (by earth pressure). How far the renewed increase in strength at high temperatures results from a temperature dependence of the solubility of the admixture cannot yet be decided. In any case, initially different crystals can be brought to a com-parable state by suitable annealing. Natural crystals of rock salt of the most diverse origin give a value of about 40 g./mm.^{2} for the critical shear stress of the dodecahedral glide system after annealing at 600° C. for 6 hours. In Table XXV are given the values obtained with crystals grown from the melt under specially controlled con-ditions (De Haen material of maximum purity; the unglazed porcelain crucible was used for a single melt only).

Specially noteworthy is the value for crystals of rock salt grown from the melt. It is almost twice that of the annealed natural crystals, and it cannot be substantially reduced by a similar anneal.

Investigation of the dependence of the yield point upon tempera-ture showed a very pronounced decrease at high testing temperatures [(446), (477)]. At temperatures of about 100° C., however, there seems to be a range of slight dependence upon temperature. For the stress limit determined by X-rays which, as already mentioned, greatly exceeds the yield point, the values for sodium chloride were

TABLE XXV

Critical Shear Stresses of the Dodecahedral Glide Systems of Alkali Halide Crystals Grown from the Melt

Salt.	Total of impurities, weight, %.	Critical shear stress of the dodecahedral glide system, g./mm.².	Literature.
NaCl . .	0·030	75	(476)
KCl . .	0·016	50⎫	
KBr . .	0·030	80⎬	(466)
KI . .	0·020	70⎭	

found to fall to zero at the melting point both in tensile and compressive tests (462). This behaviour differs characteristically from that of metallic crystals (Section 47).

A comparison of the critical shear stresses of the glide systems obtained in tension or compression with those obtained with other types of loading is possible as yet only for torsion (478). The dependence upon orientation of the torsional yield point of rock salt crystals drawn from the melt is also governed in general by the shear stress law (minimum yield point for orientations close to the [101] direction); rods parallel to the cube edge cannot be twisted by dodecahedral glide. The critical shear stress was found to be 111 g./mm.². If we accept the condition for torsional yielding as discussed in Section

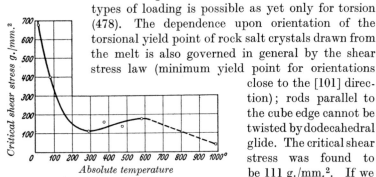

FIG. 181.—The Critical Shear Stress, under Torsion, as a Function of Temperature. NaCl Crystals (478).

41, according to which the whole section of the crystal has to be deformed, we obtain a value of 70 g./mm.², which agrees sufficiently with the value determined in tension. The influence of temperature on the critical shear stress as revealed by torsion tests is shown in Fig. 181. If the temperature is reduced from room temperature to −253° C. it increases about eightfold, but even when the temperature is raised, a slight increase in yield point is observed in the neighbourhood of 300° C. Investigations into the bending of rods of cubic orientation are discussed in [(479) to (481) and (467)].

68. *Further Course of the Glide Process*

The further course of the glide process in ionic crystals has received far less attention than the start of glide. The establishment of quantitative criteria, such as work-hardening curves, has been hampered by the fact that instead of glide taking place in one glide system only, several equivalent and intersecting systems participate in the deformation. However, the extension of sodium- and potassium-halide crystals oriented parallel to the cube edge, for example, does not proceed by the same amounts of glide along the four geometrically equivalent systems. Usually only two related dodecahedral systems become active (they belong to the same (100) zone), and it frequently happens that one of these is specially preferred. This means that the orientation of the crystal does not usually remain unchanged, as would correspond to the stable initial position; instead, where extension is large, lattice rotations up to approximately 30° occur.

It is clear from the shape of the stress–strain curves which show a rise in stress with increasing extension, that here, too, the deformation is accompanied by a *shear hardening* of the operative glide systems (cf. Fig. 176). Such an increase in stress always points to an increase in the critical shear strength of the operative system (Section 43).

The amount of extension which can be achieved under tension at room temperature does not usually exceed more than a few per cent. Extensions of more than 50 per cent. have, however, been reported for very thin annealed rods of rock salt. The increase in extension with reduction in cross-section, like the similar increase in yield point, starts at initial diameters of about 0·6 mm., at which it is very marked (472). Even when rods of sodium chloride are loaded parallel to the body diagonals, a perceptible though small extension precedes fracture. It is clear from the orientation relationship discussed above that this extension cannot result from dodecahedral glide; instead, glide occurs here on the cube planes, as is shown in (451).

As the temperature rises and the yield point falls, there is substantial increase in the amount of plastic deformation that can be obtained. Crystals of rock salt can be easily bent and twisted by hand in a flame of alcohol (482). At temperatures above 400° C. there is obvious necking which may produce local extensions of up to thirty times (462). The reduction of the yield stress with increasing temperature corresponding to varying amounts of strain is seen clearly from Fig. 182, in which the initial portions of the stress–strain curves of melt-grown crystals of rock salt are represented.

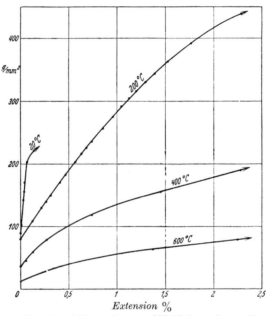

FIG. 182.—The Stress–Strain Curves of NaCl Crystals as a Function of Temperature (477).

The increase of plasticity due to increasing temperature was measured by torsion tests in (483) and (478).

Measurement of the compression of square cleaved specimens of alkali halide crystals (down to 1/10 of the original height) has led to the discovery of a very simple relationship. If s is the compressive strain referred to the final height $\left(s = \dfrac{h_0 - h}{h}\right)$, λ the " slenderness " of the specimens ($\lambda = h_0/d$; $d = $ length of edge of the square section), then for not too low pressures p (from about 2000 kg./cm.^{2} upwards) $s = b\lambda p$, where b represents a characteristic constant of the material. Table XXVI, which contains the values which have so far been determined for b, reveals a relationship between these and the lattice constants.   Where the cation is the same, b increases approximately linearly with the lattice constant. At the high pressures employed, the origin and previous history of the crystals are no longer found to exert an important influence [(484), (485), (486)].

TABLE XXVI

Compression of Alkali Halide Crystals (486)

Material.	$b.$ cm.^{2}/kg.	Lattice constant, Å.
NaCl	0·0020	5·628
NaBr	0·0052	5·96$_2$
NaI	0·0086	6·46$_2$
KCl	0·0044	6·27$_7$
KBr	0·0070	6·58$_6$
KI	0·0094	7·05$_2$

Mention may finally be made of experiments with load reversal. It was observed long ago that the hardening of rock salt crystals in bending tests is directional (487).   *Inhomogeneously* stressed crystals therefore show the phenomenon which came to be known later as the *Bauschinger effect* [cf. also (488)].   Similarly, *hysteresis* and *after-effect* occur in plastically bent crystals.   The cause of this behaviour should be sought in the internal stresses stored in the crystal as a result of the inhomogeneous stressing, the existence of which stresses can be made directly visible by the movements which occur after removal of the surface layers by partial dissolution (481).

69. Fracture of Crystals.   Sohncke's Normal Stress Law

As mentioned in Section 53, the first systematic investigations into the influence of the crystal orientation upon the tensile strength of sodium chloride crystals led to the formulation of the *Normal Stress Law* (489).   For all orientations ([100], [110], [120] and [111]) of tensile direction that have been examined the cube plane appeared as fracture plane; as the angle between fracture plane and the direction of pull was reduced, the tensile strength of the specimen increased to an extent roughly consistent with the existence of a critical normal stress of the cube plane [equation (40/2)].   In view of subsequent investigations which revealed that the nature of the planes forming the surface of the specimen influences the magnitude of the tensile strength, the normal stress law was rejected (461), and not until it had been rediscovered many years later on metal crystals was it also possible to demonstrate its validity for crystals of rock salt (490). It was found that the experiments which had led to the rejection of the law could also be regarded as its confirmation, provided there was one pair of surface planes common to the variously oriented rods. The use of cylindrical specimens avoids the surface effect, which can perhaps be regarded as an increase of the tensile strength due to the preparation of the specimen. Whereas theoretically the tensile strengths to be expected for the [100], [110] and [111] directions should be proportional to 1 : 2 : 3, the ratios actually observed are 1 : 2·24 : 3·32 when based on maximum values, and 1 : 2·22 : 3·51 for mean values. It is a striking fact, however, that the values for a [112] direction do not fit into this picture : instead of the (theoretically) expected factor of 1·5, an observed strength of about double that amount is obtained [(452); cf. also (469)]. These tests show that constancy either of the normal dilatation perpendicular to the cubic fracture plane, or of the elastic energy, is excluded as a condition of fracture [(491), (492)].

So far the normal stress law has not been confirmed for melt-grown crystals of KCl and KBr. The tensile strengths in this case are so widely scattered that the scatter range has been resolved into several maxima of frequency. Nevertheless, the values obtained for tensile directions lying in a cube plane do not appear to conflict with the assumption of a constant normal strength of the cube plane of fracture. No influence of the orientation of the surface planes was observed [(466), (470)].

A model showing the relationship between orientation and tensile strength of cubic crystals which fracture along the cube plane according to the normal stress law is illustrated in Fig. 183. The minimum strength corresponds to the direction of the edge of the cube; the maximum strength, which is three times as great, to the space diagonal.

The tensile strength values for the cube plane of rock salt crystals, like the values for the yield point, fluctuate within wide limits, usually between 200 and 600 g./mm.^{2}, according to origin and history. Sometimes values up to 1800 g./mm.^{2} are obtained (464). For the *elastic dilatation* perpendicular to the cube plane at the moment of fracture the order of magnitude of 10^{-4} is again obtained; owing to the dependence of tensile strength and modulus of elasticity upon orientation, the normal dilatation in tension parallel to the space diagonal is about three times as high as in tension parallel to the edge of the cube.

The tensile strength of sodium chloride crystals, like the yield point and elongation, also shows dependence on the *cross-sectional area* of the specimen. This, too, becomes observable only from very small diameters (about 0·8 mm.) downwards; increases up to twenty times the normal strength have been determined (472).   In the case of thin specimens produced by dissolution in water (493)

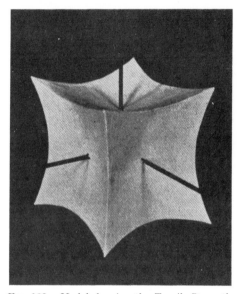

Fig. 183.—Model showing the Tensile Strength of Cubic Crystals which Fracture along the Cube Plane (452).

instead of in the melt, the objection could be raised that the water might exert a hardening effect [(480); cf. also Section 72].

The tensile strength of rock salt is influenced by *additions* in much the same way as the yield point.   Tensile strengthening and shear hardening proceed along parallel lines, as is seen from Fig. 180.   If two added constituents which form solid solutions are present simultaneously in small quantities, the strengthening effect of the single constituents becomes additive (476).   Formally, therefore, the tensile strengthening caused by the additions can be expressed in the same way as the shear hardening caused by alloying (Section 45).

Annealing, too, affects tensile strength in the same way as has been described above for the yield point. Fig. 184 shows the effects obtained with natural rock salt crystals of various origins. It will again be noted that the chief result of annealing for several hours at 600° C. is to equalize substantially the strength values, which originally differed very considerably. This common strength value (approx. 170 g./mm.²) represents a minimum which is reached after

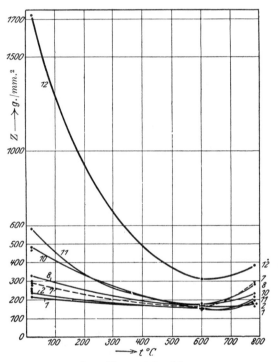

Annealing period : 6 hours

Fig. 184.—Effect of Annealing on the Tensile Strength of Natural NaCl Crystals of Various Origins (464).

the hardening arising from the previous history of the specimens has been eliminated. If the temperature of annealing is raised still further, there will be a slight increase in strength, no doubt owing to the presence of impurities.

Table XXVII summarizes the values for the normal strength of fracture planes of ionic crystals. It contains in the first place the values for melt-grown crystals of the NaCl type and produced from

De Haen preparations of maximum purity. Fluorspar was examined as a natural crystal, while strontium chloride, which also crystallizes in the fluorspar lattice, was investigated as a melt-grown crystal. Assuming the validity of the normal stress law, the extreme values for the tensile strength of *octahedral fracture planes* vary in the proportion of 3 : 1. Whereas for cube cleavage the strength maximum is found parallel to the space diagonal, and the minimum parallel to the edge of the cube, the exact opposite is true for octahedral cleavage.

It must be emphasized here that the low tensile strengths observed and their marked dependence on the orientation are closely related to the appearance of smooth fracture planes. If these are absent,

TABLE XXVII

Normal Strength Values of the Fracture Planes of Ionic Crystals

Salt.	Fracture plane.	Normal strength of fracture plane, g./mm.².	Reference.
NaCl . .	(100)	220	(476)
KCl . .		200	
KBr . .		250	(466)
KI . .		240	
CaF$_2$ . .	(111)	1550–2430 *	(495)
SrCl$_2$ . .		approx. 1100	(496)

* Cf. also (494), in which differences in strength in the proportion of 1 : 3 were observed on crystals of different origin.

or if the fracture proceeds more or less at random through the crystal, higher strength values and an appreciably lower anisotropy are obtained.

In regard to the *temperature dependence* of the tensile strength, most of the available results relate to rock salt crystals. It is necessary in this case to distinguish between static tests and those in which the load is increased rapidly. In the first case the crystal stretches considerably before it fractures, especially at high temperatures; in the second, plastic extension of the crystal is largely prevented by the rapid increase in load, with the result that the tensile strength can be determined without being affected by tensile strengthening. Natural crystals exhibited a virtually constant tensile strength of about 450 g./mm.² within the temperature range of −190° C. to + 600° C.; the scatter was of the order of only 5–10 per cent. (462).

If, on the other hand, the load is increased so slowly that marked plastic deformation can occur, appreciable tensile strengthening takes place corresponding to the extent of the deformation; this strengthening can amount to thirty times the tensile strength value at room temperature. Fig. 185 shows the increase of the tensile strength of natural rock salt crystals as a function of the reduction of area during extension. The graph gives the results obtained at high temperatures (up to 600° C.) as well as after cooling the crystals down to room temperature. The fact that all experimental results could be expressed by a smooth curve is remarkable, since it indicates that the magnitude of the tensile strength is determined solely by

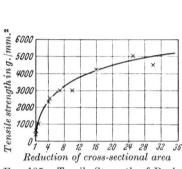

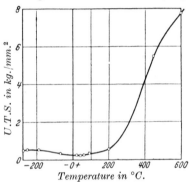

FIG. 185.—Tensile Strength of Rock Salt Crystals as a Function of Extension (Reduction of Cross-sectional Area) (462).

FIG. 186.—The Static Tensile Strength of Rock Salt Crystals as a Function of Temperature [(477), (497), (498), (499)].

the elongation, and not by the temperature at which the elongation takes place. At 600° C. certain types of rock salt exhibit tensile strengths as high as 10,000 g./mm.² (477). Fig. 186 shows the temperature dependence of the static tensile strength of melt-grown NaCl crystals (normal strength for the cubic fracture plane) within the range −269° to 600° C. At the lowest temperatures (brittle fracture) it is roughly constant; it diminishes as the temperature rises, reaches a minimum at about + 40° C. and increases rapidly at higher temperatures.

Very interesting observations have been made on the temperature dependence of the tensile strength of NaCl crystals containing various amounts of SrCl₂. The results of the experiments are given in Fig. 187. At temperatures above −100° C. the strengthening effect of the added constituent (which is also indicated by the critical shear stress of the glide planes) becomes increasingly apparent as the temperature of the test is raised.   The position is reversed for temperatures below this value.   In this case the strength of the contaminated crystal is always lower than that of the pure crystal, the difference increasing with the percentage of ingredient and with a reduction of the temperature.

It seems that both a determination of the dependence of the *strain-hardening* curve on temperature, and an investigation of the critical normal strength of the fracture plane as a function of the glide strain and the temperature, are necessary for a systematic study of the matter.

So far little is known about the mechanism of *tensile fracture*.   Clearly it is a process that takes place at a finite speed.   This fact is demonstrated by the deep cracks, parallel to the cubic cleavage plane, which are often found on tensile specimens of annealed NaCl crystals.   With crystals of potassium halides between crossed Nicols it has sometimes been possible to observe the progress of fracture across the section of the specimen, starting from one side (466).   Preliminary cinematographic tests to ascertain the duration

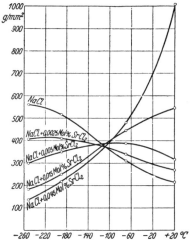

Fig. 187.—Tensile Strength of Rock Salt Crystals Containing Various Amounts of SrCl$_2$ as a Function of Temperature (498).

of the fracture of rock salt crystals led in three cases to times of 10^{-3} to 10^{-4} seconds (500).

Determinations of the *bending strength*, which for brittle fracture should coincide with the tensile strength, gave substantially different values for different orientations of NaCl crystals (501); it is remarkable, however, that these values were twice as high as the corresponding tensile strengths.[1]   This discrepancy may be due partly to the unjustified use of the theory of elasticity.   It is certain that at the moment of fracture, plastic deformations are already present in the outer parts of the bar, and that this causes appreciable deviation from the theoretical elastic stress distribution (502).   Moreover,

---

[1] Compare with the analogous case that greater force is needed to produce tensile cleavage than compressive cleavage in the bent crystal (Fig. 175).

the plastic deformation of the outer zones may have already produced tensile strengthening there. Both factors may serve to explain the high *torsional strengths* which have been obtained experimentally [about 2700 g./mm.^{2} (503)]. In this case the surface of fracture often consists of several planes. The cube plane, which is distinguished by the maximum value of the normal stress, is always very prominent in such experiments (478).

Owing to the fact that the normal stress law has so far been confirmed only for temperatures at which fracture is preceded by more or less marked deformation, this law has recently been regarded as originating from a law of strain strengthening from which no inference can be made about the behaviour of undeformed crystals (504). It is clear, in fact, from static tensile tests carried out with NaCl crystals at elevated temperatures that tensile strengthening can be very considerable. Nevertheless, in view of the extremely diverse gliding possibilities which are available in the various tensile directions (see above), it seems to us that the normal stress law, which was established at room temperature for crystals of the NaCl type, usually at small plastic deformations, should be regarded as analogous to the shear stress law, and should be applied to cases where no deformation has taken place.

70. *Hardness* (*see* 505)

Hardness, which is closely related to the plasticity and strength of crystals, is a property much more obscure than the critical shear stress of the glide systems or the normal strength of the fracture planes which have been discussed above. In reality the term embraces many properties which are distinguished according to the method by which the " hardness " is determined. As a characteristic property of materials, hardness is technically very important; a further advantage is the relative ease with which it can be measured.

The various methods for determining hardness can be summarized as follows (see 506) :

I. The penetrating body moves along the surface of the test piece :

    1. Comparison of hardness by mutual scratching.

    2. Scratching with a hard point under a controlled load :

        (*a*) measurement of the load at which the scratches become just visible;

        (*b*) measurement of the load for a given width of scratch ;

        (*c*) measurement of the width of scratch at a given load.

70. Hardness 249

3. Determination of the wear resistance of the test surface or of its resistance to penetration by a drill or a rotating disc.

II. The penetrating body moves vertically relative to the test surface:

1. Indentation tests (ball or cone indentation).
2. Impact test. A punch of a fixed shape is driven into the surface of the specimen with a given energy.

An intermediate position is taken up by a method which derives hardness from the damping of an oscillating pendulum which rests on a ball, knife edge or point on the test surface. Finally, mention should be made of a method which measures hardness from the magnitude of the rebound of a small drop-hammer.

From the above summary it is seen that, as mentioned above, hardness is not a uniquely defined quantity. Quantitative comparisons become possible only if the same test procedure is applied. To-day it is still impossible to express hardness values in terms of simple mechanical properties.

The following remarks relate to the various methods of testing and the hardness values obtained therewith.

The basis for the testing of hardness by mutual scratching is provided by Mohs' hardness scale, the standard minerals of which are shown in Table XXVIII.

TABLE XXVIII

Hardness Scale (Mohs)

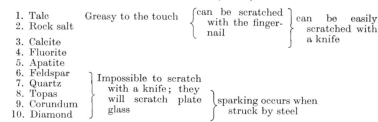

A material that is not scratched by a mineral of hardness number n, but is scratched by a mineral of number $(n + 1)$, has a hardness which lies between these two numbers and is therefore designated by $(n + 1/2)$. Much finer differences than are discernible by this method can be obtained with the aid of the methods mentioned under I. 2. These more refined procedures for determining scratch-hardness make it possible to study anisotropy of hardness in various directions on the same surface. A particularly convenient apparatus which operates on this principle is the Martens scratch-hardness testing machine (507). In Figs. 188 and 189 are shown the results of scratch-hardness determinations on the cube and octahedral planes of NaCl and fluorspar crystals, plotted in the form of curves (508). These are obtained by connecting the terminal points of lines which start from a point on the surface and whose length is proportional to the scratch-hardness measured in that direction.

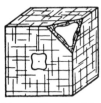

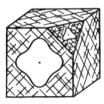

FIG. 188.—Scratch-hardness Patterns on (100) and (111) Planes of NaCl Crystals.

FIG. 189.—Scratch-hardness Patterns on (100) and (111) Planes of Fluorspar Crystals.

The relationship of the scratch-hardness to the crystal symmetry, and also to the cleavability as indicated in the figure, is seen clearly. It can be said in general that measurable differences in hardness appear especially in crystals that possess cleavability; if the surface under investigation is perpendicular to cleavage planes, the hardness parallel to the direction of cleavage is a minimum; if the surface is inclined to a cleavage plane, the hardness is usually different in opposite directions. Scratch-hardness measurements have also disclosed relationships between hardness, crystal structure and chemical constitution (506).

Drill-hardness tests serve only to determine surface hardness. The measure of drill-hardness is the number of revolutions which a drill or disc, under constant load, must perform in order to produce a hole of given depth. Hardness tests by grinding, too, yield as a rule only a measure of surface hardness. The work necessary under given conditions to grind away 1 cm.³ of the crystal parallel to the surface examined, serves as a measure of hardness. It was recently shown that this method is also sufficiently sensitive to detect anisotropy of hardness on the surfaces of crystals (509).

The indentation and impact methods, as well as the method of measuring hardness by rebound, are used mainly for metals. The Brinell Ball Hardness Number, in particular, is one of the most important technological data. This test has recently been systematically applied to alkali halide crystals. Table XXIX contains the values obtained in this way.

By means of such hardness measurements it was possible to follow the work hardening of crystals by plastic compression and their softening by recrystallization. An investigation of the plasticity of ionic crystals by the *cone*-indentation method led to a formula

TABLE XXIX

Brinell Hardness of Alkali Halide Crystals [(485), (486)] (*Based on the Diameter of the Indentation on the Cube Plane with a Load of* 5 *kg. and a Ball Diameter of* 0·71 *cm.*)

Material.	Hardness.	Material.	Hardness.	Material.	Hardness.
NaCl .	12·4	NaI	8·4	KBr	5·4
NaBr .	9·2	KCl	5·8	KI	3·2

containing five constants of the material, representing the cone-indentation hardness at different temperatures and for various periods of pressure (510). This shows clearly that hardness is by no means a simple basic property. On the idealizing assumption that the forces are transmitted to the crystal by the cone indenter (apex angle 90°) only in a direction perpendicular to the crystal face, the distribution of shear stresses in the dodecahedral glide systems in the tested surface has been calculated; it was found to agree qualitatively with the pressure figures which appear on the cubic, rhombic, dodecahedral and octahedral faces of rock salt (511). The pendulum hardness test, too, was for long restricted to polycrystalline

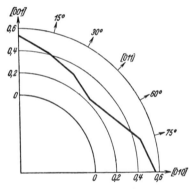

Fig. 190.—Anisotropy of the (001) Plane of NaCl Crystals as Determined by Pendulum-hardness Tests (514).

metals (512); recently, however, it has also been applied with success to the examination of ionic crystals [(513), (514)]. By way of example, Fig. 190 shows the anisotropy of hardness on the cube plane of rock salt, measured very accurately by this method.

In conclusion, it should be emphasized that we shall be unable to interpret so complicated a property as hardness until we have succeeded in relating the much simpler phenomena of crystal cleavage and deformation to the constitution and lattice structure of the

Elasticity Coefficients (s_{ik}) *of Ionic*

Material.	Crystal. Class.	Crystal. System.	s_{11}.	s_{12}.
Rock salt, NaCl . . .	⎫	{	24·3	− 5·26
	⎪		23·0	− 5·0
Sodium bromide, NaBr . .	⎪		40·0	−11·5
	⎬ O_h cubic {		27·4	− 1·38
Sylvite, KCl	⎪		29·4	− 5·3
Potassium bromide, KBr . .	⎪		31·7	− 4·7
Potassium iodide, KI . .	⎭		39·2	− 5·4
Fluorspar, CaF$_2$ . . .			6·91	− 1·49
Pyrites, FeS$_2$	T_h		2·89	+ 0·44
Sodium chlorate, NaClO$_3$ . .	T		24·6	+12·6
Beryl, Al$_2$Be$_3$(SiO$_3$)$_6$ . . .	D_{6h}	hexagonal	4·42	− 1·37
Hæmatite, Fe$_2$O$_3$ . . .	D_{3d}		4·42	− 1·02
Calc.-spar, CaCo$_3$ . . .	D_3	rhombo-	11·3	− 3·74
Quartz (rock crystal), SiO$_2$.	D_3	hedral	13·0	− 1·66
Tourmaline	C_{3v}		3·99	− 1·03
Topaz, (AlF$_2$)SiO$_4$ . . .	⎫		4·43	− 1·38
Barytes, BaSO$_4$ . . .	⎬ V_h	rhombic	16·5	− 8·97
Aragonite, CaCO$_3$ . . .	⎭		6·96	− 3·04

crystals. In addition, the quality of the surface will also need very careful consideration. Adsorption of polar molecules can cause, by the reduction of the cohesive forces near the surface by amounts corresponding to the energy of adsorption, very substantial reductions in hardness (more than 50 per cent.) (515).

71. *Effect of Cold Working on Various Properties. Recovery and Recrystallization*

We shall refrain from a detailed discussion of the anisotropy of the physical properties of ionic crystals, since the effect of cold working on these properties has not yet been studied quantitatively. We shall present only the elastic constants in Table XXX in so far as complete determinations were available.

As in the case of metal crystals, so, too, in that of the heteropolar crystals there is no evidence that *lattice dimensions* are influenced by deformation. The lattice constant of a plastically bent crystal of sylvine equals that of undeformed specimens within 0·5 per thousand. Diffusion of the K_α doublet was not observed (518). In agreement

Crystals (in dyn.$^{-1}$/cm.^{2} . 10^{-13})

$s_{44}.$	$s_{33}.$	$s_{13}.$	$s_{14}.$	$s_{22}.$	$s_{55}.$	$s_{66}.$	$s_{23}.$	Literature.
78·7	—	—	—	—	—	—	—	(516)
78·0	—	—	—	—	—	—	—	(517)
75·4	—	—	—	—	—	—	—	(517)
156·0	—	—	—	—	—	—	—	(516)
127·0	—	—	—	—	—	—	—	
161·0	—	—	—	—	—	—	—	(517)
238·0	—	—	—	—	—	—	—	
29·6	—	—	—	—	—	—	—	
9·48	—	—	—	—	—	—	—	
83·6	—	—	—	—	—	—	—	
15·3	4·71	−0·86	—	—	—	—	—	
11·9	4·44	−0·23	+0·79$_5$	—	—	—	—	(516)
40·3	17·5	−4·33	+9·15	—	—	—	—	
20·0$_5$	9·90	−1·52	−4·32	—	—	—	—	
15·1	6·15	−0·16	+0·58	—	—	—	—	
9·25	3·85	−0·86	—	3·53	7·53	7·64	−0·66	
84·0	10·7	−1·92	—	18·9	34·9	36·0	−2·51	
24·3	12·3	+0·43	—	13·2	39·0	23·5	−2·38	

with this *the changes in density* of NaCl crystals after deformation and heat treatment do not exceed 0·5 per cent. (519).

However, again in line with metal crystals, the virtual invariance of the lattice dimensions after cold working is by no means equivalent to complete intactness of the lattice. Attention has already been drawn in Section 67 to the stress birefringence which can be observed in polarized light after glide. An attempt to measure quantitatively the tensile (and compressive) stresses parallel to the operative (110) glide planes of NaCl crystals gave a value of about 2·3 kg./mm.^{2} (463). The value of these local stresses exceeds considerably not only the yield point but also the tensile strength. Their determination is based on the proportionality within the elastic range between stress and birefringence. The distribution of the stresses corresponds to a bending of the glide packets. Very much higher stresses amounting to 60 kg./mm.^{2} have been calculated from the asterism of the Laue photographs of bent NaCl crystals (520). Owing to the inhomogeneous distribution of these stresses along the operative glide planes there occurs in the vicinity a lattice distortion of not more than 1° (521).   As was observed with metal crystals (Section 59), reverse bending of the crystals is accompanied by a reversal of the bending of the lattice.

Attempts have also been made to determine the magnitude of internal stresses by measuring the absorption spectrum of deformed NaCl crystals (522).   It has been found experimentally that the absorption maximum of coloured crystals (see below) which is at 4650 Å., is displaced after deformation by about 100 Å. towards the red.   This indicates a reduction of the mean photo-electric work function for the most loosely held electron of the colour centres by $8 \cdot 10^{-14}$ erg per atom.   If this energy were available as stored elastic energy in the ionic volume, it would imply local stresses of the order of 300 kg./mm.^{2}

An attempt has been made to determine directly, on compressed crystals of synthetic KCl, the *lattice disturbances* which accompany deformation, by measuring the intensities of X-ray interferences and of the background scatter between the Debye–Scherrer circles [(518), (523)].   The disturbances accumulated in the vicinity of the operative glide planes do not merely reduce the absolute intensity in the directions of interference ;   they also modify the relative intensities of various reflexions.   In accordance with their distribution they do not influence the width of the lines (524), but they increase the diffuse scattered radiation.   The intensities of the various orders reflected from the cube plane, measured by means of an ionization spectrometer, do in fact show a reduction of the relative intensities of the higher orders in the compressed state compared with the undeformed state of the crystal.   Based on this observation, an estimate of the lattice disturbances present after 3·8 per cent. compression indicates that about 2 per cent. of the lattice points are displaced from their normal positions with a maximum displacement of about 1/8 of the identity period in the direction of glide.   Measurement of the intensity of the X-rays which had been diffusely scattered by a compressed KCl crystal (between the interference maxima) was carried out with a Geiger counter.   It showed clearly the increase in scattered intensity resulting from lattice disturbances.

Changes in the *moduli of elasticity* as a result of plastic deformation have not so far been observed.   Bending tests carried out on NaCl crystals showed that Hooke's Law was valid up to 97 per cent. of the fracture load (479).

The changes in the crystal lattice which accompany plastic deformation can be seen very clearly from their effect on the phenomena of *coloration* which occur when the crystals are irradiated with ultra-violet light, X-rays or γ-rays. Crystals of NaCl, with which these investigations have been mainly carried out, become yellow by such irradiation. This is caused by the appearance of a selective absorption band, which at room temperature lies between 3500 and 5500 Å., with a definite maximum at 4650 Å. This absorption band is probably due to single sodium *atoms* which do not belong to the lattice and which absorb visible light; rock salt that has been coloured yellow by treatment with sodium vapour followed by sufficiently rapid cooling, shows exactly the same absorption spectrum (525). The free sodium atoms arise from the absorption of a quantum of radiation by a chlorine ion, by which process its extra electron is transferred to a sodium ion (526). The intensity of the colour increases with the intensity and duration of radiation, until finally a state of equilibrium is reached between the formation of atoms and the reverse formation of ions, so that the intensity of

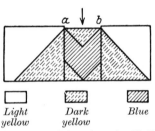

FIG. 191.—Coloration of a NaCl Crystal Compressed between *a* and *b* (529).

the colour remains constant. In certain cases impurities dissolved in the lattice may undergo a sort of coagulation, and this may lead to a new coloration—the cause of which, however, is quite different (colours due to diffraction by submicroscopic particles).

There are two ways in which plastic deformation influences coloration [(527), (528)]. First, it diminishes the intensity of the existing colour as measured by the absorption coefficient in the maximum of the band, but without changing the spectral distribution; secondly, it increases the capacity for developing a colour by irradiation, *i.e.*, the absorption of crystals that are still uncoloured (cf. Figs. 191 and 192). In this way plastically deformed zones can be made visible to the eye by a simple device. Crystals grown from the melt or annealed crystals are more susceptible to colouring than natural crystals [(530), (531), (532)].

Although the greater part of the research into the effect of deformation upon colouring has been carried out on NaCl, particulars are also available for other crystals. A detailed discussion of the work which has been done in this field so far, together with particulars of experimental coloration tests, will be found in (533) and (528). It may be mentioned as an interesting fact that the twinned parts of calcite twins react to radiation in the same way as the original crystal.

Electrolytic conductivity is another property which can be substantially influenced by deformation. It is true that the experimental data available on this subject are still somewhat conflicting, owing mainly to the difficulty of carrying out reliable measurements. At low temperatures the currents are very small; at high temperatures the effect produced by deformation is small compared with the normal conductivity. Crystal recovery and recrystallization, too, interfere with the study of the effect of deformation. It is certain, however, that with NaCl crystals a sudden increase in conductivity accompanies each increase in load within the range of plastic deformation (534). The increased conductivity then diminishes

(a) Undeformed uncoloured crystal.

(b) Uncoloured crystal, bent, and then coloured by irradiation.

(c) Undeformed crystal coloured by irradiation.

(d) Crystal coloured by irradiation and then bent.

FIG. 192 (a)–(d).—Change in the Capacity of NaCl Crystals for Coloration as a Result of Bending (527).

with time, and it is not clear whether it continues to fall until the initial value is reached, or whether an effect of the plastic deformation remains [(535), (536)]. If crystals from which the load has been removed are loaded again, the jump in conductivity does not occur until the previous maximum load has been exceeded. This hardening effect is fairly permanent at 40–50° C.; even after 24 hours no softening effects are observed. With regard to the influence of deformation, the same contrast exists between ionic conductivity (heteropolar crystals) and electronic conductivity (metals) as in the dependence of conductivity on temperature. Whereas rising temperature and plastic deformation lead to an increase in *resistivity* where conduction is by electrons, they result in an increase in *conductivity* where conduction is by ions.[1]

The *rate of solution* of deformed NaCl crystals is raised within areas of high internal stress (revealed by double diffraction) (519).

[1] Melt-grown crystals exhibit a conductivity which is many hundred times higher than that of the best solution crystals (532); impurities increase conductivity (537).

The changes which deformation produces in the properties of ionic crystals can be reversed by *crystal recovery* and *recrystallization*. These phenomena were observed even at room temperature in *compressed* crystals of NaCl and KCl, which formed the main subject of study [(538), (539)]. The most suitable method of investigation was found to be the coloration method, by means of which it was also possible to observe directly the shape of newly formed grains and their speed of growth. The grain boundaries are seen to be cube planes which push forward into the deformed material, along parallel lines, at a speed that often remains constant for days (about 0·1 mm. per day at room temperature). Sometimes the speed changes abruptly (540). The growth of a newly formed grain has been attractively demonstrated in a slow-motion film (541). The speed of recrystallization increases with the degree of deformation and the annealing temperature (542). A diagram, similar to the recrystallization diagram, showing the speed of recrystallization (reciprocal value of the time which elapses before coloured spots appear when irradiating compressed salt) as a function of the time of annealing and the percentage of reduction, is available for a certain type of rock salt (543).

This observed constancy of the rate of growth conflicts to some extent with the results of corresponding tests on metal crystals (Section 62), which revealed a reduction in speed as annealing continued. The explanation may be that recovery can be only slight in the compressed and microscopically distorted NaCl crystal.

Experiments with NaCl, KCl and KBr crystals showed that irradiation has a very marked retarding effect on the formation of recrystallization nuclei, but not on the rate of growth of the new grains. As a rough quantitative estimate it may be said that the neutralization of about 10^{-6} of the ions suffices to delay the recrystallization for many days [(543), (544)].

Detailed studies of the recovery of ionic crystals have not yet been carried out. That recovery actually takes place has already been concluded in (519) from the fact that flow occurs in bending tests at room temperature. A reduction of the rate of recrystallization of compressed crystals by a short intermediary annealing at high temperature (540) is probably also an effect of recovery.

72. *The Effect of a Solvent on the Mechanical Properties.*
Joffé Effect

The diversity of the plasticity and strength phenomena shown by ionic crystals increases greatly if the crystal is exposed to a solvent while it is subjected to stressing. In recent years experiments of this type have been very numerous, owing to their importance for the theory of crystalline strength. The material used was generally rock salt.

The solvent has a twofold effect : it causes an appreciable increase not only of the plasticity but also of the strength.

It has long been known in salt mines that the normally brittle crystals can be bent and twisted in warm water. Systematic bending tests on rock salt prisms confirmed not only the increased plasticity of the crystals under water, but also showed that the maximum tensile stress in crystals deformed in the wet state can greatly exceed the bending strength of dry crystals (545). Since for ionic crystals, too, the dynamics of the glide process should be determined by the initial critical shear stress of the glide system, and by the work-hardening curve, the effect of the solvent must be described in terms of its influence on these properties. The following observations have been made of the start of plastic deformation in crystals which are being simultaneously dissolved. Initially there is a very perceptible increase in the rate of flow of wetted crystals in the bend test [(547), (549)]; the limits of plasticity, however, remain unchanged (550). Invariance of the yield point was observed under tensile stress also (552). In particular, the shear stress law still applied in cases where the crystals were dissolved as uniformly as possible, and the value of the critical shear stress agreed within the limits of error with the value obtained in the dry test (553). If the direction of tension is in the vicinity of a body diagonal, then, in view of the dependence of the yield point upon orientation, extension will take place by glide in a cube plane. The critical shear stress for this glide system amounts at room temperature to 238 g./mm.^{2}. That is more than three times the value for the principal (dodecahedral) system (554).

Unfortunately, it is still not possible to describe the effect of water in terms of its influence upon the strain hardening. We do not yet know the strain hardening curve appropriate even to the normal tensile test. The results so far indicate that if deformation and dissolution both take place at the same time, the increase in shear stress is much less than in the dry test. The shear hardening for a given deformation is greatly reduced by the solvent (550). Consequently, if a crystal under tension is wetted, its rate of flow is increased. It is also remarkable that if rock salt is subjected to a bending test, the (shear) softening influence of the water takes almost full effect if only the compressed side is wetted, but is almost absent if dissolution occurs on the tensile side only (547).   Since very high deformations can be achieved in water, the glide bands of square specimens which have been fractured under water can become visible by ordinary transmitted light (Fig. 193).

The effect of solution on the tensile strength of rock salt crystals was revealed by the Joffé experiment (546), in which crystals are subjected to a given load and then surrounded with water up to a certain level, the water being removed at the moment of fracture. The values obtained in this way, referred to the final cross-section, greatly exceed the normal strength of rock salt; they amount to as much as 160 kg./mm.^{2}.   Although such exceptionally high values could not be obtained when the tests were repeated, it is certain that the strength can be increased up to 25 times of the value in the

Fig. 193.—Glide Bands on a NaCl Crystal which has been Fractured under Water (552).

dry state (*i.e.*, to about 10 kg./mm.^{2}).   An element of uncertainty is introduced in these experiments by the circumstance that the tension could not have been uniaxial in the narrowest cross-section of the constricted specimen (see Fig. 194).   When the conditions of solution were changed so as to reduce the cross-section along the whole crystal [periodic raising and lowering of the water level (555); rotation of the crystal placed horizontally in the solvent (553)], the tensile strength values could be noticeably increased, and fracture was preceded by elongations of up to 45 per cent.

The most important facts about the Joffé effect may be summarized as follows :

1. It is not absolutely essential for the attainment of higher strength that solution should take place while the specimen is loaded. Unloaded crystals which have been subjected to solvents and tested immediately after drying exhibit values which are just as high, providing solution has proceeded sufficiently far.   In general, where

the initial cross-section was the same, the greater the degree of solution the higher the strength [(548), (551), 553)].

2. It was found that the orientation of the crystal had no influence. Those specimens in particular which were oriented parallel

(a) Full view.

(b) Appearance of the upper portion.

(c) Upper portion of a crystal dissolved in hot water.

Fig. 194 (a)–(c).—Shape of Cylindrical NaCl Bars after Fracturing under Water (551).

to a body diagonal (slight cubic glide in place of extensive dodeca-hedral glide; cf. Section 67) exhibited the effect to the same extent (551). Consequently, the normal stress law is valid also for wetted crystals (552).

3. Fracture in water is always preceded by plastic deformation.

This is shown by direct measurements of length and by observing the changes in orientation (inclinations of up to 30° were observed for the cubic fracture plane). The same applies to crystals wetted without load (cf. Fig. 195). Extension occurs also in specimens with the axis parallel to one of the body diagonals, but on a smaller scale, presumably by cubic glide [(551), (552), (554)]. The fractured rods can be more easily coloured by ultra-violet light as a consequence of the preceding deformation (556). A tensile test in which the load is increased rapidly with the object of minimizing extension so far as possible is described in (548).

4. Tempered crystals whose strength in the dry state had been reduced almost by half, attain in the Joffé tensile test the same high

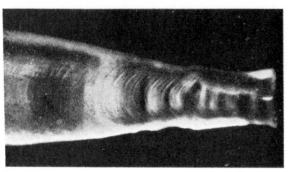

Fig. 195.—Glide Bands on a Wetted NaCl Crystal without Load after Subsequent Fracture in the Dry State (551).

values as are exhibited by material which has not been previously treated [(551), (552)].

5. High tensile strengths are exhibited by rock salt crystals even when dissolved in more or less saturated solutions of NaCl (investigated in up to 80 per cent. saturated solutions) [(551), (559)]. The effect is said to be absent if the solution is saturated (546). The same mechanical properties as in water were also obtained in concentrated sulphuric acid and in a 25 per cent. solution of SO_3 in H_2SO_4 in which NaCl is decomposed (557).

6. The Joffé effect was found with crystals of potassium chloride and potassium iodide immersed in water, and with potassium iodide immersed in anhydrous methyl alcohol [(558), (557)].

There is still some uncertainty as to the relation between the increase of strength on the one hand, and the initial stress, degree of solution and the deformation which precedes fracture, on the other.

The experimental results differ in regard to the after-effect of solution. Whereas in the bending test the increase in plasticity ceases as soon as the water is removed (547), in the tensile test the normal strength of the dry state is not regained until after a few days (548). The time required for this after-effect varies with the solvent (559).

With NaCl specimens oriented parallel to the cube edge it is very noticeable that down to a definite initial stress of 216 g./mm.^{2} no increase in strength takes place (551). In this case the crystals do not fracture where the cross-section is smallest, but in the zone that has remained dry, about 0·1–0·5 mm. above the water level. The tensile values are therefore identical with the initial stress, which is about 50 per cent. lower than the normal dry strength. These tests give no indication as to the strength of the portion which has been exposed to the solvent. Where the initial stresses are small there is appreciable solution, fracture takes place in general at the narrowest part of the crystal, and high tensile strengths are obtained. Dependence of the strength upon the magnitude of the applied stress, if it exists at all within this range of initial stress, merely takes the form of a slight increase of strength as the initial stress is reduced. Complete independence of the initial stress would involve the independence of the resultant strength from the degree of solution. This conclusion is not borne out by the tests carried out with wetted specimens under no load, mentioned under (1).

Owing to the shape of the wetted crystals the recognition of a relationship between the increased tensile strength and previous deformation is difficult (cf. Fig. 194). The extension is by no means distributed evenly over the wetted portion of the crystal. In any case, so far all tests devised to reveal a clear connection between strength and the degree of deformation have failed. Even specimens with axis parallel to a body diagonal show the same increase in strength, although the extension is very slight.

There is still no satisfactory explanation of the solution effect. The explanations attempted hitherto differ in their basic assumptions.

According to the interpretation suggested by the discoverer of the effect, the cause of the increase of strength is to be sought in the elimination of surface cracks which, by their notch effect, tend to lower the strength (546). The inherent strength of the *crystal* is not raised; the *technical* strength obtained in the normal tensile test is less than that of the true strength of the crystal owing to the notch effect of the fine cracks which are always present. In support of the view that stresses of the order of 100 kg./mm.^{2} can actually occur in the interior of crystals, tests have been carried out with spheres of rock salt whose temperature had been suddenly raised from the temperature of liquid air (560). The stress produced within the specimen on immersion in a lead bath at 600° C., which was calculated at 70 kg./mm.^{2} and which did not lead to fracture, was regarded as a confirmation of the high inherent strength of the crystal.

An objection to this interpretation is that it assumes a very great depth for the surface cracks, since no increase in strength is produced until a large part (more than 50 per cent.) of the section has been removed. The " sphere " experiment has also been objected to on the grounds that plastic deformation occurs in the surface layers, thus preventing the development of high stresses in the interior (561).

Recently this view has been discussed again with the necessary addition that the depth of the crack depends on the dimensions of the crystal (562); in this way an explanation can also be found for the increase of strength observed with very thin crystals in the dry test (563).

According to a second explanation of the Joffé effect, the increase in strength is to be ascribed to the preceding deformation. It is assumed that the strength of the undeformed crystal is small and that it rises only as a result of deformation to the high values which are theoretically required (561). The effect of the water is believed to consist in the removal of impediments from the surface, thus assisting the crystal to deform (547).

This interpretation assumes the existence of a definite relationship between the observed strength and the degree of deformation which, however, has so far not been proved. In this connection attention is again drawn to the results which were obtained with specimens parallel to the body diagonals and which are difficult to reconcile with this assumption.

A combination of the two conceptions is discussed briefly in (553), (564). The high strength is attributed exclusively to the effects of work hardening. The requisite high degree of plasticity exhibited by the wetted crystals is believed to be due to the removal by solution of cracks which develop in the course of gliding, and which in the dry test lead to fracture.

Further attempts at elucidation assume that water penetrates into the crystal. While on the one hand the crystal is presumed to acquire in this way a very great capacity for strain strengthening (" Reissverfestigung ") (565), it is also believed, especially in view of experiments on the dependence of the Joffé effect upon orientation, that changes in the interior of the crystal due to the penetration of water are mainly responsible for the high strengths obtained (551). That plastic deformation is not an indispensable condition for the change in the mechanical properties of wetted crystals is suggested by the increased scratch hardness of crystals subjected to extensive solution (566); however, in later tests in which solution was less severe this increase was not found (567). An increase in the ionic conductivity also points to an internal effect of the water (565).

Although it was not possible to obtain a direct verification of the penetration of the water by measuring the lattice constants and the density [with an accuracy of about 0·1 per thousand (566)], this proof has been supplied for wetted crystals by observing in them the ultra-red absorption typical of water (568).

With the object of disproving the theory that penetrating water is responsible for the increase in strength, experiments were carried out in which narrow strips of crystal were protected against solution by a coating of vaseline; the observed strengths agreed with those obtained in the dry test (569).

From what has been said it will be seen that a satisfactory explanation of the Joffé effect has still to be found. An important contribution to the solution of this problem would probably result from tensile tests carried out on crystals in varying stages of dissolution, and in which plastic deformation during the tests had been minimized (low temperature, short distance between the clamps).

INTERPRETATION OF THE BEHAVIOUR OF SINGLE CRYSTALS
AND CRYSTAL AGGREGATES

CHAPTER VIII

THEORIES OF CRYSTAL PLASTICITY AND CRYSTAL STRENGTH

The deformation and fracture of crystals obeys a number of laws. Although the criterion of mechanical twinning has not yet been established, the most important of the deformation mechanisms (gliding) has already been largely elucidated. The principal facts are given in Section 50; they can be recapitulated as follows :

1. Shear stress law.

2. Low critical shear stress which, down to the lowest temperatures, is only slightly dependent on temperature.

3. Increase in the shear strength with increasing glide; work -hardening.

4. Dependence of work hardening on temperature (and speed of deformation). This is insignificant at the lowest temperatures and close to the melting point, but substantial at intermediate temperatures.

A simple law has also been discovered governing the cleavage fracture of crystals (Sohncke's normal stress law). The values for the critical normal stresses are as low as for the critical shear stresses for glide.

A physical theory should have for its object the formulation of a conception of the structure of the solid which explains the empirical laws, and the quantitative derivation therefrom of the observed values of the mechanical properties. In addition, the processes of recovery and recrystallization which are observed in plastically deformed crystals (Section 49 and Chapter VI, G) have also to be explained.

It is naturally a condition of the theoretical interpretation of the tensile and shear strengths of crystals that the effective binding forces and the laws by which they are governed should be known. It is precisely in this field, however, that our knowledge is limited. According to whether the lattice particles are ions, or atoms showing no polarity, a distinction is made between polar and non-polar binding. Theoretically the position is clearest in the case of the polar (heteropolar) crystals, such as salts. Here electrostatic forces are responsible for holding the lattice together; the law assumed for these has been generally confirmed. To the crystals of non-polar binding belong, first of all, the metals. According to present conceptions a metal crystal is held together by positively charged atom cores embedded between the free metal electrons. However, there is still no satisfactory theory on the subject.[1] Other non-polar crystals are the molecule lattices (H_2,N_2,CO and many organic compounds) and the crystals of the rare gases. In this case it has been possible, with the aid of quantum mechanics, to account for the attractive forces which hold the lattice together by attributing them to the deformability of the molecules or atoms. The forces decrease with the inverse seventh power of the distance. Special difficulties arise with a series of non-polar crystals, such as materials of the diamond type with tetrahedral bonds, benzene, etc.

73. Theoretical Tensile Strength

The mechanical property which, so far, has been examined theoretically in greatest detail is the tensile strength. The only calculations carried out strictly on the basis of the lattice theory relate to the tensile strength of rock salt, that is of an ionic crystal. This is because the ionic crystals best illustrate the laws of the interatomic forces. Two forces are mainly effective in these heteropolar crystals : electrostatic (Coulomb) forces between the ions, and a repulsive force which prevents the ions from penetrating into each other. The electrostatic forces are governed by Coulomb's law, repulsion by the interaction of the electron clouds of the ions. In view of the existence of a stable equilibrium, repulsion is bound to diminish much more rapidly with the distance than the attraction. At a given distance, represented by the lattice constant, attraction and repulsion are in equilibrium. From this we obtain for the force (K) the equation (570)—

$$K = \frac{e^2}{\rho^2} - \frac{b}{\rho^n + 1} \ . \quad . \quad . \quad . \quad (73/1)$$

in which e and ρ represent the charge and the distance between the ions, while b and n are constants (cf. Fig. 196). $n \simeq 9$ was determined for the alkali halides from the lattice constants and the compressibilities. In order to arrive at the tensile strength it is necessary to know also the " lattice energy " ϕ, i.e., the energy which must be employed in order to dissociate completely 1 g.

---

[1] *Translator's Note.* This statement is no longer true : see the works of Mott and Jones, Brouillin, Seitz, Hume-Rothery and others.

molecule of the crystal (to place infinite distance between the ions).

If now a crystal of rock salt is stressed parallel to a cube axis, then the cubic lattice (with the lattice constants, a_0) is transformed into a tetragonal lattice (with the constants a and h). In this

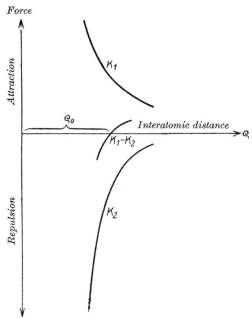

Force

Attraction

K_1

ϱ_0

Interatomic distance → ϱ

K_1-K_2

Repulsion

K_2

FIG. 196.—Overlapping of the Coulomb Attractive and Repulsive Forces in Ionic Crystals.

process, of course, the lattice energy of the crystal depends on h. But since Poisson's ratio results solely from elongation, then

$$\left(\frac{\partial\phi(a, h)}{\partial a}\right)_h = 0$$

from which the functional connection of a and h is obtained. If inserted in the expression for ϕ, then the lattice energy is obtained solely as a function of h. If the tensile stress K is necessary in order to increase h by dh, then clearly $K \cdot dh$ is equal to the reduction of the lattice energy $d\phi$:

$$K = -\frac{d\phi}{dh}. \quad \cdot \quad \cdot \quad \cdot \quad \cdot \quad \cdot \quad (73/2)$$

If h increases from a_0 onwards, then initially the repulsive force increases also in an attempt to restore the original undeformed condition. This increase in strength, however, does not persist at greater deformations. After a maximum has been exceeded the energy soon becomes negligibly small as dilatation continues. The maximum K_z corresponds to the tensile strength. It can be calculated from (73/2) for a constant h characterized by $\dfrac{d^2\phi}{dh^2} = 0$. The calculated tensile strength for cubic specimens of rock salt (571) was 200 kg./mm.[2] with an elastic elongation of 14 per cent.[1] Similar elongations before fracture are also to be expected for other ionic crystals, so that about one-tenth of the modulus of elasticity in the direction of stress may be regarded as a rough indication of the theoretical tensile strength.

The same order of magnitude for the tensile strength of crystals was derived from energy considerations (573) which, owing to their independence of the valid law of energy, are by no means limited to ionic crystals. The basic idea is as follows. The additional surface energy which results from the development of the two fracture planes must at the moment of fracture be available in the crystal as elastic energy of deformation. In order that the surface energy may be fully effective, it is necessary that the distance between the two fracture planes shall be greater than the range of action of the molecular forces. Let Δl be the elastic extension of the specimen immediately prior to fracture. If the grips are held rigidly in position, this amount is also the distance between the two fracture planes. If now α is the free surface energy and Z the tensile strength, then

$$Z \cdot \Delta l > 2\alpha \quad \cdots \quad \cdots \quad (73/3)$$

since the elastic stress energy is certainly smaller than $Z \cdot \Delta l$. With the aid of the values for α and Z it is therefore possible to calculate a lower limit for the range of molecular forces. For rock salt (α NaCl ~ 150 dyn/cm., $Z = 2 \cdot 2 \cdot 10^7$ dyn/cm.[2] for fused crystals) the value is ~ 1400 Å. Undoubtedly this value is unduly large since we know from other sources that the radius of action of the lattice forces is no greater than the distance between the lattice points (amounting to a few Å.). This contradiction could be resolved only by assuming a tensile strength which was 100–1000 times greater than that experimentally determined.

[1] Recently in place of (73/1) a law of energy was quoted in which repulsion is represented by an experimental function and in which allowance is made for the van der Waals forces of attraction which are always present (572). A new calculation of tensile strength has not yet been undertaken. It is true, however, that the lattice energy will vary by no more than 0·2 per cent. if rock salt is used.

If this method of estimating the theoretical tensile strength is applied to metallic crystals, values are obtained which again exceed the test values by many orders of magnitude. In zinc crystals, for instance, Δl is ~ 9000 Å. when fracturing about a basal plane disposed obliquely to the tensile direction ($a_{zn\ \text{fused}} \sim 800$ dyn/cm.* $Z_{(0001)} = 1 \cdot 8 \cdot 10^7$ dyn/cm.^{2}).

The main conclusion to be drawn from these considerations is that the observed tensile strengths of crystals cannot be regarded as due to a breakdown of the lattice forces over the whole of the fracture area. The lattice strengths are higher than the technical strengths by 2–3 orders of magnitude. Independently of any calculation this is readily shown by the fact that the modulus of elasticity is constant nearly to the point of fracture (574). If the forces which hold the crystal together were, in fact, overcome at the moment of fracture, then this would correspond to a modulus of elasticity of zero, parallel to the tensile direction, and thus appreciable reductions in the modulus should be observed at an earlier stage.

74. *Calculations of the Theoretical Shear Strength*

The conception underlying a (lattice) theoretical calculation of the shear strength of glide planes is that the forces which are effective between the lattice planes can be represented by a model resembling a file (575). A reciprocal displacement in a longitudinal direction of two files in direct contact is possible only if they are slightly separated from each other before each stage of glide begins. Normal dilatation perpendicular to the glide planes is also stated to be a condition of glide in crystals. Before glide can take place, therefore, it is necessary to reduce the cohesive forces. A mathematical analysis of this model in terms of the lattice, making certain assumptions for the sake of simplicity, gives the same order of magnitude for the relationship between shear strength and shear modulus (1 : 10) as was found above for the relationship between tensile strength and modulus of elasticity. Just as the elastic extension should amount to about 10 per cent. at the point of fracture, so, too, the elastic shear should be about 10 per cent. at the start of glide. The theoretical shear strengths, like the tensile strengths, amount to several hundred kg./mm.^{2}. This gives for the magnitude of normal dilatation perpendicular to the glide plane a value of 1–2 per

---

* The (unknown) surface energy of the crystal is greater; it implies therefore an even greater value of Δl.

cent. of the tangential displacement. Fig. 197 shows clearly that glide cannot proceed along these lines. It shows the movement perpendicular to the glide plane which is exhibited at the yield point by zinc crystals of various orientations. Whereas normal *dilatation*

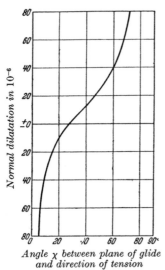

Angle χ between plane of glide and direction of tension

FIG. 197.—Normal Dilatation (ϵ_z) Perpendicular to the Glide Plane at the Yield Point of Zn Crystals.'

$$\epsilon_z = S_{33}\,\mathrm{tg}\,\chi + S_{13}\,\mathrm{cotg}\,\chi.$$

occurs when the basal plane makes a large angle with the crystal axis, glide is accompanied by normal *contraction* where the initial position of the basal plane in relation to the longitudinal direction exceeds 27° 30'.

The same order of magnitude for the theoretical shear strength of glide systems is derived from a rough estimate which assumes that the two halves of a crystal would have to be elastically displaced with reference to each other by about half a glide unit, if they are to pass into the new position without additional stress (576).

Actual shear stress values are smaller than these high values by more than three orders of magnitude (about 100 g./mm.²). Here, too, then, the lattice forces are by no means overcome—a fact which appears clearly from the low shear values (about 10⁻⁴) which prevail at the yield point. Just as crystals fracture long before the tensile strength of the lattice is reached, so, too, glide, and probably mechanical twinning, occur long before the shear strength of the lattice is attained.

75. Attempts to Resolve the Difference between the Theoretical and Technical Values

Although it is generally impossible to realize in practice the tensile shear strengths which have been determined theoretically, it is, nevertheless, quite obvious that the lattice strengths must exceed the experimental values by several orders of magnitude. The failure of the crystals under mechanical stressing (start of gliding and twinning; fracture) is premature and takes place long before the lattice forces have been overcome. In an attempt to account for this great difference, Voigt drew attention to the significance of structural and thermal inhomogeneities in the actual crystals (577).

Both of these ideas have been further developed. The structural inhomogeneities provide the basis of Smekal's theory of vacant lattice sites [(578), (579)]. According to this theory, real crystals differ fundamentally from the ideal crystal by reason of the defects in their lattice (holes, defective orientation, foreign atoms). In the case of transparent crystals these defects can be experimentally demonstrated in a number of ways (*e.g.*, coloration, absorption). Some idea of the prevalence of lattice faults can be gained from an estimate which gives one fault for every 10,000 atoms built perfectly into the lattice. Since the faults originate while the crystal is growing, their frequency and their nature depend both on the material itself and on the conditions under which it crystallizes. The properties of the crystal which are decisively influenced by these faults (the structurally sensitive properties) contrast with the structurally insensitive properties which in the main are determined by the ideal lattice units distributed between the faults (cf. Section 60).

The significance of the faults in the lattice for bridging the large gap between technical and lattice strengths is to be sought in their notch effect. This is due to the fact that, when cracks are present in an elastically stretched solid, appreciable changes in the stress distribution will occur in the vicinity of the cracks. Initially, elastic deformation will increase with stress, yet only up to the point at which either the maximum stress achieved is equal to the lattice strength, or until less energy is required (surface energy) to enlarge the crack than to increase the stress. The stress attained at this point represents the technical tensile strength.

In the case of a plate with an elliptical hole (radii a and b) which is subjected to a tensile stress σ in the undisturbed zone, in the plane perpendicular to the major axis, a maximum stress will occur at the edge of the hole

$$\sigma_{\text{max.}} = \sigma\left(1 + \frac{2a}{b}\right) \quad . \quad . \quad . \quad . \quad (75/1)$$

at the ends of the major axis, while at the ends of the minor axis there will be a compression stress of $-\sigma$ (580). If $\left(\rho = \dfrac{b^2}{a}\right)$ represents the radius of curvature at the apex of the ellipse, then the expression for the maximum stress becomes

$$\sigma_{\text{max.}} = \sigma\sqrt{\frac{a}{\rho}} \quad . \quad . \quad . \quad . \quad (75/1a)$$

In order to calculate, on an energy basis, the technical strength required to extend a crack, it is first necessary to ascertain the elastic energy inherent in a thin plate containing a slot (581). It will be found that in a plate (thickness I) subjected to stress σ the stored elastic energy is less by

$$A_e = \frac{I}{E}\pi a^2 \sigma^2$$

(E = modulus of elasticity, a = half the length of the slot) than in a plate that is intact. Also there is in the plate with the crack an amount of energy equal to $4a\alpha$ (α = surface energy) which originates from the surfaces of the crack. Fracture of the crystal will occur without the addition of further energy if the elastic energy which has been acquired by widening the crack just suffices to supply the surface energy needed to enlarge the surface. Therefore

$$\frac{\partial}{\partial a}\left(\frac{I}{E}\pi a^2\sigma^2\right) = \frac{\partial}{\partial a}(4a\alpha)$$

is the equation for determining the tensile strength σ_1; it results in the expression

$$\sigma_1 = \sqrt{\frac{2E\alpha}{\pi a}} \quad \cdot \quad \cdot \quad \cdot \quad \cdot \quad \cdot \quad (75/2)$$

by means of which the tensile strength of an elastic solid can be calculated from its shape, that is, from the length of the existing cracks and from its physical properties. This theory has been tested and confirmed on scratched glass and quartz.

The notch effect associated with the microscopic cracks discussed above is also influenced by the faults in the crystal referred to earlier in this section. With the aid of the formula (75/2), and if the tensile strength, modulus of elasticity and surface energy are known, then a length of crack corresponding to the faults can be calculated (582). For example, the following values were obtained for rock salt and zinc : rock salt ($\sigma_1 = 2\cdot2 \cdot 10^7$ dyn/cm.^{2}; $\alpha = 150$ dyn/cm.; $E = 4\cdot9 \cdot 10^{11}$ dyn/cm.^{2}) $2a = 0\cdot2$ cm.; zinc (fracture along the basal plane; $\sigma = 1\cdot8 \cdot 10^7$ dyn/cm.^{2}; $\alpha = 800$ dyn/cm.; $E = 3\cdot5 \cdot 10^{11}$ dyn/cm.^{2}) $2a = 1\cdot1$ cm. Naturally, no real significance attaches to such large cracks. Formula (75/2), of course, can be used only if the maximum tensile stresses at the edge of the notch do not exceed the strength of the lattice. But for the cracks whose lengths have already been obtained from the plausible assumption that the radius of curvature at the base of the notch is of the same order of magnitude as the lattice constants ($5 \cdot 10^{-8}$ cm.), this has already occurred. (75/1a) gives for rock salt $\sigma_{\text{max.}} = 630$ kg./mm.^{2}, for zinc $\sigma_{max.} = 1200$ kg./mm.^{2}, which values are about three times higher than the lattice strengths recorded in Section 73. $(75/1a)$ shows that cracks ten times as short would suffice to overcome the lattice strength at the base of the notch. Even in that case, however, the cracks would still be of macroscopic dimensions, which is contrary to the prevailing conception of the nature of a fault in a crystal. An attempt to postulate even shorter cracks with the aid of plastic glide will be found in (583).

Although we have been unable to account quantitatively for the low mechanical properties of crystals along these lines, qualitatively the concept employed leads to a plausible interpretation of a series of phenomena (578). The selection of fracture planes would be decisively influenced by the spatial distribution of crystal faults, for instance, along planes of minimum surface energy. The Sohncke normal stress law is derived from the fact that the maximum stress at the base of the notch results from the stress components perpendicular to the crack. Likewise the small effect of temperature on the critical normal stress is also explained. Further, the increase in the tensile strength of polycrystalline specimens with decreasing grain size (582) has been accounted for by assuming that the cracks are halted initially at the grain boundaries and do not penetrate into the adjacent grain until the external stresses have been increased (584). In order to avoid the long cracks which, in thin crystals at least, are obviously impossible, attention has been drawn on the one hand to the reciprocal intensifying effect of neighbouring cracks (578). On the other hand, it was assumed (cf. Section 72) that the crack length ceases to be a constant of the material if the dimensions of the solid come within its order of magnitude (584).

The effect of surface cracks on the tensile strength of mica sheets is made very clear in (585). By removing the load from the edges of the specimens (the width of the specimens was greater than the width of the grips) the tensile strength was increased tenfold. Consequently, it amounted to about one-tenth of the theoretical value, which shows that in this instance the effect of *internal* faults was no longer decisive. This can be explained in terms of the notch theory by assuming that, in the mica under investigation, the structural faults were disposed mainly parallel to the cleavage plane, *i.e.*, parallel to the direction of pull.

It was only natural that the notch effect of cracks and faults should be adduced to account also for the low shear strength of crystals. To facilitate calculation an extended ellipse was again substituted for the crack, the stress distribution being determined along the boundary of the ellipse [(586), (587)]. The result is wholly in line with that obtained in the tensile test. The decisive factors are the length of the crack and the radius of curvature at its end. Stress concentration again occurs at the ends, and the direction of maximum shear stress coincides with the direction of the crack itself. In spite of lower total effective shear stress, it is there that the shear stress of the lattice is said to be reached and that glide begins. According to this conception, therefore, the glide elements of the crystals would be decisively influenced not only by the anisotropy of the lattice strength but also by the distribution of the faults.[1]

The formula for the shear strength τ_1, which, like formula (75/2), is obtained by balancing the energy, is as follows :

$$\tau_1 = 2\sqrt{\frac{G \cdot \alpha}{\pi a(1 - \mu)}} \quad . \quad . \quad . \quad (75/3)$$

in which G represents the modulus of shear and μ Poisson's ratio (587). Cracks of the same order of magnitude as those which have already been derived from the formula for tensile strength are obtained.

A theory which avoids these large cracks can be developed if the following conception of the mechanism of glide is adopted (589). Glide starts locally at random points in the crystal as a result of " dislocations " (deviations from the strictly geometrical lattice structure; caused probably by thermal movement) which become separated from each other, under the influence of the local shear stress, by migration parallel to the direction of gliding. In the ideal crystal the migration of the first few " dislocations ", or in other words the plastic deformation, would begin at minimum shear stresses. It is fundamental to this theory that these dislocations do not migrate through the whole crystal, but very soon meet with obstacles which prevent them from scattering beyond a mean distance L. It is mainly from this picture that the slope of the hardening curve is derived. From a comparison between observed and calculated curves a value of about 10^{-4} cm. is obtained for the length L.

In order to explain the experimentally determined final values for yield point, reference is made to the mosaic structure which is

---

[1] The *weakening* effect of the notches described above has nothing to do with the shear *hardening* which is produced in plastic crystals by subsequently perforating or notching them (Fig. 198). In this case it is a question of producing lattice disturbances which impede glide in the vicinity of the notches.

often observed in natural crystals. The crystal is composed of lattice blocks, having roughly the same linear dimensions as L, which are rotated from each other about small angles of approximately 1′ (590). In consequence, internal stresses develop within the crystal. An estimate gives shear stresses approximately one ten-thousandth of the shear modulus. Not until these stress limits have been overcome can the dislocations start to migrate in the crystal mosaic, and plastic strain occur. It will be seen that the order of magnitude of the experimentally determined critical shear stresses has in this way been correctly given.

Dislocations are also assumed to be nuclear points of glide in (591) and in (592) (in the former they are illustrated by the picture of a vernier).

Hitherto, low strength was always attributed to faults which, while their orientation corresponds to the anisotropy of the crystal, are statistically distributed throughout it. A theory of a more sweeping nature assumes the existence of regularly distributed

(a) Tensile specimens of a perforated Mg crystal.

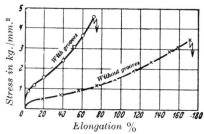

(b) Stress–strain curves of magnesium crystals with and without longitudinal grooves.

FIG. 198 (a) and (b).—Shear Hardening of a Crystal Caused by Subsequent Drilling and Notching (588).

inhomogeneities, and connects these with a reduction in the potential energy of the crystal and consequently with the creation of a more stable condition than that represented by the normal lattice (593). The normal lattice structure is said to be overlaid by a " secondary " structure, characterized by the periodical occurrence, at distances of a multiple of the lattice constants, of planes which are abnormally close packed. According to (594), (595), however, the calculation carried out for the crystals of alkali halides cannot be regarded as valid for a less symmetrical lattice of higher stability. Consequently, although there is much experimental data to support the theory of the construction of actual crystals from the lattice blocks, it is unlikely that an explanation along these lines will be possible. Moreover, it has been shown that the mosaic texture is not a general but an individual property which is appreciably dependent on the conditions of growth and deformation (596). How far this applies to the adsorption of small additions into the internal surfaces of crystals [for ionic crystals cf. (597); for Bi (598)] cannot yet be decided. An attempt to account for the great difference between the theoretical and experimental values on the basis of a secondary structure would therefore appear to be inadmissible.

In this connection, attention is drawn to the very different function performed by the system of internal surfaces in the glide theory based on the migration of " dislocations " which has been outlined above. In this theory the " dislocations " serve solely to restrict migration; the process of deformation takes place in the interior of the blocks.

The theories we have been discussing assume that structural inhomogeneities are the cause of the premature failure of actual crystals. The second type of inhomogeneity to which reference has already been made, namely thermal, originates in the thermal movement of the atoms. Becker has very carefully studied their effect on glide [(599), (600)]. In this case, however, it was not, as hitherto, the onset of plastic flow that was calculated, but the speed of flow (u) as a function of the applied shear stress (S) and of the absolute temperature (T). The line of thought is roughly as follows : irregular thermal movement in the neighbourhood of the glide planes causes stress fluctuations to be superimposed on the applied shear stress. After short periods the shear stress increases to the value of the lattice shear strength (S^*) : the crystal glides by discreet amounts. The increase in length of the crystal per second (the speed of flow) is obtained from the product of Z, the number of glide-plane sections involved in a unit of glide, the average length Δl by which the specimen is extended as a result of such a unit of glide and the probability W that such glide will occur :

$$u = Z . \Delta l . W \quad . \quad . \quad . \quad . \quad (75/4)$$

For the probability W, *i.e.*, the frequency with which the limiting stress S^* is exceeded in the vicinity of the glide plane, within a volume V we obtain

$$W = e - \frac{V(S^* - S)^2}{2GkT} . \quad . \quad . \quad . \quad (75/4a)$$

in which G represents the modulus of shear and k the Boltzmann constant $(1\cdot37_2 . 10^{16}$ erg/degree). In experiments with tungsten crystals (599) and polycrystalline copper (601) the marked dependence of the speed of flow upon temperature (the speed being doubled when temperature rises by $10°$), which follows from formula (75/4), was very well confirmed.

However, an attempt to account for the low experimental shear strengths exclusively in terms of thermal fluctuations meets with difficulties owing to the low values for critical shear stress which have also been observed at very low temperatures. The observed dependence of the flow speed upon temperature can also be regarded as an effect of crystal recovery, which works in opposition to the work hardening which accompanies increasing deformation (Section 49). Since recovery depends on the effect of thermal fluctuations (cf. Section 77), it should be easy to interpret the results of the flow experiments (576). On the other hand, in the case of the amorphous solids, plasticity seems, in fact, to result exclusively from thermal fluctuations (cf. also Section 77).

In (592) the effects of both structural and thermal inhomogeneities are combined to explain the low shear stresses of crystals. Notches are held mainly responsible for the wide discrepancy between theoretical and experimental shear strengths; they cause stress concentrations amounting to about one-third of the theoretical shear strength. Nevertheless, thermal fluctuations are said to impose characteristic features on crystal glide. If certain allowances are made, the slight dependence of the critical shear stress of zinc and cadmium crystals upon temperature also agrees in general with the theory. On the other hand, the marked dependence upon temperature of the flow speed of these metals, on which the application of the thermal theory rests, has still to be proved.

76. *Theory of the Work Hardening of Crystals*

It has been shown that so far no reliable explanation of the low tensile and shear strength of crystals has been forthcoming. Even less is known of the phenomenon of work hardening, which consists in an increase in the mechanical properties with increasing deformation. This subject still remains within the realm of speculation.

Recent theories of the notch effect naturally assume that hardening results from changes in the strength-reducing notches. The shear and tensile strengths of the lattice are assumed to be fixed; the experimentally determined values can only approximate to them; they can never exceed the lattice strength; a hardening of the lattice is impossible. Several opinions have been expressed as to the kind of changes which must occur in the notches in order to reduce their effect. For instance, as a result of deformation, the cracks associated with the faults in the structure of the crystal could be shortened. It is therefore reasonable to suppose that the dangers to which they give rise are reduced, where simple glide is involved, for the *latent* planes, while in the case of complex glide the relief is more general. Owing to the movement of the glide " packets " the cracks are further subdivided, and the individual portions displaced stepwise in relation to each other (584). The relatively slight changes in the tensile strength of metals with unique glide planes (Sections 53 and 54) are also in agreement with this conception. In regard to the shear hardening of the *operative* glide plane, it has been pointed out that glide occurs at an increasing number of points as deformation proceeds. Whereas the stress concentration at the ends of the cracks which first become effective is very high, by reason of the fact that the load has been removed from large areas in the vicinity, the areas which remain undeformed diminish as elongation proceeds. In terms of the notch-effect theory it follows that the new cracks must be increasingly short, and their stress concentration correspondingly low : thus, the glide plane hardens (586). According to the more recent conception of glide as a migration of dislocations (589) the shear hardening is similarly interpreted as an increase in the number of dislocations. As the distance between the dislocations is reduced (perpendicular to the plane of gliding) the shear strength increases. This theory yields the following expression for the connection between shear strength (S) and gliding (a), *i.e.*, for the shear-hardening curve

$$S = \kappa \, . \, G \sqrt{\lambda/L} \, . \, \sqrt{a} \quad . \quad . \quad . \quad (76/1)$$

in which G represents the modulus of shear, λ the identity period in the glide direction, L the distance of the free migration of the dislocations (size of the lattice blocks; approx. 10^{-4} cm.) and κ a constant (approx. 0·2). The shear-hardening curves are thus represented by parabolas, and there is, in fact, close agreement with the experimental results obtained with cubic metals.

In this connection mention must be made of a theory which associates strain hardening with increasing disorientation along the glide planes (578). This is related to the rotation of portions of the lattice in the vicinity of the glide planes (cf. Section 59), which has been theoretically deduced and experimentally demonstrated from the stress distribution at the crack [(587), (586)]. As strain along the operative glide planes is intensified the single crystal is said to disintegrate into lattice blocks of increasingly variable orientation. This not only impedes further glide (shear hardening), but it also renders more difficult the formation of smooth fracture planes permeating the entire crystal (tensile hardening). It should be noted, however, that if the direction of stress is reversed the degree of disorientation will be reduced, but the hardening will continue to increase (cf. Sections 59 and 61).

The hypotheses of the nature of work hardening so far described relate to changes which take place, during gliding, in the strength-reducing faults present in the crystal. The tendency of these changes is to increase the effective lattice strength with the percentage of working. According to other hypotheses, which take no account of the discrepancy between the theoretical and actual strengths of crystals, the phenomenon of hardening is due to modifications of the crystal lattice. These include the earlier modification hypotheses, together with the version in which they are best known : the amorphous layer hypothesis [(602), (603)] which assumes the formation of a brittle and amorphous layer between the gliding lamellæ, and at the grain boundaries, as a result of friction. Apart from the fact that no proof could be adduced for the occurrence of such layers, there are thermodynamic grounds for rejecting this assumption (604). In this context mention should also be made of the displacement hypothesis (experimentally disproved) (605), which assumes a lattice displacement, to the point of complete destruction, in the course of deformation.

According to the interference hypothesis (606) *local* disturbances of the crystal lattice, and consequently of the interatomic forces, are regarded not only as an important cause of hardening by cold working, but also of hardening through alloying. A more precise picture of local lattice disturbances of this type is obtained if the glide lamellæ are visualized as being elastically bent (Fig. 199). In the case of glide accompanied by bending [(607) and especially (608)] the lamellæ are not displaced parallel to each other; instead, in the course of gliding they curve about an axis which in the glide plane is perpendicular to the glide direction. Compared with its original intact condition, the lattice in the bent crystal is changed. In the first place the boundary planes of the bent lamellæ represent "internal separation planes", while, secondly, elastic stresses (tensile or compressive stresses on the convex or concave side of the lamellæ) are distributed inhomogeneously in the crystal. In the

FIG. 199.—Sketch to Show the Process of Glide during Bending (608).

present case the model of the macroscopically bent crystal does not limit application of the general theory, since undulations of the glide planes have also been determined by X-ray diffraction in crystals which have been extended uniformly. Hardening, especially hardening of the latent glide systems, is now attributed to the internal separation planes. These prevent glide planes which intersect the first planes from becoming effective. Hardening of the primary system, together with that of other systems, occurs from the start of deformation, although these act only in a subordinate capacity.

The conditions occurring on the internal separation planes are subjected to mathematical analysis in (609). It is shown there that the disorientation of a single atom in a series of atoms is unstable and that the atom must revert of its own accord to its position in the lattice. On the other hand, linear groups of disoriented atoms become stabilized. Consequently the resultant stresses remain even after the externally applied stress has been removed. A model illustrating very clearly the stability of these groups has already been shown in (610). The relationship between the two forces operating on irregular atoms (elastic energy exerted by one-half of the lattice on an atom belonging to it, but which has been forced out of equilibrium; energy which is present in the opposite half of the lattice and which changes periodically with the displacement) is assumed to be such that at least two stable equilibria are available for the displaced atoms. The transition from one position of equilibrium to the other proceeds by jumps when a given relative displacement of the lattice halves has been achieved. If the direction of stress is reversed, then in order to restore the original state it will be necessary to remove the load to a point at which the

deformation is smaller than that which is necessary to cause instability at that load.   Thus, dependent upon the previous history,
two different states of deformation can result from one and the same
effective stress.   This would explain elastic hysteresis in single
crystals (cf. Section 55).   The cause of the appearance and per-
sistence of closed loops under alternating stress should be sought,
not in the crystallographic mechanism of deformation, but in the
mechanically reversible atomic movement between two adjacent
stable positions (611).

The model described above has also been used to explain the low
elastic after-effect of crystals.   Thermal energy, and not the supply
of elastic stress energy, is held to account for the jump from one
state of equilibrium into another.   The requisite amounts for overcoming the shear stress of the lattice are thus supplied locally in
exactly the same way as was assumed in the theory of crystal flow
by gliding (cf. Section 75).

According to a third group of hypotheses the cause of hardening
must be sought in the atoms themselves.   Control of the speed of
solution, and the change which is produced in the colour of alloys
by cold working, have been accounted for in this way (604).   The
circumstance that the lattice distortions which are revealed by
X-ray photographs do not entirely correspond to the changes in
the mechanical properties and the electrical resistance, and that the
effect of hardening and recovery varies for different properties of
metals of the same lattice type, has been put forward as evidence
that a deformation of the electron shells of the atoms is the primary
cause of the work hardening effects.   These atomic deformations
can be accompanied by slight disturbances of the crystal lattice, but
they need not be so accompanied [(612), (613)].   Without denying
the possibility of such atomic deformations, it should be pointed out
that they ought rather to be regarded as a secondary effect of the
primary lattice distortions (614).

It cannot be said at present which of these theories of work hardening will survive when our knowledge of the subject is more complete. It is, however, an advantage of the first group of hypotheses that
they approach work hardening phenomena from the same angle as the
equally unsolved problem of the low mechanical properties of crystals.

77. *The Theory of Recrystallization—Atomic Migration Plasticity*

The phenomena and principles of recrystallization (and crystal
recovery) have been discussed in previous sections (49, 61–65 and

71). The process underlying this phenomenon is the *migration* of the atoms brought about by the increased thermal movement incidental to heat treatment. We shall discuss theories which enable us to understand the recrystallization temperature and the *shape* of the recrystallization diagrams. Then we shall describe a type of plasticity which, while it occurs with crystalline material, does not proceed in a regular crystallographic fashion but is also probably due to atomic migration.

There are two explanations for the existence of the temperature of recrystallization. The first, which applies to cubic crystals, relates to the thermal migration of atoms while also describing the diffusion processes (614a). A condition of atomic migration is that the energy of the atoms in question must be greater than a given limiting value (E), and this applies to that part of the atomic array which is represented by $(e^{-\frac{E}{kT}})$. With the aid of the specific atomic frequency (ν) both the number of migrations per second and the time required until all the atoms have migrated can be calculated. The deformed material is distinguished from the undeformed by an increase in energy (ΔE), which is characterized by the change in the specific resistance. The probability of migrations is thereby increased, since the quantity of energy to be produced by the thermal movement of the atoms is now only $E - \Delta E$; the time required for the migration of all atoms is reduced. An important prerequisite for the occurrence of recrystallization is the *unsymmetrical* distribution of energy accumulations in the deformed material, which otherwise would be indistinguishable from material at high temperature.

The points at which maximum accumulation of energy occurs determine the course of recrystallization. In order that the whole specimen may recrystallize, it is necessary that at several places all the atoms shall migrate during heating. The time of heating must therefore be stated in order that the temperature of recrystallization can be fixed. The connection established in this way, at a constant percentage of working, between recrystallization temperature $T_{R(abs)}$, time of heating t and atomic frequency ν, is as follows :

$$T_R = \frac{\text{const}}{\ln \nu t} \quad . \quad . \quad . \quad . \quad (77/1)$$

If, then, two related values of T_R and t are known, the heating period corresponding to another temperature of recrystallization can be calculated (thus an increase of T_R by 1 per cent. reduces the time of heating by 35 per cent). This formula has been tested experimentally on various materials, and the results have confirmed theoretical expectations.

According to the second explanation the temperature of recrystallization is said to be that at which the speed of recrystallization changes abruptly (609). This phenomenon can be best explained with reference to the groups of disoriented atoms which occur in hardened crystals and which have already been described. The state of metastable equilibrium in which the disoriented atoms find themselves at the boundaries of the glide lamellæ increases with the number of atoms that are interlocked; the limiting energy required to overcome this equilibrium is therefore correspondingly great. If we heat a hardened crystal in which such interlocked atoms are dispersed, then, in accordance with Maxwell's Law of the distribution of energy, only a few interlocked atoms will have sufficient energy at low temperatures to free themselves from the bond. At such temperatures very long heating periods will be necessary to attain the final recrystallized state of equilibrium. If, however, the heat treatment supplies sufficient energy within a short time to a large number of interlocked atoms, then, as the mathematical analysis shows, all interlockings are dissolved simultaneously owing to the reduced stability. The temperature at which this occurs in a short time is the temperature of recrystallization.[1]

The shape of the *recrystallization diagrams* [coarse grain size after low percentage of working and high annealing temperatures; fine grain size after heating at low temperature specimens which had been heavily worked (cf. Fig. 169)] was derived qualitatively from thermodynamic considerations which avoid the still unknown details of atomic processes (600). The atomic migrations which take place at a temperature above that of recrystallization increase in frequency with the growing confusion of the atomic arrangement which augments with the percentage of working (increased hardening). If, as a result of this migration, a number of atoms arrive in positions which are crystallographically suitable, they will remain in these positions for a much longer period : a crystal nucleus will form. This means that the frequency of the migrations will decisively influence the formation and subsequent growth of crystal nuclei. In the first place it will be observed that not every new grouping of atoms is capable of serving as a nucleus. For instance, if the

---

[1] However, it is difficult to reconcile with this explanation the fact that the temperature of recrystallization falls as the percentage of working increases, while the number of interlockings, far from being reduced, probably grows as hardening proceeds.

number of related atoms is too restricted and the fragment of crystal too small, its stability will not exceed that of the deformed material in the immediate vicinity. If the nucleus remains below a certain critical size it will disappear again, since its vapour pressure exceeds that of its surroundings. The problem is, therefore, to represent the vapour pressure of the basic material as a function of the percentage of working and of the temperature of annealing, and that of the newly formed nuclei as a function of size and temperature. From an equalization of the two we obtain the minimum size (r) of the stable nucleus as a function of the percentage of working [measured by the heat of recrystallization (δ) and temperature (T)]. The expression for r is as follows :

$$\frac{1}{r} = \frac{\delta(T_s - T)}{T_s} \cdot \frac{d}{2\sigma M} \quad \cdot \quad \cdot \quad \cdot \quad \cdot \quad (77/2)$$

in which T_s represents the melting point, d and σ the density and surface tension of the crystal, M the molecular weight of the vapour. The dependence of the nuclear size (r) upon δ and T which results from (77/2) agrees with the shape of the recrystallization diagrams.

The foregoing discussion has dealt with the average grain size of the recrystallized structure. Observations relating to the distribution of grain sizes will be found in (615). In (616) grain-size distribution has been calculated on the basis of an assumed constant frequency of nucleus formation and constant linear speed of growth. This distribution should be the same both for the cast and recrystallized structures, but apparently it does not entirely conform to actual experience. To assume a constant speed of nucleus formation is to simplify the problem unwarrantably, at least so far as recrystallization is concerned.

In connection with this discussion of the recrystallization of hardened crystalline materials, attention should also be drawn to another consequence of the atomic migration phenomena. There is a type of plasticity which accompanies recrystallization (and phase transformations) in the course of stressing, and which strictly speaking cannot be regarded as a movement due to crystallographic glide. On the other hand, it can easily be explained as a result of atomic migration, since under stress those migrations are naturally preferred which lead to stress relief or to deformation consistent with the stress. This type of plasticity has therefore been designated " *amorphous plasticity* " because migration is the mechanism by which an amorphous vitreous material deforms. In this way it may happen that at temperatures above recrystallization a recrystallized material, which has been annealed and softened, will be stronger than work-hardened unannealed material in which vigorous migration takes place as heating proceeds (600). Experiments with tungsten coils (617) and with copper and aluminium wires (618) reveal, in fact, that pronounced flow accompanies recrystallization. In Fig. 200, by way of example, flow curves are shown which were obtained in a filament-stretching apparatus with hard and soft copper wires treated at 600° C.; in every case the hardened recrystallizing wire flows more readily than the one which has been previously annealed.

Similar results have also been obtained when investigating the

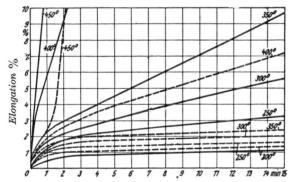

FIG. 200.—Flow Curves of Hard and Soft Copper Wires at Various Temperatures (618).

————— : Hard.   – – – – : Soft.

" creep " strength of iron and nickel and of some of their alloys at elevated temperatures (619). The extensions in this case were not determined by mirror reading, but were compensated for by changing the temperature. It was again found that material which had been pre-heated was very much stronger (in terms of creep strength) at temperatures above the range of recrystallization. The technical significance of these facts is obvious. If the observed phenomenon is generally valid, then metals used within the temperature range in which recrystallization and recovery occur could in many cases be better employed in the heat-treated than in the cold-worked state.

In view of these facts it was natural to expect a reduction in flow resistance in cases of phase changes also. Here, too, atomic migration supplies the mechanism by which the lattice is reconstructed.

Detailed investigation of the γ–α phase change of nickel steel (30 per cent. nickel) has confirmed this expectation (620).   If the phase change occurred while the wire specimen was being stressed, then extensions of more than 10 per cent. were observed to take place at the same time.   Since the specimen had been merely cooled, it follows that in this case it cannot have been a question of migration of thermal origin.   It is therefore the occurrence of active atomic migration, and not the manner in which this migration is brought about, which is the essential feature of the mechanical weakness.

CHAPTER IX

THE PROPERTIES OF POLYCRYSTALLINE TECHNICAL MATERIALS IN RELATION TO THE BEHAVIOUR OF THE SINGLE CRYSTAL

Having discussed the phenomena of the plastic deformation of single crystals, we saw in the previous chapter how far we still are from a satisfactory theoretical interpretation of the observed facts. The present and last chapter will be devoted to an examination of the problem—scientifically less impressive but technically very important—of the relationship between the properties of poly-crystalline aggregates and the properties and arrangement of the individual grains (texture). An understanding of this relationship will enable us not only to explain the behaviour of aggregates, but also to estimate what properties can be obtained in the material under optimum conditions. A calculation of the properties of the material from the behaviour of the single crystal and the arrangement of the grains will be particularly successful if the effect of the grain boundaries does not make itself felt in the polycrystal.

In the first place it is the properties which are structure insensitive (cf. Section 60) for which crystalline behaviour and structure are the sole determining factors (*e.g.*, the elastic properties, thermal expansion). But even in the large group of plastic properties which are structure sensitive, we may expect to find a manifestation of polycrystalline behaviour which is at least qualitatively correct, owing to the directionality of these properties being frequently very marked. In this way we gain a clearer insight into a number of technological problems and so are enabled to make more effective use of the material. But in so far as the influence of the grain boundaries predominates, the method of approach outlined above becomes less applicable. Properties which are based on inter-crystalline processes occurring at the grain boundaries, such as hot shortness caused by melting of a eutectic, intercrystalline disintegration due to corrosion—do not, of course, come within the scope of the present discussion.

Before dealing with the mathematical side of the problem, and before noting examples from the technology of metals, we will discuss the methods in use for determining the crystalline arrangement in polycrystalline material, and we will describe and trace to their origin the textures which are produced in metals by the various methods of working. We have omitted rock structures from this discussion. These have been widely investigated in recent years, and the subject is treated exhaustively in the monographs [(621), (622)].

78. *Determination and Description of the Textures* (*see* 623)

In principle, the same methods are used for determining the distribution of orientations of the grains of a polycrystalline aggregate as for determining the orientation of single crystals (cf. Chapter IV). The superiority of the X-ray method has been clearly demonstrated, especially where fine-grained and intricate textures are involved, and it is this method which we propose to discuss almost exclusively.[1]

Supplementing the information contained in Chapter IV, we give below, briefly, the solution of the following two problems.

1. Determination of the orientation of crystal grains relative to a *direction* imposed by the shape or previous history of the specimen (*e.g.*, direction perpendicular to the cooling surface in cast material,[2] and to the longitudinal axis in drawn, rolled or recrystallized wires and extruded bars).

2. Determination of the texture relative to a *co-ordinate system of axes* suggested by shape or previous history (*e.g.*, the rolling, transverse and normal direction in sheets).

If, in the first case, one and the same lattice direction in each crystal grain coincides with the imposed direction (axis of the fibre), then the simplest type of crystallite arrangement is present and we obtain a simple fibre texture (625).[3] In this case a monochromatic X-ray diagram perpendicular to the working direction coincides with the diagram of a crystal which has been rotated about the relevant lattice direction. In the specimen all those crystalline positions are spatially distributed which appear successively as the crystal is rotated. The

---

[1] If a quick and rough estimate of the anisotropy of rolled sheet is required, the Chladni resonance figures can be used (624).

[2] Fig. 201 shows the structure of a technical casting. Several different zones can be clearly distinguished. At the outer edge, close to the wall of the die, there is a thin layer of fine-grained equiaxed crystals, which is followed by a fairly coarse-grained layer of columnar crystals, the longitudinal direction of which is perpendicular to the cooling surface. The interior of the bar is made up again of smaller crystals of irregular shape. Deviations from the random arrangement of the crystals are to be expected mainly in the columnar zone. The prevailing texture in this layer is known as the casting texture. When determining the texture by X-ray methods it is advisable that the test bars taken from the casting should have their axis parallel to the direction of growth of the crystals.

[3] Fibre—the name originates from the discovery of this type of texture in natural fibres (626).

interference patches are located on the layer lines, and it is from the distance of the latter that, knowing the crystal structure, and using formula (21/2a), the identity period along the fibre axis and the

FIG. 201.—Cast Structure.   Section through a
Copper Ingot.

crystallographic nature of the texture can be determined.   The nature of the fibre axis can be checked by examining the distribution of the interferences on the layer lines, which, as mentioned in Section 21, is entirely regular.   Examples of a cast specimen and of

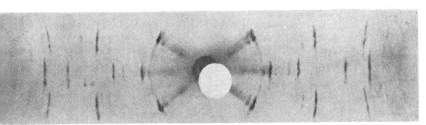

FIG. 202.—Pattern of a Cast Texture.   Al (627); Axis of the Fibre ‖ [100].

a recrystallized wire are shown in Figs. 202 and 203, which were obtained by irradiation perpendicular to the axis of the fibre. Fig. 204, a and b, reveals the presence of a so-called " double fibre texture " represented by the superimposition of two simple fibre textures with a fibre axis common to both.   In this case there are two groups of crystallites, each of which is characterized by a definite crystal direction parallel to the axis of the fibre.

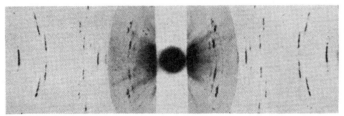

FIG. 203.—Fibre Pattern : Recrystallized Aluminium Wire (628) ; Axis of the Fibre ‖ [111].

As an example of the determination of the orientation of the crystallites in relation to a three-axial rectangular system of co-

(a) Texture pattern.

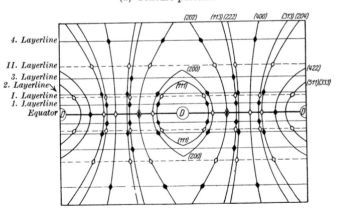

(b) Theoretical diagram of a double fibre-texture ; fibre axis ‖ [111] and ‖ [100]. The layer lines through the texture with Arabic numerals refer to [1̄11], those with Roman numerals to [100].

FIG. 204 (a) and (b).—Drawing Texture : Cu (629).

ordinates, attention is drawn to the so-called pole figure method which was applied to metals for the first time in (631). This does not merely consist in giving the crystallographic directions which lie parallel to the three axes in the individual crystals, but also—and this is the main advantage—in a representation, characterized by pole figures of the more important crystal faces, of the whole distribution of orientations present in the specimen. In order to obtain these pole figures the interferences on the Debye–Scherrer circles are plotted in the stereographic representation of the normals of the reflecting planes (cf. Section 24, where this plot is shown for the Laue interferences). This method is used especially for defining the textures of sheets, and it is to such an application that the following remarks apply. If the distribution of orientations is to be fully accounted for, a single diagram taken approximately perpendicular to the plane of the sheet will by no means suffice. If the specimen is not moved, then reflections can be obtained only from those crystal planes which, according to Bragg's formula, are in an exact position to reflect. If therefore a complete picture is required, it will be necessary to irradiate the sheet obliquely in various directions to produce a series of photographs. It is preferable so to arrange the directions of radiation that they lie in the planes : sheet normal—rolling direction and sheet normal—transverse direction. In such oblique photographs the primary ray is no longer perpendicular to the rolling plane of the sheet, that is, to the projection plane. It will now be a question of transferring the interference positions of the oblique photographs to the stereographic projection, and so to a representation of the reflecting lattice planes. This is achieved by projecting the interferences first of all in the same way as the photographs which have been taken perpendicularly. The primary ray is then perpendicular to the projection plane. The plane of the sheet, however, does not coincide with the projection plane, as would be necessary for a uniform representation. On the other hand, the inclination of the sheet normals to the primary ray is known. For instance, if, in a photograph, the primary ray in the plane containing the sheet normal and the rolling direction has been inclined at an angle α to the sheet normal, then the sphere of reflection, which initially was projected perpendicularly, must be rotated about the transverse direction by the angle α. In this way the rolling plane will coincide with the projection plane, while the primary ray will impinge obliquely at angle α. Thus, a number of photographs can be evaluated and the results entered on the same stereographic projection. Finally, the points obtained are reproduced in accordance with the existing symmetry elements of the texture. With a view to a more exact definition, the density of distribution in the pole figure is estimated from the intensity of the reflections. The diagram is prepared either in terms of the different degrees of density (see Fig. 205) or by suitably shading the pole figure (see Fig. 206).

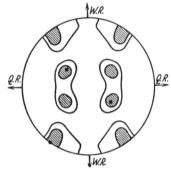

(a) Pole figure of the (100) plane.

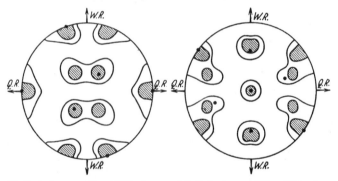

(b) Pole figure of the (111) plane.   (c) Pole figure of the (110) plane.

Fig. 205 (a)–(c).—Rolling Texture of α-brass (632).

WR = direction of rolling.   QR = transverse direction.

Table XXXI contains particulars of the casting textures of a series of metals and solid-solution alloys based on the *results of texture determinations* [in particular the determinations in (627)]. In the columnar zones there is a simple fibre texture : an important lattice direction coincides more or less in the fibre axis of all crystal dendrites. Zinc and cadmium are the only exceptions. Here a " circular " fibre texture is present (634); in all crystals the longitudinal direction is distinguished only in so far as it lies in the basal

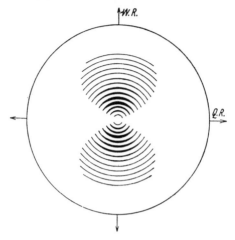

(a) Pole figure of the (0001) plane (basal).

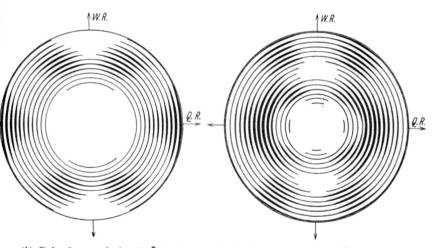

(b) Pole figure of the (10$\bar{1}$0) plane (prism type I, order 1).

(c) Pole figure of the (10$\bar{1}$1) plane (pyramid type I, order 1).

Fig. 206 (a)–(c).—Rolling Texture of Zinc (633).

WR = direction of rolling.   QR = transverse direction.

plane; otherwise it presents a different direction from crystal to crystal.[1]

The deformation textures, which are distinguished according to the method of working (drawing textures, rolling textures, etc.), cannot as a rule be fully described by indicating the principal crystal orientations. With rolling textures the pole figure is usually employed. This indicates the degree of scattering, and so affords a better idea of the distribution of the crystal orientations present.

TABLE XXXI

Cast Textures

Metal.	Parallel to the longitudinal direction of the crystals.
Al . . . . . Cu . . . . . Ag . . . . . Au . . . . . Pb . . . . . α-Brass . . .	[100]
α-Fe . . . . β-Brass . . .	[100]
β-Sn	[110]
Mg . . . . . Zn . . . . . Cd	[11$\bar{2}$0] [0001] Perpendicular
Bi	[111]

But other refinements in the structure of the deformation textures must also be taken into account. They relate mainly to the inhomogeneity of the texture with regard to type and scatter in the different layers of the material. Where a double fibre structure is present it will also be necessary to indicate the frequency with which the individual crystal orientations occur. In Fig. 207 are found four diagrams which have been obtained from copper wire etched to different depths. It will be seen immediately from the varying

---

[1] Regular textures also occur, under suitable conditions, in metals which have been electrolytically deposited from aqueous solutions; they are simple fibre textures with the direction of the current flow serving as the fibre axis. A whole series of factors is responsible for the selection of this axis—solvents, ions in solution, amperage, the material of the cathode. For a summary of the observed textures, together with particulars of the working conditions, see (623).

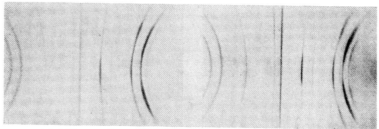

(a) 1·75 mm. diameter.   (b) The same wire etched down to
                                    1·3 mm.

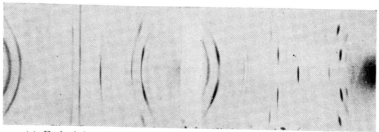

(c) Etched down to 1·0 mm.   (d) Etched down to 0·4 mm.

FIG. 207 (a)–(d).—Texture Present in Different Layers of Drawn Copper
Wire (629).

length of the interference arcs that the pattern is much sharper in
the centre of the wire than in the peripheral zones. That this
discrepancy in scatter is not due to differences in the thickness of
the wire will be clear from Fig.
208, which illustrates drawn,
unetched wire 0·05 mm. in diameter. In addition to the
sharpness of the pattern, inhomogeneity affects also the
texture (note the absence of
symmetry in Figs. 207, a–c, and
208).

The real drawing texture resulting from uni-axial tension can
be described as follows in the case
of cubic face-centred and prob-

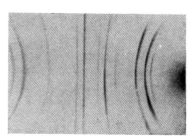

FIG. 208.—Texture Pattern of Copper
Wire which has been Substantially
Hard Drawn (629). Diameter : 0·05
mm.

ably also cubic body-centred metals : in the interior of the wire there is
a normal fibre texture ; towards the exterior it merges gradually into

a simple " conical fibre texture ", which is characterized by the fact that the preferred crystal axes constitute the generators of a simple circular cone about the central axis of the wire.   Direction and

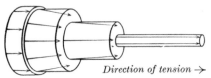

Direction of tension →

FIG. 209.—Sketch showing the Texture of a Hard Drawn Wire (629); cf. also (630).   The Direction and Length of the Arrows Indicate the Direction, and the Degree of Uniformity of Direction, of the Fibre Axis.

counter-direction of the wire axis are not equivalent, and the plane perpendicular to the axis of the wire is not a plane of symmetry for the texture.   The apex angle of the cone increases with the distance

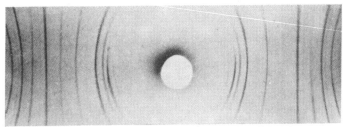

(a) Wire 1·1 mm. in diameter.   ↑ direction of drawing.

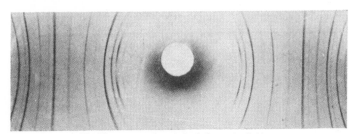

(b) The same wire etched down to 0·4 mm.

FIG. 210 (a) and (b).—Inhomogeneity of the Drawing Texture of Zinc (635).

of the reference point from the axis of the wire, attaining approximately the angle of inclination of the die just below the surface of the wire (see Fig. 209).

This same inhomogeneity is also observed in the drawing textures of the hexagonal metals, as will be apparent from Fig. 210.   Whereas in the interior of zinc wires there is a double conical fibre or spiral fibre texture,[1] the surface of the wire (unsymmetrical diagram owing

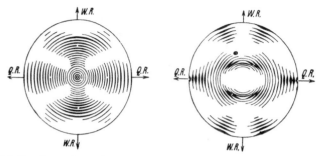

(a) For the surface of the sheet.   (b) For the centre layer of the sheet.

Fig. 211 (a) and (b).—Inhomogeneity of the Rolling Texture of Al (636). Pole Figure of the Octahedral Plane.

WR = rolling direction. QR = transverse direction.

to absence of the single basal interference) exhibits also only a simple conical fibre texture.

Since inhomogeneity of texture, in the case of drawn wires, results from inhomogeneity of flow, analogous results might be expected to follow from other methods of deformation. In fact, careful study of the rolling texture of aluminium and zinc has revealed substantial differences in the orientation of the crystals in the outer and centre layers of the sheets. This is shown by the pole figure of the octahedral planes of a 5-mm.-thick sheet of aluminium in Fig. 211, and of zinc in Fig. 212 (to be compared with Fig. 206a).

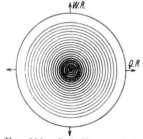

Fig. 212.—Pole Figure of the Basal Plane in the Surface Layer of a Zn Sheet (637).

The intricate nature of the real texture of cold-worked materials is shown clearly by these examples. In addition, the conditions of fabrication, such as the percentage and speed of working, reduction per pass, die angle and diameter of the roll, clearly affect the texture

[1] In the case of a spiral texture the preferred crystal axes constitute at each point surveyed the generators of a double cone ; in the case of a double conical texture all the preferred crystal axes lie on a double cone about the central axis of the wire.

of the peripheral zones of the specimens even if not that of the interior. The particulars of the preferred orientations contained in Tables XXXII–XXXIV can therefore serve only as a first rough

TABLE XXXII

Drawing Textures

Metal.		Parallel to the axis of the wire.	
		1.	2.
Al		[111]	—
Cu			
Au	}	[111]	[100]
Ni			
Pd			
Ag		[100]	[111]
a-Fe			
W	}	[110]	—
Mo			
Mg : drawn . . .		(0001) [10Ī0]	—
scraped . . .			—
Zn		(0001) Inclined at an angle of 18°	—

TABLE XXXIII

Compressive and Torsional Textures

Metal.	Compression. Parallel to the direction of compression.	Torsion. Parallel to the longitudinal direction.
Al	} [110]	[111]
Cu		
a-Fe	[111]	1. [110] 2. [112]
Mg	[0001]	—

approximation. In regard to Table XXXII (drawing textures) it should be observed that, despite the different method of fabrication, the same texture appears also in the interior of fracture cones of tensile bars, as well as in the interior of cold-rolled wires (copper, aluminium, β-brass). It would therefore appear as though the type of deformation employed, rather than the method of applying the force and the actual stress distribution, is mainly responsible for the development of the structure [(638), (639); cf. also Section 80].

It is not easy to generalize regarding the textures which result from the heat treatment of cold-worked metals—recrystallization structures (cf. Sections 63 and 64). In this case, history, degree of purity and type of heat treatment profoundly influence the behaviour

TABLE XXXIV

Rolling Textures

Metal.	Parallel to the rolling	
	Direction.	Plane.
Al Ag α-Brass . (Pt)	[112]	(110)
Cu Ni (Au)	1. [112] 2. [111]	1. (110) 2. (112)
α-Fe Mo	[110]	(001)
Mg	—	(0001)
Zn	[1120] Inclined at an angle of 20°	(0001) Inclined at an angle of 20°
Cd	—	(0001) Inclined at an angle of 30°

of the material. Three main types of recrystallization can be distinguished in rolled sheets (640). First, the orientation of the new grains can be entirely random; secondly, a regular texture can appear at the onset of recrystallization, but it will be unstable, and a random distribution of orientations will be produced as treatment continues at higher temperatures; in the third case, the regular texture may persist at the highest annealing temperatures. As already mentioned, however, it is impossible to distribute the various metals accurately among these three groups. Aluminium, for instance, was at one time placed in the first group (640), but subsequent experiments revealed the presence of a regular recrystallization texture (641). In silver foil, which is representative of the second group, the recrystallization texture was found to disappear at 750° C. (642); but another type of silver foil retained the texture even at high temperatures of heat treatment (632). Recrystallized copper foil, in particular, was considered by various authorities to

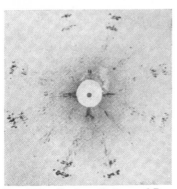

FIG. 213.—Texture Pattern of Recrystallized Copper Sheet (632). Cube Position.

belong unquestionably to the third group. In this case the very simple recrystallization position shown in Figs. 213 and 214 persists up to the highest temperatures. The hexagonal metals probably belong to this group also. The rolling texture of these metals has been observed always to correspond to the recrystallization texture [(645), (646)].

The recrystallization textures of rolled sheets as defined by their main positions are shown in Table XXXV. The pole figures obtained with iron can be represented as three superimposed preferred positions, the first two occurring with nearly the same frequency, while the third is much more rare (647).

TABLE XXXV

Recrystallization Textures of Rolled Sheets

Metal.	Parallel to the rolling	
	Direction.	Plane.
Ag . . . . . a-Brass . . . . Bronze (5% Sn) . .	[112]	(311)
Al . . . . . Cu . . . . . Ni . . . . . Au	[100]	(001)
a-Iron	[110] [112] [110]	(001) (111) (112)

In annealed drawn wires the recrystallization texture at low temperatures is usually observed to be identical with that of the drawing texture; the drawing texture still persists to some extent, however, in material which has been annealed at high temperatures. In pure aluminium wire the recrystallized texture is even more marked than the drawing texture (cf. Fig. 203). The recrystalliza-

FIG. 214.—Etch Pits in Recrystallized
Cu Sheet (643).

tion texture of copper wire which has been annealed at high temperatures differs from the drawing texture, being a simple fibre texture with [112] parallel to the direction of the wire.

79. *Behaviour under Strain of a Grain Embedded in a Polycrystalline Aggregate*

Before we discuss the development of textures we propose to examine in the present section, with reference to the formation of deformation textures, the behaviour under plastic strain of a single grain embedded in a polycrystalline aggregate, since the deformation of the aggregate also usually takes the form of a stretching of the individual grains and not of their displacement along the grain boundaries [cf. Fig. 215 and also (649)]. It is, of course, a fact that continuity on all sides with neighbouring crystals of different orientation results in considerable difference between the behaviour of the aggregate and that of the deformed, free, single crystal.

In order to make this difference clear, let us examine the case of a single crystal. If, for instance, two tin crystals are joined together along a short boundary, and if the crystals are then stretched, it will be noticed that the unjoined parts elongate with the typical formation of glide bands, while the region of the junction remains unchanged (650). It is obvious that the point of coalescence (the grain boundary) has a restrictive effect on elongation. If the stress were increased, the bi-crystal, too, would deform, but this would be accompanied by very substantial strains and distortions along the

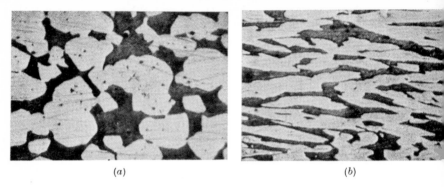

(a)   (b)

(<i>a</i>) and (<i>b</i>) (<i>a</i> + <i>β</i>)-brass before and after deformation (648).

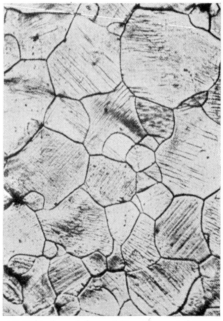

(<i>c</i>) Glide in a deformed Fe polycrystal (644).

Fɪɢ. 215 (<i>a</i>)–(<i>c</i>).—Grain Deformation Resulting from the Stretching of a
Polycrystal.

grain boundary. These can develop to a point at which they cause
grain boundaries to split, as has already been demonstrated in the

case of bi-crystals of bismuth. Indirectly the high strains along the grain boundaries are rendered visible by a greatly increased local capacity for recrystallization. Under annealing conditions which, in the elongated single crystal of tin, produce only a few large grains, the elongated bi-crystal changes into a fine crystalline structure.

These facts become intelligible when it is realized that the coalesced surface has to accommodate itself to the deformation of *both* crystals. This requires, however, that the crystals shall glide along the grain boundary, which would be impossible without overcoming the intercrystalline forces. If the stress is insufficiently

Fig. 216.—Slightly Bent Zn Sheet, with Chains of Deformation Twins (651).

raised, or if fracture takes place earlier, then the crystals deform in a way less suited to their inherent mode of deformation. Pronounced distortions of the crystal lattice occur, especially in the vicinity of the grain boundary, accompanied by such phenomena as greater work hardening and an increased capacity for recrystallization.

The grain boundary effect is even more considerable in polycrystalline aggregates than in bi-crystals. The single grain is much more restricted in its deformation by glide, being stressed along the whole of its surface by forces which distort it and impose upon it a very general modification of shape which cannot be achieved by simple glide alone. But if only a single glide plane is available, as in the case of the hexagonal metals, then the individual grains are but insufficiently equipped to bring about a general change in shape. In this case heavy additional distortions will be necessary, and in particular twinning, which results in greater hardening of polycrystalline aggregates than of the freely stretched single crystal for the same percentages of working. Fig. 216 shows by way of example a coarse-grained zinc sheet which has been slightly bent, and which exhibits deformation twins in numerous grains. It should be noted that the twinning lamellæ often originate at the point of contact of three crystals and that connected chains form across the grain boundaries. Consequently, any portion of a grain boundary on which a deformation twin impinges is liable, owing to the very marked concentration of stress at this point, to give rise to another twin in a neighbouring crystal.

The position is different for cubic crystals, where there are many possibilities for glide. Face-centred crystals possess already twelve crystallographically equivalent octahedral glide systems. Consequently in their case deformation is by no means restricted to the system which is initially the most favourable, and multiple glide makes it much easier for the crystal to adapt itself to the imposed change in shape. Mathematical analysis has shown that any desired shape can be obtained if glide takes place simultaneously on five different glide systems (652). In such cases, therefore, much smaller differences may be expected in the hardening of single and poly-crystals. This is confirmed by the stress–strain curves of single and polycrystals of magnesium and aluminium which are shown in

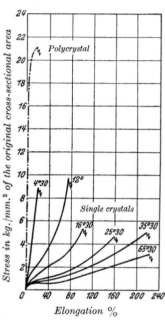

FIG. 217.—Stress–Strain Curves of Mg Crystals of Various Orientations, and of a Mg Polycrystal (655). The Initial Angles of the Basal Plane are Included for the Curves of the Single Crystals.

Figs. 217 and 90a (page 121). It will be seen that, for the magnesium polycrystals, the stress applied is greatly increased and the ductility reduced compared with the single crystal; while for aluminium the stress–strain curve of the polycrystal falls within the limits of the curves for the single crystal [(653), (654)].

A similar contrast between the behaviour of hexagonal and cubic metals is observed under dynamic stressing. While with magnesium polycrystals subjected to alternating torsional stress fracture starts as a rule along the grain boundaries (high stress concentration caused by restricted ductility), with aluminium poly-crystals the course of fracture is mainly transcrystalline [(656), (657), (658)].

The interaction between the crystals becomes increasingly important with decreasing grain size. Similarly, work hardening is observed to increase with decreasing grain size for the same amount of deformation (659). Hardness, too, increases with decreasing grain size (cf. 660). It is possible that the experiments which revealed an exceptional increase in the tensile strength of zinc sheets with increasing fineness of the grain admit of a similar explanation (661). This would mean that, despite the low temperature of the test (−185° C.), the results could be interpreted in terms of tensile hardening dependent upon grain size (compare with this the very different interpretation in Section 75).

80. *On the Origin of Textures*

The cast structure usually results from the location of the longi-tudinal axes of the columnar crystals at right angles to the cooling surface (wall of the mould). This also holds for relatively intricate moulds (662). A principle of selection which explains the columnar arrangement and the existence of an orderly structure in terms of geometry and crystallography can be obtained from Fig. 16 (page 28). Having regard to what has already been said in Section 13, the fibre axis which corresponds to the longitudinal direction of the columnar crystals should be identified with the direction of maximum speed of growth in the crystal. According to this conception the crystal directions given in Table XXXI must also represent the directions of maximum speed of crystallization.[1]

The reorientations experienced by the individual grains in the course of working are closely related to deformation textures. This has been strikingly confirmed by the interpretation of the deforma-tion textures of hexagonal metals, and by the great difference in behaviour of zinc and cadmium on the one hand and magnesium on the other (663). The mechanism which is mainly responsible for the change of shape, namely basal glide, makes the basal plane approach a position parallel to the direction of deformation. In the case of magnesium it is impossible for twinning to follow gliding, since with the basal plane in the above position, twinning would be accompanied by a contraction in the tensile or rolling direction

---

[1] It is probable that the electrolytic textures owe their origin to a similar selection in growth.

(cf. Section 31, and especially Section 39).   The process of stretching therefore results mainly in primal basal glide.   The most frequent end position is one in which the basal plane lies in the tensile direction or rolling plane.   The compression texture observed in magnesium is also what would be expected from the behaviour of the single crystal.   In zinc the formation is somewhat more complicated. Tensile tests on single crystals have already shown that in this case primary basal glide is followed by mechanical twinning, which, unlike the case of magnesium, results in a lengthening of the crystal.   The angle of inclination between the basal plane and direction of tension at which the new type of deformation becomes effective is between 8° and 16° at room temperature.   Twinning brings the basal plane into a position about 60° to the tensile direction, where it is again favourably oriented for further glide, with the result that very considerable secondary glide takes place in the twin lamellæ.   It is in the texture of rolled sheet that all the stages of this tensile deformation are present simultaneously (cf. Fig. 206a).   The most frequent final position of the basal plane is found experimentally to be about 10–20° to the direction of deformation (rolling direction).   In many grains, however, the basal planes are inclined at steeper angles, since twinning leads continually to a reorientation of the basal plane followed by further gliding.   The absence of basal-plane normals in the plane containing the sheet normal and transverse direction is readily explained by the fact that grains which are oriented in this way cannot undergo any lengthening in the rolling direction, and so must be brought, initially by twinning, into an orientation which is more suitable for subsequent deformation.[1]

Difficulties often arise, however, when such a direct interpretation of polycrystalline textures is attempted for cubic metals.   For instance, the tensile and drawing structure of cubic face-centred metals is a double fibre texture having the [111] and [100] directions parallel to the tensile direction.   In the case of stretched single crystals the [112] direction finally coincides with this direction. In spite of numerous attempts, no one has yet quite succeeded in deducing, quantitatively, the orientations of the crystallites in deformed components.   The various assumptions can be grouped as follows :   a mechanical principle assumes that the deformation texture is characterized by maximum strength.   As cold working

[1] In heavily rolled sheets of a 99 per cent. zinc alloy a further restricted area has been found about the rolling direction in addition to the areas already described for the hexagonal axis.   The stability of this orientation is understandable, since neither by gliding nor by twinning can deformation be achieved in the rolling process (664).

proceeds, the crystal elements are supposed to rotate gradually into an end position which is symmetrical to the main directions of deformation, and which requires a (relatively) maximum of externally applied force if further deformation is to take place (665). However, as is shown by the dependence of the yield point on orientation, the general validity of this principle is not proven. The other hypotheses relate primarily to the behaviour of the individual grain in the polycrystalline aggregate which is being deformed in tension. In (666) the point of departure is the assumption that the glide planes are bent about an axis perpendicular to the direction of gliding (direction of curvature). The reorientation of the grain embedded in the polycrystalline aggregate is said to correspond to this bending. The final position of the lattice is attained by activating two glide systems at the most. Although in certain cases the correct textures can be inferred from these assumptions, this is not always so. In particular, the case of the rolling textures of cubic metals remains unexplained.

The third group of hypotheses deals with the differences in the behaviour of extended free crystals and crystals in an aggregate which were described in the previous section. In (667) it is assumed that, if the imposed general change in shape is to be explained satisfactorily, then all the crystallographically equivalent glide systems must become active within the extended grain in the polycrystal. It has been found that, if certain allowances are made in regard to capacity for glide of glide planes and directions, the observed textures can be conceived as stable final configurations. The employment of three glide systems for the deformation of grains in cubic metals is indicated in (668) and discussed in detail in (669). Since the glide system which operates is determined by the shear stress, the magnitude of the prevailing shear stress is utilized for selecting the effective glide elements, and that final orientation is determined which, for the prescribed deformation, remains resistant to glide in the three most favourable glide systems. It was along these lines that the well-known drawing and compression textures of cubic metals could be interpreted. The main position of the rolling textures appears as a superimposition of compression and drawing textures. It is revealed as that final orientation which remains stable when compressed parallel to the sheet normals and elongated in the direction of rolling.

Another point of view from which to interpret the deformation textures is provided by the observation that different types of stress giving identical deformation yield identical textures. We have seen, for instance, that the cold rolling of wires (copper and aluminium) produces the same texture as pulling or drawing them (639), and that steel sheet drawn through flat dies exhibits the same texture as rolled sheet (638). This emphasizes clearly the importance of the change in shape for the development of the textures. It therefore appears that the symmetry of the directions of flow in the metal is a more influential factor than the symmetry of the applied forces.[1] We again observe the importance of the direction of flow in the case of the inhomogeneous texture of wires drawn in one direction. In the centre of the wire the direction of flow coincides with the direction of drawing. Here the undisturbed texture can develop. At the surface, on the other hand, the material is compelled by the conical die wall to flow towards the centre. The axes of the crystals are rotated relative to the longitudinal direction, the amount of the rotation corresponding roughly to the taper of the die.

We are still unable to account fully for the origin of recrystallization textures—which, as already stated, are largely influenced by the conditions under which they are produced. However, it is reasonable to suppose [cf. especially (670)] that, also in the case of deformed polycrystals, recrystallization proceeds in two phases, namely by nuclear formation and grain growth—a mechanism which can be inferred especially from experiments which have been carried out with deformed aluminium crystals (cf. Section 63 and also 77). Opposed to this view is the " single-phase " conception of the recrystallization process (671). According to this theory those parts of the lattice which are least deformed serve as nuclei for the new lattice; removal of the distortions which are present in the deformed state enables these nuclei to grow. Consequently, over moderately large lattice areas only a *single* lattice orientation is present—although in various conditions, according to the number of distortions (" *Gehalt an Verhakungen* ").

81. *Calculation of the Properties of Quasi-isotropic Polycrystalline Aggregates*

As already pointed out, effective calculation of the behaviour of crystal aggregates, based on the behaviour of the single crystal,

---

[1] In this connection it might be possible to interpret the deformation textures by determining the stable end positions of the lattice at which those three glide systems become active which give maximum extensions in the preferred directions of flow for the same amount of glide (minimum energy of deformation). Since for small amounts of glide the strain is obtained by multiplying the amount of glide with the same factor ($\sin \chi \cos \lambda$) as when calculating the shear stress from the tensile stress (see Section 26), the same choice of glide systems becomes available as under the condition of maximum shear stress.

should be possible especially for those properties for which grain boundary interferences can be disregarded.

Hitherto, calculations have been carried out principally for the quasi-isotropic polycrystal (random orientation). Common to all determinations of average values is the assumption that the grains are large with respect to the range of the binding forces, but small in relation to the dimensions of the specimens, and that they completely occupy the space.

For the *elastic properties* this averaging (*Mittelung*) was first undertaken in (672) at a time when experimental material for testing the formula was scarce. The adhesion of the grains during elastic deformation is assured if one assumes that the stresses and strains are continuous across the grain boundaries. The averaging of the stress components (equation 7/1) for the polycrystal of random orientation is then based on the mean value of the elastic parameters C_{ik}. Integration of these expressions over the total range of orientation gives the elastic parameters of the polycrystal (in accordance with the system in Section 8) from which the moduli of elasticity and torsion are then derived.

It has been shown in (673), as the result of tests carried out on a number of metals, that while this method of calculation yields fairly accurate values for crystal material which is slightly anisotropic, the margin of error increases with increasing anisotropy. This is attributed to the conditions which exist at grain boundaries. Since the principle of action and reaction requires that the three stress components perpendicular to the boundary plane in adjacent crystals shall be equal to each other, it will usually be possible for only three, and not for all six, elastic deformations to be equivalent to each other. Very great difficulties are encountered when arriving at averages under the new limiting conditions. It becomes necessary to assume a special lamellar type of structure in the polycrystal. Such calculations yield a noticeably closer approximation to the values of the moduli as determined experimentally. The first process mentioned above is found to be an approximation where only slight anisotropy exists.

Underlying all calculations of averages is the assumption of a state of uniform stress in all grains (674). The new calculation proceeds along analogous lines (672), except that in accordance with the changed assumption the average is reached by way of S_{ik} [equation (7/2)]. This method of calculation yields values which are persistently lower than the observed results.

Finally, in (675) averaging is performed by integrating over the whole range of orientation the expressions for the moduli of elasticity and torsion, as given by the theory of crystal elasticity, for a specimen taken in any desired direction relative to the crystallographic axes [formulas (10/1, 2) and (10/4, 5)]. Special limiting conditions governing the constancy of the components of stress or deformation become superfluous at the grain boundaries. This is justified because the cohesion of the grains is assured by the distortions which take place in the outer layers. This can, in fact, be directly observed in the case of the much larger stresses which lead to plastic deformation. Integration over the total range of orientation (corresponding to the random texture of the polycrystal) results in finite expressions for the moduli of hexagonal crystals, and in rapidly converging series for those of cubic and tetragonal crystals (676). This method of averaging reproduces the observed facts with at least the same degree of accuracy as that described in (673). A statistical comparison with the results obtained in practice will be found below in Table XXXVI (678).

TABLE XXXVI

Calculated and Experimentally Determined Properties of Polycrystalline Metals (Random Orientation)

Metal.	Modulus of elasticity, kg./mm.².		Modulus of shear, kg./mm.².		Thermal expansion, 10^{-6}.		Specific electrical resistance, 10^{-6} Ω/cm.		
	Calculated.[1]	Observed.[2]	Calculated.	Observed.[2]	20–100° C. calculated.	0–100° C. observed.[3]	Calculated (81/3).	Calculated (81/2).	Observed.[3]
Aluminium .	7,170	7,200	2,660	2,700					
Copper . .	11,950	12,100	4,280	4,400					
Silver . .	7,500	8,000	2,640	2,700	Isotropic.				
Gold . .	7,750	8,100	2,650	2,800					
α-Brass (72% Cu)	10,500	10,200	3,550	4,100					
α-Iron . .	20,700	21,400	7,770	8,400					
Magnesium .	4,510	4,500	1,770	1,800	25·9	26·0	4·29	4·32	4·4
Zinc . .	10,040	10,000	3,620	3,700	30·7	30·0	5·89	5·91	6·0
Cadmium . .	6,110	5,100	2,130	2,200	31·8	31·6	7·30	7·37	7·4
β-Tin . .	4,480	4,650	1,570	1,700	20·5 [4]	23·0	11·0	11·4	11·1

[1] All calculations based on the properties of crystals recorded in Tables XVIII and XX.
[2] According to Landolt, Bornstein, Roth and Scheel, *Physico-chemical Tables*, 5th edition.
[3] According to F. Kohlrausch, *Manual of Practical Physics*, 16th edition.
[4] At approximately 20° C.

In this context, mention may be made of Poisson's equation (cf. Section 9). It has been found that when applying the averaging process according to (672), the validity of the Cauchy relations for the single crystal leads inevitably to the Poisson equation for the ideal random polycrystal. This relationship might be expected to emerge from other methods of averaging also. The discrepancy in the Poisson's ratio, as determined on the polycrystal, of 0·25, makes it possible to gauge the extent to which the Cauchy relations have been satisfied in respect of the single crystal (cf. Table III).

Averages for the *magnetic properties* will be found in (677). Here it is a question of the relative change in length $\left(\dfrac{\delta l}{l}\right)$ of the crystal parallel to the magnetization vector (magnetostriction), the relative change in the specific resistance $\left(\dfrac{\delta \rho}{\rho}\right)$ resulting from magnetization, and finally of the energy $\sigma = \displaystyle\int_0^j H_p dJ$ which is necessary for magnetization, in which H_p represents the components of the outer magnetic field parallel to the vector of magnetization. In the event of saturation, the difference in the properties of ferro-magnetic cubic crystals in the direction (S) under investigation and about the cube edge [100]

$$\left(\frac{\delta l}{l}\right)_S - \left(\frac{\delta l}{l}\right)_{[100]}, \quad \left(\frac{\delta \rho}{\rho}\right)_S - \left(\frac{\sigma \rho}{\rho}\right)_{[100]} \quad \text{and} \quad \sigma S - \sigma_{[100]}$$

are proportional to the expression $\gamma_1{}^2 + \gamma_2{}^2 + \gamma_2{}^2 \gamma_3{}^2 + \gamma_3{}^2 \gamma_1{}^2$ [which is also important for the moduli of elasticity and torsion (Fig. 154a)], where γ_i is the cos of the angles of the magnetization vector in relation to the cubic axes. The mean value $\bar{f}$ of the three properties of a quasi-isotropic polycrystal is obtained from integration over the entire range of orientation

$$\bar{f} = \tfrac{1}{5}(2f_{[100]} + 3f_{[111]}) \quad . \quad . \quad . \quad . \quad (81/1)$$

in which $f_{[100]}$ and $f_{[111]}$ represent the values for cube edges and the body diagonals. There is close agreement between the observed and calculated results of the change in the electrical resistance of iron.

The same principle of averaging was also applied to the *thermal expansion* (α) and the specific resistance (ρ) of the tetragonal and hexagonal crystals (678). Integration of the expression representing dependence upon orientation (58/1) gives

$$\bar{f} = \tfrac{1}{3}f\| + \tfrac{2}{3}f\bot \quad . \quad . \quad . \quad . \quad (81/2)$$

in which $f\|$ and $f\bot$ represent the values parallel and perpendicular to the main axis. So far as *specific resistance* is concerned, this result conflicts with the values obtained earlier. It was pointed out in (672) that it was preferable to obtain averages for the constants of conductivity rather than for those of specific resistance, and the expression for the specific resistance of the random oriented polycrystal was stated to be

$$\frac{1}{\rho} = \frac{1}{3\rho\|} + \frac{2}{3\rho\bot} \quad . \quad . \quad . \quad . \quad (81/3)$$

The conductivity $\left(\frac{1}{\rho}\right)$ has been averaged over the whole range of orientation in (679), from which it follows that for $\rho > \rho\bot$

$$\rho = \sqrt{\rho\bot(\rho - \rho\bot)} \arctan \sqrt{\frac{\rho\| - \rho\bot}{\rho\bot}} \quad . \quad (81/4a)$$

and for $\rho\| \angle \rho\bot$

$$\rho = 2 \frac{\sqrt{\rho\bot(\rho\bot - \rho\|)}}{\ln\dfrac{\sqrt{\rho\bot} + \sqrt{\rho\bot - \rho\|}}{\sqrt{\rho\bot} - \sqrt{\rho\bot - \rho\|}}} \quad . \quad . \quad (81/4b)$$

A comparison of the methods of calculation will be found in Table **XXXVI**, which also compares the mean calculated values of some other properties with those determined experimentally.

The great difficulty presented by a comparison of this kind lies in the selection of the observed values. Only in the rarest cases has the texture of the polycrystalline specimen investigated also been determined. The assumed random orientation of the crystals on which the calculations are based is therefore by no means warranted.[1] In the case of specific resistance there is a further element of uncertainty in so far as the determinations on the single crystals and polycrystals were not always carried out on the same type of material, so that the degree of purity may have been different. We believe, however, that the tabulated figures are fairly accurate.

The values obtained by averaging approximate closely to the observed values and so appear to justify the averaging methods

---

[1] The random orientation of the crystals can be estimated if the cubic compressibility (K) of the material, or Poisson's ratio (μ), of the specimen is known in addition to the elastic moduli (673). The theory of elasticity yields for the isotropic solid the relationships $K = \dfrac{9}{E} - \dfrac{3}{G}$ and $\mu = \dfrac{E}{2G} - 1$. Satisfaction of these equations is therefore a necessary although inadequate indication of quasi-isotropy.

employed. This should be particularly emphasized in respect of the coefficient of expansion, which in the case of zinc exhibits such exceptional anisotropy $\left(\dfrac{\alpha_\|}{\alpha_\perp} = 4 \cdot 51 \right)$. The distortions at the grain boundaries, which, as the temperature changes, are necessary for the cohesion of the polycrystal, appear therefore to be confined to such small areas that their effect can be only slight. A decision regarding the average values for specific resistance [the values calculated according to (81/4, a and b) lie between those given in the table] cannot be reached on the basis of results so far obtained, especially in view of what has just been said regarding the uncertainty of the comparative figures.

In (680) Sachs has furnished an important clue for calculating the *plastic properties* of quasi-isotropic polycrystals from the behaviour of the single crystal. These are arrived at by determining the ratio of the tensile and torsional yield points of the cubic face-centred metals. The shear stress law (cf. Section 40) provides the basis for this calculation. By taking the mean values from graphs it will be found that the maximum shear stress at the start of deformation of the polycrystal is, in the tensile test, greater by 12 per cent., and in the torsional test greater by 29 per cent., than the critical shear strength of the octahedral glide system. From this it follows that the ratio of the torsional yield point to the tensile yield point of the polycrystal (measured by the maximum shear stress) is 1·15. This figure agrees well with the direct experimental determinations made on copper and nickel, which showed the effect of the mean principal stress, and gave for the above ratio an average figure of 1·125 [(681), (682)]. Previously this could be explained only on the assumption that the energy of deformation represented a measure of the probability of flow (683).

An attempt to calculate the yield point of cubic and hexagonal crystals on the basis of an assumed condition of flow of the single crystal, equivalent to a square function of the stresses, will be found in (674). (For objections to this type of flow condition in the single crystal, see Section 40.)

82. *Interpretation of the Properties of Technical Components*

The ideal random arrangement of the crystals, which served as a basis for discussion in the previous section, represents an extreme case which will not normally be present in technical components. As a rule these will be characterized according to their previous history by a more or less pronounced directionality of the grains.

It is on this texture that an exact calculation of the properties of polycrystals must be based, and it is owing to the complicated nature of such actual textures (described in Section 78) that the calculations are difficult to perform. That is why on the whole we still have to be content with a qualitative interpretation of polycrystalline behaviour. However, that this method has added greatly to our knowledge of materials will be apparent from the following examples.

Before we examine the individual results, however, let us glance at a group of polycrystalline properties which can be plausibly interpreted only on the basis of the general behaviour of single crystals [cf., for instance, (684)]. The properties in question are the elastic after-effect, hysteresis and the Bauschinger effect. The occurrence of these phenomena in the case of single crystals of metals has been described in Section 55, while a very tentative explanation of them will be found in Section 76. The reason for their occurrence in the polycrystal is thought to be completely different from that in the single crystal. The different resistance to flow exhibited by the individual grains furnishes a clue to an understanding of the phenomenon. If a polycrystalline specimen is subjected to increasing stress, the yield point will not be exceeded simultaneously in all grains. Whereas those crystals which are very favourably oriented for deformation are soon plastically deformed, the grains which are less favourably placed remain within the range of purely elastic deformation. Thus the individual crystallites do not uniformly resist the externally applied force. Whereas the soft grains, by deforming plastically, are not heavily stressed and are subjected to a load which only slightly exceeds the yield point, the strength of the material resides principally in the grains which have not yet been plastically deformed. In these grains considerable elastic stresses predominate. Even when the stress has been increased to such an extent that the strongest grains become plastically deformed, differences still persist in the stress content of the various crystals. If the load is removed from the solid, the average value of the stress reverts to zero, but not so the stress in the individual grains. In order that the crystals which have already experienced purely elastic deformation can become completely unstressed they must first induce reverse deformation in their plastically deformed neighbours. This is possible only to a limited extent. Consequently, after the removal of the load, there will remain in the grains which resisted the previous load a stress appropriate to this load, while in the plastically deformed crystals there will be a stress of opposite sign. It is in this way that Heyn's concealed elastic stresses (685) can be explained structurally (686).

Since the crystals which have already been deformed plastically continue to be the source of further deformation, it follows from the stress distribution in the unloaded specimen that there will be a reduction in the resistance to a stress applied in the reverse direction to that of the original stress (Bauschinger *effect*) and an increase in the resistance to deformation in the same direction (*work hardening*).

In view of the plastic inhomogeneity of the crystal, the *elastic*

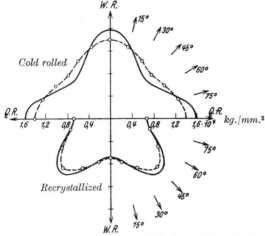

FIG. 218.—The Modulus of Elasticity in Relation to Direction in Cold-rolled and Recrystallized Cu Sheet (690).

WR = rolling direction. QR = transverse direction.

————— : Calculated.   O– – –O : Measured.

after-effect, i.e., the gradual emergence of a condition of equilibrium after application or removal of the load, can now be explained satisfactorily as a *plastic* process. This phenomenon, which normally is absent in the single crystal, consists in an equalization, over a period of time, of the various stresses present in the grains—a process which, as already stated, is possible only with the aid of plastic deformation (687). The quantitative development of this theory of after-effect, and its combination with the earlier phenomenological theories, will be found in (688).

In accordance with this conception the dependence of the shape, which a specimen must possess in order to achieve a certain condition of stress, upon the path which has brought about this condition—a phenomenon known as *hysteresis*—can also be attributed to plastic deformation in those individual crystals which are favourably placed. In this connection the appearance of glide bands in individual crystals at stresses far below the limit of proportionality is highly important (689).

In order to illustrate the connection between the properties of the polycrystal, the texture and the behaviour of the single crystal, diagrams are reproduced in Figs. 218 and 219, which show the

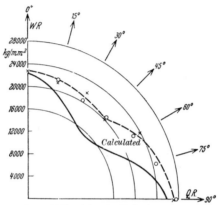

FIG. 219.—The Modulus of Elasticity in Relation to Direction in Cold-rolled
Fe Sheet (691).

WR = rolling direction.   QR = transverse direction.

dependence upon direction of the *modulus* of elasticity of copper and iron sheets. The curve which has been calculated on the basis

TABLE XXXVII

*Dependence of the Strength and Elongation of Recrystallized Sheets
upon Direction*

Metal.	Parallel to the rolling		Strength (kg./mm.²). Angle to the rolling direction			Elongation (%). Angle to the rolling direction.		
	Direction.	Plane.	0°.	45°.	90°.	0°.	45°.	90°.
Aluminium (641) . Copper (692) . .	[100]	(001)	(5·8 =) 1 (18·5 =) 1	1·13 1·08	1·03 0·99	(6·1 =) 1 (20 =) 1	1·97 3·40	1·21 1·15
Calculated . .			1	1·06	1·00	1	2·30	1·00
Silver (693) . . α-Brass (72% Cu) (693)	[112]	(311)	(21 =) 1 (39·5 =) 1	0·92 0·95	0·90 0·94	(44 =) 1 (50 =) 1	1·00 1·09	0·96 1·12
Calculated . .			1	0·92	0·89	1	1·10	1·12

of the principal positions of the texture agrees closely with the experimentally determined curve for recrystallized copper sheet, while for the cold-rolled sheets also the approximation to the observed results is good. No doubt the agreement would be still more marked if closer regard could be paid to the actual textures [cf., in particular, the detailed discussion in (689a)].

Table XXXVII contains the *mechanical properties* of recrystallized sheets of cubic face-centred metals. The observed dependence upon direction is contrasted with a calculated figure derived from the properties of correspondingly oriented aluminium crystals (cf. Fig. 99). Consequently, although in this case it is a question of transferring to other metals results already obtained with aluminium, and although only the main position of the recrystallized texture has been considered, the calculated values correspond in general quite closely to the behaviour of the sheets. In this table the low maximum for strength, the high maximum for the elongation in the 45° direction of sheets with the cube texture, and in the case of sheets with the [112]—(311) texture, the slight decrease in strength and increase in elongation with increasing angle to the rolling direction, are shown very clearly. The graph shown in Fig. 220 contains in addition the yield point and fatigue strength of recrystallized copper sheet. The dependence of these properties upon orientation corresponds substantially to that of the static tensile strength. It seems probable, therefore, that fatigue strength, like the tensile strength, is related to the orientation of the crystals. It is thus essential that the texture should be taken into account not only in the case of the static properties but also when assessing the fatigue strength.

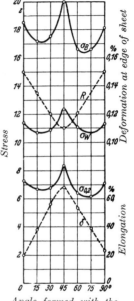

Angle formed with the rolling direction

FIG. 220.—Anisotropy of the Static and Dynamic Mechanical Properties of Annealed Cu Sheet (692).

σ_B = U.T.S. ;
$\sigma_{0.2}$ = yield point ;
δ = elongation ;
σ_W = bending fatigue strength (10^6 stress reversals) ;
R = edge deformation.

The *deep-drawing properties* of sheets are also largely influenced by the texture; in fact, the cupping test is merely an intricate tensile test in which the fibres are stretched in all possible directions (694). In the deep drawing of hollow components the anisotropy of the sheet will reveal itself in " earing " (" Zipfelbildung ") and in a reduced capacity for deep drawing [aluminium (632), copper (695), (693), iron (696)].

The zonal type of structure which is characteristic of cold-drawn wires (cf. Fig. 209) suggests that in different layers of the same wire

TABLE XXXVIII

Directionality of Certain Properties of Zinc Sheets. [*Taken from (637)*]

Property.	Sheet thickness, mm.	Parallel to the :		
		Rolling direction.	Transverse direction.	Normal direction.
Coefficient of expansion $\times$ 10^6 between 30° and 50° C.	0·65 2·27	21·0 30·5	14·1 18·7	36·7
Modulus of elasticity in kg./ mm.^{2} [taken from (701)]	0·28 2·27	9180 8200	10110 10100	
Poisson's ratio [taken from (701)]	2·27	0·299	0·226	0·320
Yield point (0·2%) in kg./ mm.^{2}	0·65 2·27	9·7 14·8	13·1 18·8	
Ultimate tensile strength in kg./mm.^{2}	0·28 0·65 2·27	24·4 20·4 28·5	31·6 27·2 35·9	
Elongation, %	0·28 0·65 2·27	2 10 12	~2·5 7 3	

there will be differences in those properties which are dependent on orientation. Such differences were, in fact, observed in the tensile strength. Tensile tests on etched copper wires revealed that the tensile strength of the core zone is about 10 per cent. higher than the average value for the original wire. In spite of the greater distortion caused by deformation in the surface zone, the work hardening of the more heavily deformed core sections predominates (629). Compression tests on drawn aluminium bars also revealed this difference between the core and surface zones (697).

Attention has been drawn in (696) to the connection between the anisotropy of the *magnetization* of iron and nickel sheets on the one hand, and the texture and crystalline properties on the other. The significance of texture in the *corrosion resistance* of sheets is clearly shown in the case of copper with random and regular recrystallized texture (698).

In regard to the *specific resistance* and the *coefficient of expansion* we will examine in the first place the changes which accompany the plastic working of hexagonal metals and tin.   In every case the changes could be related to reorientations of the crystals, whether by gliding or by the development of deformation twins [(679), (699), (700)].   The coefficient of expansion of zinc and the hexagonal zinc alloys in the rolling, transverse and normal directions, is examined with particular care in (645). The maximum value for thermal anisotropy (the maximum differential expansion between the rolling and transverse directions) was found to occur at rolling percentages of about 20–40 per cent.; beyond 80 per cent. the anisotropy dis-

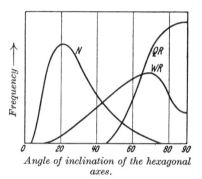

Angle of inclination of the hexagonal axes.

Fig. 221.—Distribution of the Angles of the Hexagonal Axes for the Rolling, Transverse and Normal Directions in Zinc Sheets.

WR = Rolling direction.
QR = Transverse direction.
N = Normal direction.

appears.   This behaviour is explained by the opposing effects of gliding and twinning (twinning increases, basal glide diminishes anisotropy in the rolling plane).

Table XXXVIII summarizes the directionality of properties of zinc sheets of various thickness.   In view of the substantial range of scatter (Fig. 206*a*) the results cannot be explained in terms of the rolling structure if only one main position of the hexagonal axis is used.   Fig. 221 gives for the rolling, transverse and normal directions an approximate picture of the distribution of the angles at which the hexagonal axes are located in the sheet under investigation. These distributions must be taken into account when deriving the properties of the sheet from the behaviour of the single crystal and the texture.   On a qualitative basis this operation leads to very satisfactory agreement in the case of the coefficient of expansion and modulus of elasticity.   The absence of anisotropy in specific resistance in the plane of the sheet can be readily understood

(accuracy of measurement 1 per cent.).   For the mechanical properties the comparison is less accurate; the reason is that hexagonal
crystals glide on the basal plane only; thus they are quite unable to
cope with the deformation demanded by the crystal aggregate (cf.
Section 79).   Nevertheless, if the behaviour of the single crystal is
taken into account (Figs. 87 and 96) and if allowance is made for the
diversity of orientation in the sheet, it is possible to account roughly
for the observed anisotropy of the yield point, ultimate tensile
strength, elongation and bending capacity (702).

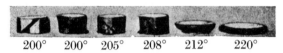

200°   200°   205°   208°   212°   220°

FIG. 222.—Sudden Change in the Plastic Properties of a
Magnesium Alloy (AZM) at about 210° C. (703).

We will mention in conclusion the example of an Elektron alloy
(AZM—6·3 per cent. Al, 1·0 per cent. Zn, remainder magnesium).
It will be seen from Fig. 222 that a sudden change takes place in the
plasticity of the technical alloy at a working temperature of about

TABLE XXXIX

*Tensile and Comprehensive Yield Point of a Deformed Elektron
Alloy (AZM).   [Taken from (703)]*

Component.	Yield point (0·2%), kg./mm.².	Compressive yield point (−0·2%), kg./mm.².
Extruded rod.	23·0	13·0
Forged Billet.	12·0	18·0
	18·0	12·0
	8·0	8·0

210° C.   This behaviour results from the activity of new glide
elements at higher temperatures in the hexagonal solid solutions of
magnesium (cf. Table VI), as a result of which the material deforms
more readily.

A very characteristic effect of the mechanisms by which hexagonal crystals deform (gliding *and* twinning) can also be observed in the behaviour of the tensile and compressive yield points of worked material. Table **XXXIX** contains particulars of bar material and forged discs. The appreciable discrepancies can be readily explained in terms of the deformation textures (hexagonal axis normal to the longitudinal direction of the bar, or in the forging direction). Basal glide can occur only in specimens taken at an angle of 45° to the forging direction. In other cases deformation must normally start by twinning. This leads, however, to a very definite change of shape (Section 39). If this change is in the direction of the external stress, the flow resistance will be low, but if it is in the opposite direction to the imposed deformation, then the resistance will be high. Where basal glide is possible, deformation takes place in both directions at the same low stress. By producing a specially fine grain in the extruded rods, and by frequently changing the direction of forging and so avoiding formation of too pronounced a texture, it is possible to eliminate plastic anisotropy—usually a technically undesirable property (703).

TABLE XL

Crystalline Lattice Structure of the Elements

Legend:

□ Cubic face centred.
⊡ Cubic body centred.
⊠ Cubic Diamond.
○ Hexagonal close packed.
△ Rhombohedral.

1 H hex.															2 He □
3 Li ⊡	4 Be ○										5 B	6 C ⊠ hex.		7 N kub.hex.	10 Ne □
11 Na ⊡	12 Mg ○										13 Al □	14 Si ⊠	15 P kub. △	8 O rhomb.hex.	18 Ar □
19 K ⊡	20 Ca ○ α ⊡ β	21 Sc	22 Ti ○	23 V ⊡	24 Cr ⊡ kub.kub. α γ β	25 Mn kub.tetr.	26 Fe ⊡ □ γ	27 Co α β γ	28 Ni □(○)	29 Cu □	30 Zn ○	31 Ga rhomb.	32 Ge ⊠	33 As △	16 S rhomb.	36 Kr □
37 Rb ⊡	38 Sr □	39 Y ○	40 Zr ○ α β	41 Nb ⊡	42 Mo ⊡	43 Ma	44 Ru ○	45 Rh kub.	46 Pd □	47 Ag □	48 Cd ○	49 In tetr.	50 Sn ⊠ tetr. α β	51 Sb △	34 Se hex.mono-klin	54 X □
55 Cs ⊡	56 Ba ⊡	57–71 Selt. Erden	72 Hf ○	73 Ta ⊡	74 W ⊡ kub. α β	75 Re ○	76 Os ○	77 Ir □	78 Pt □	79 Au □	80 Hg △	81 Tl ○ □ α β	82 Pb □	83 Bi △	52 Te hex.	53 J rhomb.
87	88 Ra ⊡	89 Ac □	90 Th □	91 Pa	92 U ⊡										84 Po	86 Em
															35 Br	9 F
															85	17 Cl

322

Table XLI

Type and Dimensions of the Lattices of the Important Metals

Metal.	Lattice type.	Parameter,* Å.	Metal.	Lattice type.	Parameter,* Å.	
					a.	c.
Al	Cubic face-centred A1 †	4·040	Be	Hexagonal close packed A3	2·281	3·577
γ-Fe ‡		3·56	Mg		3·202	5·199
β-Co ‡		3·54	α-Co		2·51	4·07
Ni		3·517	Zn		2·659	4·935
Pd		3·881	Cd		2·974	5·606
Pt		3·915				
Cu		3·608	α-Sn	Cubic face-centred (diamond) A4	6·46	—
Ag		4·078				
Au		4·070				
Pb		4·939	β-Sn	Tetragonal body-centred A5	5·819	3·175
Ta	Cubic body-centred A2	3·29	Bi	Rhombohedral body-centred A7	4·736	ω = 57° 16′ u = 0·474
Mo		3·140	Sb		4·50	ω = 57° 5′ u = 0·466
W		3·158				
α-Fe		2·861	Te	Hexagonal A8	4·445	5·912 u = 0·269

* Measurements taken from M. C. Neuburger, *Z. Kristallogr.*, Vol. 86 (1933), p. 395.
† Designation of the lattice types according to P. P. Ewald and C. Hermann, *Strukturbericht*, Leipzig, 1931.
‡ Extrapolated at room temperature.

Table XLII

Type and Dimensions of the Lattices of some Ionic Crystals *

Salt.	Lattice type.	Parameter, Å.	Salt.	Lattice type.	Parameter, Å.	
					a.	c.
			β-ZnS	Wurtzite B4 (hexagonal)	3·84	6·28
			CdS		4·14	6·72
NaF	Rock salt B1 (cubic)	4·62	BeO		2·69	4·37
NaCl		5·628	ZnO		3·24	5·18
NaBr		5·96				
NaI		6·46	NiAs	Nickel Arsenide B8 (hexagonal)	3·61	5·0₃
KF		5·33	NiS		3·42	5·3₀
KCl		6·28				
KBr		6·59	CaF₂	Fluorspar C1 (cubic)	5·45₁	—
KI		7·05				
RbCl		6·54	TiO₂	Rutile, C4 (tetragonal)	4·58	2·95
NH₄I		7·24				
MgO		4·20				
PbS		5·9	CdI₂	Cadmium iodide C6 (hexagonal)	4·2₄	6·8₄
			Mg(OH)₂		3·1₂	4·7₃
CsCl	Cæsium chloride B2 (cubic)	4·11	α-Al₂O₃	Corundum D51 (rhombohedral)	5·12	a = 55° 17′
NH₄Cl		3·86				
NH₄Br		4·05	CaCO₃	Calcite G1 (rhombohedral)	6·361	a = 46° 7′
			MgCO₃		5·61	48° 1′₀
α-ZnS	Zinc blende B3 (cubic)	5·42	ZnCO₃		5·62	48° 2′₀
CuCl		5·41	MnCO₃		5·84	47° 2′₀
CuI		6·05	FeCO₃		5·82	47° 4′₅
			NaNO₃		6·32	47° 1′₅

* According to P. P. Ewald and C. Hermann, *Strukturbericht*, Leipzig, 1931. Selection from Table XXIV.

BIBLIOGRAPHY

I. SOME FUNDAMENTALS OF CRYSTALLOGRAPHY

For detailed studies consult :
P. Niggli, " Lehrbuch der Mineralogie ". Berlin : 1924, I.
H. Mark, " Die Verwendung der Röntgenstrahlen in Chemie und Technik ". Leipzig : 1926.
P. P. Ewald, " Handbuch der Physik ". Berlin : 1933, **23/2**.

II. CRYSTAL ELASTICITY

General Treatises

W. Voigt, " Lehrbuch der Kristallphysik ".
J. W. Geckeler, " Handbuch der Physik ". Berlin : 1928, **6**.
M. Born and O. F. Bollnow, " Handbuch der Physik ". Berlin : 1927, **24**.
 1. P. P. Ewald, " Cauchy Relations for Metal and Ionic Crystals ". Müller-Pouillet's, " Lehrbuch der Physik ". 11th ed., **1/2**, 925.

Determination of the Elastic Parameters

 2. P. W. Bridgman, " Alkali Halides ", *Proc. Amer. Acad. Arts Sci.*, 1929, **64**, 19.
 3. K. Försterling, " Sylvin ", *Z. Physik*, 1920, **2**, 172.
 4. W. Voigt, " Fluorspar and Pyrite ", *Wiedemanns Ann.*, 1888, **35**, 642.
 5. W. Voigt, " Sodium Chlorate ", *Wiedemanns Ann.*, 1893, **49**, 719.
 6. E. Goens, " Gold ", *Naturwiss.*, 1929, **17**, 180—Al. *Ann. Phys.*, [5], 1933, **17**, 233.
 7. E. Goens (and J. Weerts), " Copper ", *Z. Instrumentenkunde*, 1932, **52**, 167.
 8. H. Röhl, " Silver, Gold ", *Ann. Physik*, [5], 1933, **16**, 887.
 9. M. Masima and G. Sachs, " α-Brass ", *Z. Physik*, 1928, **50**, 161.
10. E. Goens and E. Schmid, " α-Iron ", *Naturwiss.*, 1931, **19**, 520 ; *Z. Elektrochem.*, 1931, **37**, 539.
11. P. W. Bridgman, " Tungsten, Zinc, Cadmium, Bismuth, Antimony, β-Tin ", *Proc. Amer. Acad. Arts Sci.*, 1925, **60**, 305.
12. S. J. Wright, " Tungsten ", *Proc. Roy. Soc.*, 1930, **126**, 613.
13. E. Goens and E. Schmid, " Magnesium ", *Naturwiss.*, 1931, **19**, 376.
14. P. W. Bridgman, " Magnesium ", *Phys. Rev.*, 1931, **37**, 460 ; *Proc. Amer. Acad. Arts Sci.*, 1932, **67**, 29.
15. E. Grüneisen and E. Goens, " Cadmium ", *Physikal. Z.*, 1923, **24**, 506. E. Goens, " Zinc ", *Ann. Physik*, [5], 1933, **16**, 793.
16. E. Grüneisen and E. Goens, " Determination of the Elastic Parameters by Transverse Vibrations ", *Z. Physik*, 1924, **26**, 235.
17. E. Goens, " Determination of the Elastic Parameters by Torsional Vibrations ", *Ann. Physik*, [5], 1930, **4**, 733.
18. E. Goens, " Influence of the Thickness of the Test Piece on the Frequency of the Bending Vibrations ", *Ann. Physik*, [5], 1931, **11**, 649.
19. E. Goens, " Effect of Torsion (Bending) which is Usually Produced when Bending (Twisting) a Crystal Rod ", *Ann. Physik*, [5], 1932, **15**, 455.

III. PRODUCTION OF CRYSTALS

For short summaries of the processes for producing metal crystals see :
H. C. H. Carpenter, *J. Inst. Metals*, 1926, **35**, 409.
J. A. M. van Liempt, *Amer. Inst. Min. Met. Engr. Techn. Publ.* No. 15, 1927.
20. J. Czochralski, " Absence of Recrystallization in the Cast Structure ", *Internat. Z. Metallogr.*, 1916, **8**, 1.

21. W. Fraenkel, "Absence of Recrystallization in the Cast Structure", Z. anorg. Chem., 1922, **122**, 295.
22. J. Czochralski, "Crystal Production by Work Recrystallization". In "Moderne Metallkunde in Theorie und Praxis". Berlin : 1924.
23. H. C. H. Carpenter and C. F. Elam, "Crystal Production by Work Recrystallization", Proc. Roy. Soc., 1921, **100**, 329.
24. A. Sauveur, "Crystal Production by Work Recrystallization", Proc. Intern. Assoc. Test. Mat., 1912, **2**, 11. Paper No. II, 6.
25. H. C. H. Carpenter and S. Tamura, "Cu Crystals Produced by Work Recrystallization", Proc. Roy. Soc., 1927, **113**, 28.
26. S. Takejama, "Cu Crystals Produced by Work Recrystallization", Mem. Col. Sci. Kyōtō Imp. Univ., 1930, **13**, 355, 363.
27. F. Koref, "Production of Lamp Filaments from W", Z. techn. Physik., 1926, **7**, 544.

Crystal Production by Recrystallization after Critical Cold Working

28. E. Schmid and G. Siebel, "Mg", Z. Elektrochem., 1931, **37**, 447.
29. E. Schmid and H. Seliger, "Mg and Solid Solutions", Metallwirtschaft, 1932, **11**, 409.
30. C. F. Elam, "Zn–Al Solid Solutions", Proc. Roy. Soc., 1925, **109**, 143.
31. C. F. Elam, "Zn–Al Solid Solutions", Proc. Roy. Soc., 1927, **115**, 133.
32. R. Karnop and G. Sachs, "Cu–Al Solid Solutions", Z. Physik, 1928, **49**, 480.
33. C. A. Edwards and L. B. Pfeil, "a-Fe", J. Iron Steel Inst., 1924, **109**, 129; 1925, **112**, 79, 113.
34. K. Honda and S. Kaya, "a-Fe", Sci. Rep. Tôhoku Univ., 1926, **15**, 721.
35. W. Fahrenhorst and E. Schmid, "a-Fe", Z. Physik, 1932, **78**, 383.
36. G. I. Sizoo, "a-Fe", Z. Physik, 1929, **56**, 649.
37. German Patent 291,994 (1913) : "W Crystals (Pintsch Wires)".
38. W. Böttger, "W Crystals by the Pintsch Process" (Schaller and Orbig). Z. Elektrochem., 1917, **23**, 121.
39. H. Alterthum, "W Crystals by Collective Crystallization", Z. physikal. Chem., 1924, **110**, 1.
40. Z. Jeffries, "Collective Crystallization of W : Significance of Admixtures", J. Inst. Metals, 1918, **20**, 109.
41. C. J. Smithells, "Collective Crystallization of W. Significance of Admixtures", J. Inst. Metals, 1922, **27**, 107.

Crystal Production by Crystallization in the Crucible

42. G. Tamann, "Bi. Lehrbuch der Metallographie". 3rd ed. Leipzig : 1923.
43. Obreimow and L. W. Schubnikoff, "Various Metals", Z. Physik, 1924, **25**, 31.
44. German Patent 442,085 : "Lowering of the Melting Crucible through a Vertical Tubular Furnace".
45. E. Schmid and G. Wassermann, "Te", Z. Physik, 1927, **46**, 653.
46. H. Seng, "Introduction of the Melt into the Solidification Vessel", Mitt. a.d. Mat.-Prüf.-Amt and Mitt. Kais.-Wilh.-Inst. Metallforsch., N.F., 1927, H. 5, 105.
47. G. I. Sizoo and C. Zwikker, "Raising of the Melt by Suction into a Quartz Tube, Ni", Z. Metallkunde, 1929, **21**, 125.
48. P. W. Bridgman, "Various Low Melting Metals", Proc. Amer. Acad. Arts Sci., 1925, **60**, 305.
49. R. Gross and H. Möller, "Selective Crystal Growth Determined by Speed of Growth", Z. Physik, 1923, **19**, 375.
50. S. Kaya, "Vacuum Furnace for the Production of Metal Crystals", Sci. Rep. Tôhoku Univ., 1928, **17**, 639, 1157.
51. G. Sachs and J. Weerts, "Vacuum Furnace (according to von Göler) for the Production of Metal Crystals", Z. Physik, 1930, **62**, 473.

52. K. W. Hausser and P. Scholz, " Solidification of the Entire Contents of a Crucible as One Crystal ", *Wiss. Veröff. Siemens-Konz.*, 1927, **5**, 144.
53. R. Glocker and L. Graf, " Vacuum Furnace for Crystal Production ", *Z. anorg. Chem.*, 1930, **188**, 232.
54. L. Graf, " Vacuum Furnace for Crystal Production ", *Z. Physik*, 1931, **67**, 388.
55. O. Sckell, " Production of Hg Crystals ", *Ann. Physik*, [5], 1930, **6**, 932.
56. H. Reddemann, " Production of Hg Crystals ", *Ann. Physik*, [5], 1932, **14**, 139.
57. P. Kapitza, " Crystal Production from the Molten State, Taking Precautions to Ensure Undamaged Crystals. Bi ", *Proc. Roy. Soc.*, 1928, **119**, 358.
58. A. Goetz, " Crystal Production from the Molten State, Taking Precautions to Ensure Undamaged Crystals, Bi ", *Phys. Rev.*, 1930, **35**, 193.
59. P. P. Ewald, " Production of Sn Crystals According to (57) ". Private communication.
60. F. Stöber, " Crystal Production by Slow Solidification ", *Z. Kristallogr.*, 1925, **61**, 299.
61. I. Strong, " Crystal Production by Slow Solidification ", *Phys. Rev.*, 1930, **36**, 1663.
62. S. Kyropoulos, " Method of Crystal Formation, Especially for Ionic Crystals ", *Z. anorg. Chem.*, 1926, **154**, 308.
63. J. Czochralski, " Production of Crystals by Drawing from the Melt ", *Z. physikal Chem.*, 1917, **92**, 219.
64. E. v. Gomperz, " Production of Crystals by the Drawing Method ", *Z. Physik*, 1922, **8**, 184.
65. H. Mark, M. Polanyi and E. Schmid, " Production of Crystals by the Drawing Method ", *Z. Physik*, 1922, **12**, 58.
66. A. G. Hoyem and E. P. T. Tyndall, " Change of Orientation in Crystals Produced by the Drawing Method ; Significance of the Temperature Gradient ", *Phys. Rev.*, 1929, **33**, 81.
67. A. G. Hoyem, " Significance of the Temperature Gradient for the Orientation of Crystals Produced by the Drawing Method ", *Phys. Rev.*, 1931, **38**, 1357.
68. F. Blank and F. Urbach, " Drawing Method Applied to Salt Crystals ", *Wien. Ber.*, IIa, 1929, **138**, 701.
69. E. Schiebold and G. Siebel, " Production of Mg Crystals from the Melt ", *Z. Physik*, 1931, **69**, 458.
70. F. Koref, " W Crystals from WCl₆ Vapour ", *Z. Elektrochem.*, 1922, **28**, 511.
71. A. E. van Arkel, " W Crystals from WCl₆ Vapour ", *Physica*, 1922, **2**, 56.
72. F. Koref and H. Fischvoigt, " Production of Various Metal Crystals from Vapour ", *Z. techn. Physik*, 1925, **6**, 296.
73. M. Straumanis, " Zn and Cd Crystals from Vapour ", *Z. physikal. Chem.*, 1931, **13**, 316; 1932, **19**, 63.
74. J. A. M. van Liempt, " Electrolytic Preparation of W Crystals ", *Z. Elektrochem.*, 1925, **31**, 249.
75. W. Eitel, " Crystal Formation from Solution ", " Handbuch der Arbeitsmethoden in der anorganischen Chemie ", **4/2**, 463.

IV. DETERMINATION OF THE ORIENTATION OF CRYSTALS

A brief account of the diffraction of X-rays by crystals will be found in :

R. Glocker, " Materialprüfung mit Röntgenstrahlen ". Berlin : 1927.
H. Mark, " Die Verwendung der Röntgenstrahlen in Chemie und Technik ". Leipzig : 1926.
P. P. Ewald, " Handbuch der Physik ". Berlin : 1933, **23/2**.
H. Ott, " Handbuch der Experimentalphysik ". Leipzig : 1928, **7/2**.

76. G. Tammann and A. Müller : " Determination of Crystal Orientation from Percussion Figures ", *Z. Metallkunde*, 1926, **18**, 69.
77. Cf., *e.g.*, W. Köster, " Determination of Crystal Orientation from Shape of the Etch Pits ", *Z. Metallkunde*, 1926, **18**, 112, 219.

78. G. Tammann, "Maximum Lustre Method ", Z. anorg. Chem., 1925, **148**, 293.
79. G. Tammann and H. H. Meyer, "Maximum Lustre Method ", Z. Metallkunde, 1926, **18**, 176.
80. J. Czochralski, "Determination of Orientation by Maximum Intensity of the Reflected Light ", Z. anorg. Chem., 1925, **144**, 131; Naturwiss., 1925, **12**, 425, 455.
81. P. W. Bridgman, "Determination of Orientation by Reflected Light ", Proc. Amer. Acad. Arts Sci., 1925, **60**, 305.
82. J. Weerts, "Correction to be Applied to Bridgman's Determination of Orientation, Use of Polarized Light ", Z. techn. Physik, 1928, **9**, 126.
83. M. Polanyi, "Theory of the Rotating-crystal Method ", Z. Physik, 1921, **7**, 149.
84. M. Polanyi and K. Weissenberg, "Theory of the Rotating-crystal Method ", Z. Physik, 1922, **9**, 123.
85. H. Mark, M. Polanyi and E. Schmid, "Determination of Orientation by the Rotating-crystal Method ", Z. Physik, 1922, **12**, 58.
86. K. Weissenberg, "X-ray Goniometer ", Z. Physik, 1924. **13**, 229.
87. W. F. Dawson, "X-ray Goniometer ", Phil. Mag., 1928, **5**, 756.
88. O. Kratky, "X-ray Goniometer ", Z. Krist., 1930, **72**, 529.
89. E. Schiebold and G. Sachs, "Determination of Orientation by Laue-photographs ", Z. Krist., 1926, **63**, 34.
90. E. Schiebold and G. Siebel, "Determination of Orientation by Laue-photographs ", Z. Physik, 1931, **69**, 458.
91. F. Rinne and E. Schiebold, "Polar Reflexion-chart for the Stereographic Determination of the Laue-photographs ", 'Ber. sächs. Akad. Wiss., Math.-physik. Kl., 1915, **68**, 11.
92. W. Boas and E. Schmid, "Laue Back Reflexion Diagrams ", Metallwirtschaft, 1931, **10**, 917.
93. M. Majima and S. Togino, "Laue Photographs Obtained by Changing the Direction of the Incident Beam ", Sci. Pap. Inst. Phys. Chem. Research (Tokyo), 1927, **7**, 75, 259.

V. GEOMETRY OF THE MECHANICS OF CRYSTAL DEFORMATION

94. H. Mark, M. Polanyi and E. Schmid, "Model of Simple Glide. Extension Formula ", Z. Physik, 1922, **12**, 58.
95. E. N. da C. Andrade, "Slip Bands with Hg, Pb, Sn ", Phil. Mag., 1914, **27**, 869.
96. C. Benedicks, "Band Formation with Extension of Zn Crystals ", Jb. Radioakt. Elektr., 1916, **13**, 351.
97. R. Karnop and G. Sachs, "Cristallographic Glide Strain ", Z. Physik, 1927, **41**, 116.
98. E. Schmid, "Crystallographic Glide Strain ", Z. Physik, 1926, **40**, 54.
99. v. Göler and G. Sachs, "Transformation of the Co-ordinates and Change in the Cross-section during Glide. Glide Strain with Double Glide ", Z. Physik, 1927, **41**, 103.
100. G. I. Taylor, "Simple and Double Glide in Compression ", Proc. Roy. Soc., 1927, **116**, 16.
101. G. I. Taylor, "Simple and Double Glide in Compression ", Proc. Roy. Soc., 1927, **116**, 39.
102. Y. Kidani, "Plastic Bending of Crystals ", J. Fac. Eng. Imp. Univ. Tokyo, 1930, **19**, 1.
103. P. Niggli, "Transformation of Indices in Glide ", Z. Krist., 1929, **71**, 413.
103a. M. Masima and G. Sachs, "Double Glide in α-Brass ", Z. Physik, 1928, **50**, 161.
104. v. Göler and G. Sachs, "Double Glide ", Z. techn. Physik, 1927, **8**, 586.
105. P. P. Ewald, "Double Glide ", Müller-Pouillet's " Lehrbuch der Physik ". 11th ed., **1/2**, 971.

106. We are indebted to Mr. M. A. Valouch for this representation of double glide.
107. R. Karnop and G. Sachs, " Glide Strain in Double Glide ", Z. Physik, 1927, 42, 283.
108. A. Johnsen, " Geometrical Treatment of Mechanical Twinning ", Jb. Radioakt. Elektr., 1914, 11, 226.
109. E. Schmidt and G. Wassermann, " Geometrical Treatment of Mechanical Twinning ", Z. Physik, 1928, 48, 370.
110. F. Wallerant, " Conditions for Twinning ", Bull. Soc. franç. Mineral., 1904, 27, 169.
111. A. Johnsen, " Conditions for Twinning ", Zentr. Mineral., 1916, 121.
112. O. Mügge, " Formulæ for the Transformation of Indices in Simple Glide ", Neues Jahrb. Mineral. Geol., 1889, 11, 98.

VI.   PLASTICITY AND STRENGTH OF METAL CRYSTALS

A. ELEMENTS OF GLIDING AND TWINNING

113. E. Schmid and G. Wassermann, " Determination of T from Laue Photographs ", Z. Physik, 1928, 48, 370.
114. M. Polanyi and E. Schmid, " Determination of t for β-tin from the Layer-line Diagram ", Z. Physik, 1925, 32, 684.
115. G. I. Taylor and W. S. Farren, " Analysis of the Compression of Crystals ", Proc. Roy. Soc., 1926, 111, 529.

Glide Elements of Metal Crystals

116. G. I. Taylor and C. F. Elam, " Plastic Deformation of Crystals with the Aid of the ' Unstretched Cone ' Method ; Al, 20° C.", Proc. Roy. Soc., 1923, 102, 643 ; 1925, 108, 28.
117. W. Boas and E. Schmid, " Al, 450° C., Laue and Rotating-crystal Photographs ", Z. Physik, 1931, 71, 703.
118. O. Mügge, " Cu, Ag, Au ", Götting. Nachr., 1899, 56.
119. C. F. Elam, " Cu, Ag, Au ", Proc. Roy. Soc., 1926, 112, 289.
120. C. W. Humfrey, " T of Pb ", Phil. Trans. Roy. Soc., 1903, A 200, 225.
121. G. I. Taylor and C. F. Elam, " a-Fe. Analysis of the Extension of Crystals ", Proc. Roy. Soc., 1926, 112, 337.
122. H. J. Gough, " a-Fe ", Proc. Roy. Soc., 1928, 118, 498.
123. W. Fahrenhorst and E. Schmid, " a-Fe ", Z. Physik, 1932, 78, 383.
124. F. Sauerwald and H. G. Sossinka, " a-Fe ", Z. Physik, 1933, 82, 634.
125. G. J. Taylor, " β-brass ", Proc. Roy. Soc., 1928, 118, 1.
126. F. S. Goucher, " W ", Phil. Mag., 1924, 48, 229, 800.
127. E. Schiebold and G. Siebel, " Mg ", Z. Physik, 1931, 69, 458.
128. E. Schmid and G. Wassermann, " Mg ", " Handbuch der physik. und techn. Mechanik ", 1931, 4/2, 319.
129. E. Schmid, " Mg (above 225° C.) Determination of t from the Lattice Rotation in Extension ; Basal Glide with 2 Glide Directions ", Z. Elektrochem., 1931, 37, 447.
130. H. Mark, M. Polanyi and E. Schmid, " Zn ", Z. Physik, 1922, 12, 58.
131. E. Schmid and E. Sutter (unpublished tests) ; W. Boas and E. Schmid, " Cd ", Z. Physik, 1929, 54, 16.
132. H. Mark and M. Polanyi, " β-Tin ", Z. Physik, 1923, 18, 75.
133. I. Obinata and E. Schmid, " β-Tin ", Z. Physik, 1933, 82, 224.
134. G. Masing and M. Polanyi, " Bi (according to M. Polanyi and E. Schmid) ", Erg. exakt. Naturwiss., 1923, 2, 177.
135. M. Georgieff and E. Schmid, " Bi ", Z. Physik, 1926, 36, 759.
136. E. Schmid and G. Wassermann, " Te ", Z. Physik, 1927, 46, 653.
137. M. Polanyi, " Density of Packing and Capacity for Glide ", Z. Physik, 1923, 17, 42.
138. H. G. Sossinka, B. Schmidt and F. Sauerwald, " Selection of Glide and Shear Elements in Terms of the Lattice ", Z. Physik, 1933, 85, 761.

139. E. Schmid, "Capacity for Glide and Modulus of Shear", Internat. Confer. Physic., London, 1934.
140. H. J. Gough, D. Hanson and S. J. Wright, "Glide Elements in Dynamic and Alternating Stressing", Aeronaut. Research Cttee. R. and M., 1924, No. 995.
141. R. Karnop and G. Sachs, "Deviations from the Normal Case of Octahedral Glide (Al) ", Z. Physik, 1927, **41**, 116.
142. C. F. Elam (and G. I. Taylor), "Division into Two Portions of an Al Crystal of Special Orientation ", Proc. Roy. Soc., 1928, **121**, 237.
143. R. Karnop and G. Sachs. "Deviations from the Normal Case of Octahedral Glide ", Z. Physik, 1927, **42**, 283.
144. G. I. Taylor, " Selection of the Glide System in Compression ", Proc. Roy. Soc., 1927, **116**, 16.

Twinning Elements in Metal Crystals

145. J. Leonhardt, "Meteoritic Iron ", Neues Jahrb. Mineral. Geol., 1928, Suppl. **58**, 153.
146. C. H. Mathewson and G. H. Edmunds, " α-Fe ", Amer. Inst. Min. Met. Eng., 1928, Techn. Publ. No. 139.
147. K. Tanaka and K. Kamio, " β-Tin and Zinc ", Mem. Coll. Sci. Kyōtō Imp. Univ., 1931, **14**, 79.
148. C. H. Mathewson and A. J. Phillips, " Be, Mg, Zn, Cd, Secondary Glide in Twinning Lamellæ ", Amer. Inst. Min. Met. Eng., 1928, Tech. Publ. No. 53.
149. E. Schmid and G. Siebel, " Mg ". Unpublished work.
150. O. Mügge, " α-Fe ", Neues Jahrb. Mineral. Geol., 1899, **II**, 55.
151. E. Schiebold and G. Siebel, " Mg ", Z. Physik, 1931, **69**, 458.
152. E. Schmid and G. Wassermann, " Zn Movement of the Lattice Points; Secondary Glide in the Twin ", Z. Physik, 1928, **48**, 370.
153. O. Mügge, " β-Tin ", Zentr. Mineral., 1917, 233; Z. Kristallogr., 1927, **65**, 603.
154. O. Mügge, " As ", Tsch. Min. Petr. Mitt., 1900, **19**, 102.
155. O. Mügge, " Sb ", Neues Jahrb. Mineral. Geol., 1884, **II**, 40.
156. O. Mügge, " Sb and Bi ", Neues Jahrb. Mineral. Geol., 1886, **I**, 183.
157. G. Rose, " Rose's Channels in Multiple Twinning ", Berl. Akad.-Ber., 1868, 57.
158. O. Mügge, " Rose's Channels in Multiple Twinning ", Neues Jahrb. Mineral. Geol., 1901, Suppl. **14**, 246.
159. O. Mügge, " Increase in Volume of α-Fe in Multiple Twinning ", Z. anorg. Chem., 1922, **121**, 68.
159a. J. Czochralski, " Deformation Twins in Cubic Face-centred Metals ", " Moderne Metallkunde ", Berlin: 1924.
160. C. H. Mathewson, " Movement of the Lattice Points in the Twinning Plane. Zn ", Proc. Amer. Inst. Min. Met. Eng., Met. Div., 1928, 1.
161. A. Grühn and A. Johnsen, " Sn Lattice Cannot be Displaced ", Zentr. Mineral., 1917, 370.
162. A. Johnsen, " Atomic Displacement in Bi ", Zentr. Mineral., 1916, 385.
163. E. Schmid and G. Wassermann, " Twinning of Zn-Orientation Changes in Extension and Compression ", Metallwirtschaft, 1930, **9**, 698.

B. The Dynamics of Glide

164. M. Polanyi, "Filament Extension Apparatus ", Z. techn. Physik, 1925, **6**, 121.
165. W. Boas and E. Schmid, " Adaptor for Tensile Machine ", Z. Physik, 1930, **61**, 767.
166. E. Schmid, " 'Yield Point' of Crystals. Critical Shear Stress Law ", Proc. Internat. Congr. Appl. Mech., Delft 1924, 342.
167. M. Polanyi and E. Schmid, " Absence of Influence of Hydrostatic Pressure on Glide Capacity ", Z. Physik, 1923, **16**, 336.

330   Bibliography

Dependence of the Yield Point on Orientation : Critical Shear Stress

168. E. Schmid and G. Siebel, "Mg", *Z. Elektrochem.*, 1931, **37**, 447.
169. P. Rosbaud and E. Schmid, "Zn", *Z. Physik*, 1925, **32**, 197.
170. W. Boas and E. Schmid, "Cd, Influence of Annealing and Rate of Straining—Plastic Yield Surface of Hexagonal Crystals", *Z. Physik*, 1929, **54**, 16.
171. U. Dehlinger (and F. Giesen), "Al Crystals Formed by Recrystallization with, Cast Crystals without, Clearly Defined Yield Point", *Physikal. Z.*, 1933, **34**, 836.
172. R. Karnop and G. Sachs, "Cu–Al Alloy", *Z. Physik*, 1928, **49**, 480.
173. M. Masima and G. Sachs, "a-Brass", *Z. Physik*, 1928, **50**, 161.
174. G. Sachs and J. Weerts, "Ag, Au and Alloys : Cu", *Z. Physik*, 1930, **62**, 473.
175. E. Osswald, "Ni and Ni–Cu Alloys", *Z. Physik*, 1933, **83**, 55.
176. W. Fahrenhorst and E. Schmid, "a-Fe", *Z. Physik*, 1932, **78**, 383.
177. M. Georgieff and E. Schmid, "Bi", *Z. Physik*, 1926, **36**, 759.
178. I. Obinata and E. Schmid, "β-Tin; Critical Shear Stress of Four Non-equivalent Crystallographic Systems; Comparison with Shear Strength and Density of Packing", *Z. Physik*, 1933, **82**, 224.
179. R. v. Mises, "Yield Condition for Crystals", *Z. angew. Math. Mech.*, 1928, **8**, 161.
180. W. Boas and E. Schmid, "Constancy of Shear Stress as a Condition of Flow; Elastic Shear Strain at the Yield Point", *Z. Physik*, 1929, **56**, 516; 1933, **86**, 828.
181. O. Haase and E. Schmid, "Plasticity Limit of Zn Crystals (0·002 per cent. Permanent Set)", *Z. Physik*, 1925, **33**, 413.
182. J. Czochralski, "Elastic Limit of Al Crystals (0·001 per cent. Permanent Set)", "Moderne Metallkunde". Berlin : 1924.
183. E. Schmid, "Shear Stress Law at the 0·001 per cent. Limit of Al Crystals—Plastic Yield Surface of Cubic Face-centred Metal Crystals", *Z. Metallkunde*, 1927, **19**, 154.
184. A. Ono, "Dependence of Initial Extension of Al Crystals on Section", *Proc. Internat. Congr. Appl. Mechanics, Stockholm*, 1930, **2**, 230.
185. R. Karnop and G. Sachs, "Torsional Yield Point of Cu–Al–Crystals (5 per cent. Cu)", *Z. Physik*, 1929, **53**, 605.
186. H. J. Gough, S. J. Wright and D. Hanson, "Stress Analysis in the Torsion of Al Crystals", *J. Inst. Metals*, 1926, **36**, 173.
187. H. J. Gough and H. L. Cox, "Alternating Torsion of Ag Crystals", *J. Inst. Metals*, 1931, **45**, 71.
188. H. J. Gough, "Alternating Torsion of a-Fe", *Proc. Roy. Soc.*, 1928, **118**, 498.
189. H. J. Gough and H. L. Cox, "Alternating Torsion of Zn", *Proc. Roy. Soc.*, 1929, **123**, 143; 1930, **127**, 453.
190. H. Ekstein, "Stress Analysis in the Torsion of Crystals". Unpublished.
191. J. Czochralski, "Changes in Shape in the Torsion of Crystals", *Proc. Internat. Congr. Appl. Mechanics, Delft*, 1924, 67.
192. See "General Treatises on Crystal Elasticity" (Bibliography, p. 324).
192a. E. N. da C. Andrade and P. J. Hutchings, "Critical Shear Stress of Very Pure Hg Crystals 9·3 g./mm.³ at −43° C.", *Proc. Roy. Soc.*, 1935, **148**, 120.
193. E. Schmid and G. Siebel, "Comparison between Cast and Recrystallization Crystals—Mg". Unpublished work.
194. W. Fahrenhorst, "Dependence of the Critical Shear Stress on the Speed of Straining (Zn-crystals)". Unpublished work.
195. A. Smekal, "Effect of Annealing on Crystals", "Handbuch der physikal. and techn. Mechanik", 1931, **4/2**, 116.
196. E. Schmid, "Analysis of the Stress–Strain Curve", *Z. Physik*, 1924, **22**, 328.
197. E. Schmid, "Theoretical Stress–Strain Curves for Constant Shear Stress", *Metallwirtschaft*, 1928, **7**, 1011.

198. H. Mark, M. Polanyi and E. Schmid, " Work-hardening of Crystals ",
     Z. Physik, 1922, **12**, 58.
199. G. Masing and M. Polanyi, " Work-hardening of Crystals ", Erg. exakt.
     Naturwiss., 1923, **2**, 177.
200. R. Becker and E. Orowan, " Glide by Regular Jumps ", Z. Physik, 1932,
     79, 566.
201. E. Schmid and M. A. Valouch, " Stepwise Deformation of Zn Purified by
     Distillation ", Z. Physik, 1932, **75**, 531.
202. M. Polanyi and E. Schmid, " Shear Fracture—Sn ", Z. Physik, 1925, **32**,
     684.
203. R. Karnop and G. Sachs, " Co-ordinates of the Work-hardening Curve ",
     Z. Physik, 1927, **41**, 116.
204. E. Schmid, " Co-ordinates of the Work-hardening Curve ", Z. Physik,
     1926, **40**, 54.
205. G. I. Taylor, " Work-hardening Curves for the Extension and Com-
     pression of Al Crystals ", Proc. Roy. Soc., 1927, **116**, 39.

The Work-hardening Curve

206. C. F. Elam, " Cu ", Proc. Roy. Soc., 1927, **116**, 694.
207. G. Sachs and J. Weerts, " Cu, Ag, Au ", Z. Physik, 1930, **62**, 473.
208. E. Osswald, " Cu, Ni ", Z. Physik, 1933, **83**, 55.
209. R. Karnop and G. Sachs, " Al ", Z. Physik, 1927, **41**, 116.
210. E. Schmid and G. Siebel, " Mg Shear Fracture. Upper Limit of the Glide
     Strain and Work of Deformation ", Z. Elektrochem., 1931, **37**, 447.
211. E. Schmid, " Zn ", Z. Physik, 1926, **40**, 54.
212. W. Boas and E. Schmid, " Cd ", Z. Physik, 1929, **54**, 16.
213. I. Obinata and E. Schmid, " Sn ", Z. Physik, 1933, **82**, 224.
214. E. Schmid, " Glide of Cubic and Hexagonal Crystals of the Closest
     Packed Systems ", Internat. Confer. Phys., London, 1934.
215. G. I. Taylor and C. F. Elam, " Work-hardening of Latent Glide Systems ",
     Proc. Roy. Soc., 1923, **102**, 643 ; 1925, **108**, 28.
216. C. F. Elam, " Work-hardening of Latent Glide Systems ", Proc. Roy. Soc.,
     1926, **112**, 289.
217. v. Göler and G. Sachs, " Al : Equation of the Work-hardening Curve.
     Calculation of the Model of the Ultimate Tensile Stress and Extension
     Surfaces ", Z. techn. Physik, 1927, **8**, 586.
218. J. Weerts, " Al : Equation of the Work-hardening Curve ". Research
     VDI-Heft 323, 1929.
219. K. Yamaguchi, " Al : Equation of the Work-hardening Curve ", Sci.
     Papers, Inst. Phys. Chem. Research (Tokyo), 1929, **11**, No. 205, p. 223.
220. G. I. Taylor, " Theoretical Foundation of the Equation of the Work-
     hardening Curve ", Proc. Roy. Soc., 1934, **145**, 362, 388.
221. E. Schiebold and G. Siebel, " Shear Fracture Mg ", Z. Physik, 1931, **69**,
     458.
222. W. Fahrenhorst and E. Schmid, " Upper Limit of the Glide Strain and
     Work of Deformation in Zn Crystals ", Z. Physik, 1930, **64**, 845.
223. W. Boas and E. Schmid, " Upper Limit of the Glide Strain and Work of
     Deformation in Cd Crystals ", 1930, **61**, 767.
224. E. Schmid and G. Wassermann, " Model of the Extension and Ultimate
     Tensile Stress Surface of Zn ", Z. Metallkunde, 1931, **23**, 87.

Ultimate Tensile Stress and Elongation of Cubic Metal Crystals

225. J. Czochralski, " Model of Maximum Load and Extension Surface.
     Cu ", " Moderne Metallkunde ". Berlin : 1924.
226. R. Karnop and G. Sachs, " Al ", Z. Physik, 1927, **41**, 116.
227. J. Czochralski, " Cu ", Proc. Internat. Congr. Appl. Mechanics, Delft,
     1924, 67.
228. W. Fahrenhorst and E. Schmid, " α-Fe ", Z. Physik, 1932, **78**, 383.
229. F. Koref, " W ", Z. Metallkunde, 1925, **17**, 213.

Plastic Properties of Alloy Crystals

230. P. Rosbaud and E. Schmid, " Cd–Zn, Sn–Zn ", *Z. Physik*, 1925, **32**, 197.
231. E. Schmid and H. Seliger, " Al–Mg; Zn–Mg ", *Metallwirtschaft*, 1932, **11**, 409.
232. E. Schmid and G. Siebel, " Ternary Al–Zn–Mg ", *Metallwirtschaft*, 1932, **11**, 577.
233. M. Masima and G. Sachs, " a-Brass ", *Z. Physik*, 1928, **50**, 161.
234. v. Göler and G. Sachs, " a-Brass ", *Z. Physik*, 1929, **55**, 581.
235. G. Sachs and J. Weerts, " Ag–Au ", *Z. Physik*, 1930, **62**, 473.
236. E. Osswald, " Cu–Ni ", *Z. Physik*, 1933, **83**, 55.
237. G. Sachs and J. Weerts, " AuCu$_3$ ", *Z. Physik*, 1931, **67**, 507.
238. R. Karnop and G. Sachs, " Cu–Al, Heat-treated ", *Z. Physik*, 1928, **49**, 480.
239. C. F. Elam, " Zn–Al ", *Proc. Roy. Soc.*, 1927, **115**, 133.
240. C. F. Elam, " a-Brass ", *Proc. Roy. Soc.*, 1927, **115**, 148.
241. C. F. Elam, " Al–Cu ", *Proc. Roy. Soc.*, 1927, **116**, 694.
242. E. Schmid and G. Siebel : " Thermal Treatment of Al–Mg ", *Metallwirtschaft*, 1934, **13**, 765.
243. E. Schmid and M. A. Valouch, " Initial Shear Stress of Distilled Zn ", *Z. Physik*, 1932, **75**, 531.
244. W. Boas, " Solubility of Cd in Zn ", *Metallwirtschaft*, 1932, **11**, 603.
245. E. A. Owen and G. D. Preston, " Lattice Constants of Brass ", *Proc. Phys. Soc.*, 1923, **36**, 49.
246. A. Westgren and G. Phragmén, " Lattice Constants of Brass ", *Phil. Mag.*, 1925, **50**, 311.
247. G. Sachs and J. Weerts, " Lattice Constants of Au–Ag Solid Solutions ", *Z. Physik*, 1930, **60**, 481.
248. C. H. Johansson and J. O. Linde, " Arrangement of the Atoms in AuCu$_3$ ", *Ann. Physik*, [4], 1925, **78**, 439; 1927, **82**, 449.

Dependence of Plasticity on Temperature and Speed of Deformation

249. E. Schmid (and G. Siebel), " Mg ", *Z. Elektrochem.*, 1931, **37**, 447.
250. O. Haase and E. Schmid, " Zn, Sn, Bi, Crystal Recovery ", *Z. Physik*, 1925, **33**, 413.
251. M. Polanyi and E. Schmid, " Zn, Cd ", *Naturwiss.*, 1929, **17**, 301.
252. W. Boas and E. Schmid, " Zn, Cd ", *Z. Physik*, 1930, **61**, 767.
253. W. Fahrenhorst and E. Schmid, " Zn ", *Z. Physik*, 1930, **64**, 845.
254. W. Meissner, M. Polanyi and E. Schmid, " Zn, Cd ", *Z. Physik*, 1930, **66**, 477.
255. W. Boas and E. Schmid, " Cd ", *Z. Physik*, 1929, **57**, 575.
256. J. Weerts, " Al Research ", VDI-Heft 323, 1929.
257. W. Boas and E. Schmid, " Al ", *Z. Physik*, 1931, **71**, 703.
258. D. Hanson and M. A. Wheeler, " Creep Test—Al ", *J. Inst. Metals*, 1931, **45**, 229.
259. C. A. Edwards and L. B. Pfeil, " a-Fe ", *J. Iron Steel Inst.*, 1924, **109** 129; 1925, **112**, 79.
260. A. Masing and M. Polanyi, " Sn ", *Ergeb. exakt. Naturwiss.*, 1933, **2**, 177.
261. M. Polanyi and E. Schmid, " Sn; Crystal Recovery ", *Z. Physik*, 1925, **32**, 684.
262. I. Obinata and E. Schmid, " Sn ", *Z. Physik*, 1933, **82**, 224.
263. M. Georgieff and E. Schmid, " Bi ", *Z. Physik*, 1926, **36**, 759.
264. M. Polanyi and E. Schmid, " Crystal Recovery ", *Verhandl. deut. physikal. Ges.*, 1923, **4**, 27.
265. R. Gross, " Crystal Recovery ", *Z. Metallkunde*, 1924, **16**, 344.
266. F. Koref, " Crystal Recovery ", *Z. Metallkunde*, 1925, **17**, 213.
267. M. Straumanis, " Thickness of the Glide Packet—Zn ", *Z. Krist.*, 1932, **83**, 29.

C. The Dynamics of Twinning

268. H. J. Gough and H. L. Cox, " Initial Condition ", *Proc. Roy. Soc.*, 1929, **123**, 143; 1930, **127**, 453.
269. H. J. Gough and H. L. Cox, " Initial Condition ", *J. Inst. Metals*, 1932, **48**, 227.
269a. W. Fahrenhorst and E. Schmid, " At −185° C. Twinning is much Preferred to Gliding in α-Fe ", *Z. Physik*, 1932, **78**, 383.
270. W. Boas and E. Schmid, " Reciprocal Effect of Glide and Twinning with Cd ", *Z. Physik*, 1929, **54**, 16.
271. O. Haase and E. Schmid, " Reciprocal Effect of Glide and Twinning with Zn ", *Z. Physik*, 1925, **33**, 413.
272. E. Schmid and G. Wassermann, " Reciprocal Effect of Glide and Twinning with Zn ", *Z. Physik*, 1928, **48**, 370.
273. E. Schmid, " Reciprocal Effect of Glide and Twinning with Zn ", *Z. Metallkunde*, 1928, **20**, 421.
274. J. Czochralski, " Untwinning of Zn Twins ", " Moderne Metallkunde ". Berlin : 1924.
275. G. Wassermann, " Untwinning of Sb Twins ", *Z. Krist.*, 1930, **75**, 369.
276. Unpublished work (1930) by E. Schmid, assisted by O. Vaupel and especially by W. Fahrenhorst. Repeated twinning with Zn.

D. Fracture along Crystallographic Planes

277. L. Sohncke, " Normal Stress Law for NaCl ", *Poggendorfs Ann.*, 1869, **137**, 177.
278. E. Schmid, " Zn—Normal Stress Law ", *Proc. Internat. Congr. Appl. Mechanics, Delft*, 1924, 342.
279. E. Schmid, " Strain Strengthening, Strain Recovery, Zn ", *Z. Physik*, 1925, **32**, 918.
280. E. Schmid, " Zn + 0·13 per cent. Cd; Zn + 0·53 per cent. Cd ", *Z. Metallkunde*, 1927, **19**, 154.
281. W. Fahrenhorst and E. Schmid, " Zn ", *Z. Physik*, 1930, **64**, 845.
282. W. Boas and E. Schmid, " Zn Normal Dilatation Law Contradicted ", *Z. Physik*, 1929, **56**, 516.
283. E. Schmid and G. Wassermann, " Te ", *Z. Physik*, 1927, **46**, 653.
284. M. Georgieff and E. Schmid, " Bi ", *Z. Physik*, 1926, **36**, 759.
285. G. Wassermann, " Bi, Sb ", *Z. Krist.*, 1930, **75**, 369.
286. M. Polanyi, " Strain Strengthening of W and NaCl ", *Z. Elektrochem.*, 1922, **28**, 16.
287. G. Masing and M. Polanyi, " Shear Hardening Contrasted with Strain Strengthening ", *Ergeb. exakt. Naturwiss.*, 1923, **2**, 177.

E. After-effect Phenomena and Cyclic Stressing

288. H. v. Wartenberg, " No After-effect with W and Zn Crystals ", *Verhandl. deut. physikal. Ges.*, 1918, **20**, 113.
289. J. Bauschinger, " Directional Strengthening ", *Zivilingenieur*, 1881, **27**, 299.
290. G. Sachs and H. Shoji, " Bauschinger Effect with α-Brass ", *Z. Physik*, 1927, **45**, 776.
291. H. J. Gough, D. Hanson and S. J. Wright, " Al Hysteresis ", *Aeronaut. Research Ctte. R. and M.*, 1924, No. 995.
292. H. J. Gough, S. J. Wright and D. Hanson, " Al Hardness Measurements ", *J. Inst. Metals*, 1926, **36**, 173.
293. D. Hanson and M. A. Wheeler, " Al Creep Test ", *J. Inst. Metals*, 1931, **45**, 229.
294. H. J. Gough and D. G. Sopwith, " Al Corrosion Fatigue ", *Proc. Roy. Soc.*, 1932, **135**, 392.
295. H. J. Gough and H. L. Cox, " Ag ", *J. Inst. Metals*, 1931, **45**, 71.
296. E. Schmid and G. Siebel, " Mg Change of the Mechanical Properties ", *Metallwirtschaft*, 1934, **13**, 353.

297. E. Schmid, " Zn Change of the Mechanical Properties ", *Z. Metallkunde,* 1928, **20**, 69.
298. H. J. Gough and H. L. Cox, " Zn ", *Proc. Roy. Soc.,* 1929, **123**, 143; 1930, **127**, 453.
299. W. Fahrenhorst and E. Schmidt, " Zn; Change of the Mechanical Properties ", *Z. Metallkunde,* 1931, **23**, 323.
300. W. Fahrenhorst and H. Ekstein, " Cd ", *Z. Metallkunde,* 1933, **25**, 306.
301. H. J. Gough and H. L. Cox, " Sb ", *Proc. Roy. Soc.,* 1930, **127**, 431.
302. H. J. Gough and H. L. Cox, " Bi ", *J. Inst. Metals,* 1932, **48**, 227.

F. Change of the Physical and Chemical Properties by Cold Working

Anisotropy of the Physical Properties of Metal Crystals

General Treatise : W. Voigt, " Lehrbuch der Kristallphysik ". Leipzig : 1928.

Elastic parameters : see in particular Bibliography, p. 324.
303. R. Karnop and G. Sachs, " Elastic Parameters, Al + 5 per cent. Cu ", *Z. Physik,* 1929, **53**, 605.
304. E. Grüneisen and O. Sckell, " Elastic Parameters, Electrical Resistance and Thermal Expansion of Hg ", *Ann. Physik,* [5], 1934, **19**, 387.
305. E. Schmidt, " Graphical Representation of Young's Modulus for Mg and Zn ", *Z. Elektrochem.,* 1931, **37**, 447.
306. E. Goens and E. Schmid, " Electrical Resistance and Thermal Expansion of Mg ", *Naturwiss.,* 1931, **19**, 376.
307. P. W. Bridgman, " Electrical Resistance of Mg ", *Phys. Rev.,* 1931, **37**, 460; *Proc. Amer. Acad.,* 1931, **66**, 255.
308. P. W. Bridgman, " Thermal Expansion and Thermo-electric Force of Mg ", *Phys. Rev.,* 1931, **37**, 460; *Proc. Amer. Acad.,* 1932, **67**, 29.
309. E. Goens and E. Grüneisen, " Electrical Resistance and Thermal Conductivity of Zn, Cd ", *Ann. Physik,* [5], 1932, **14**, 164.
310. E. Grüneisen and E. Goens, " Thermal Expansion of Zn, Cd ", *Z. Physik,* 1924, **29**, 141.
311. P. W. Bridgman, " Electrical Resistance and Thermal Expansion of Zn, Cd, Bi, Sb, Te, Sn ", *Proc. Amer. Acad. Arts Sci.,* 1925, **60**, 305.
312. E. Grüneisen and E. Goens, " Thermo-electric Force of Zn, Cd ", *Z. Physik,* 1926, **37**, 278.
313. A. G. Hoyem, " Thomson-effect, Zn ", *Phys. Rev.,* 1931, **38**, 1357.
314. H. Verleger, " Thomson-effect (Zn) Cd ", *Ann. Physik,* [5], 1931, **9**, 366.
315. O. Sckell, " Electrical Resistance Hg ", *Ann. Physik,* [5], 1930, **6**, 932.
316. H. Reddemann, " Thermal Conductivity, Thermo-electric Force, Hg ", *Ann. Physik.,* [5], 1932, **14**, 139.
317. H. Glauner and R. Glocker, " Rate of Diffusion, Cu ", *Z. Krist.,* 1931, **80**, 377.
318. K. W. Hausser and P. Scholz, " Rate of Diffusion, Cu ", *Wiss. Veröff. Siemens-Konz.,* 1927, **5**, 144.
319. G. Tammann and F. Sartorius, " Rate of Diffusion, Cu ", *Z. anorg. Chem.,* 1928, **175**, 97.
320. E. Schiebold and G. Siebel, " Rate of Diffusion, Mg ", *Z. Physik,* 1931, **69**, 458.

Cold Deformation and the Crystal Lattice

321. E. Schmid, " Retention of the Lattice after Cold Working ", *Naturwiss.,* 1932, **20**, 530.
322. J. Czochralski, " Asterism ", " Moderne Metallkunde ". Berlin : 1924.
323. G. I. Taylor, " Lengthening of the Reflexions in Rotation Photographs ", *Trans. Faraday Soc.,* 1928, **24**, 121.
324. R. Gross, " Asterism ", *Z. Metalkunde,* 1924, **16**, 344.
325. K. Yamaguchi, " Asterism ", *Sci. Papers Inst. Phys. Chem. Research (Tokyo),* 1929, **11**, 151.

326. S. Konobojewski and I. Mirer, " Asterism ", *Z. Krist.*, 1932, **81**, 69.
327. M. Polanyi, " Streak-shaped Reflexions ", *Z. Physik*, 1921, **7**, 149.
328. F. Regler, " Asterism ", *Z. Physik*, 1931, **71**, 371.
329. W. G. Burgers and P. C. Louwerse, " Asterism ", *Z. Physik*, 1931, **67**, 605.
330. J. Czochralski, " Reduction of Asterism after Re-torsion of Al Crystals ", *Z. Metallkunde*, 1925, **17**, 1.
331. W. P. Davey, " Broadening in X-ray Diffraction Patterns ", *Gen. Elect. Rev.*, 1925, **28**, 588.
332. A. E. van Arkel, " Broadening in X-ray Diffraction Patterns ", *Physica*, 1925, **5**, 208; *Naturwiss.*, 1925, **13**, 662.
333. U. Dehlinger, " Broadening in X-ray Diffraction Patterns ", *Z. Krist.*, 1927, **65**, 615; *Z. Metallkunde*, 1931, **23**, 147.
334. V. Caglioti and G. Sachs, " Calculation of the Elastic Stress from the Broadening of X-ray Lines, Cu ", *Z. Physik*, 1932, **74**, 647.
335. W. A. Wood, " Lattice Disturbance and Degree of Working ", *Phil. Mag.*, 1932, **14**, 656.
336. L. Thomassen and J. E. Wilson, " Broadening of the Diffraction Lines in Al; Reduction of the Working Temperature ", *Phys. Rev.*, 1933, **43**, 763.
337. E. Schmid and G. Wassermann, " Broadening of the Diffraction Lines in Al-alloys ", *Metallwirtschaft*, 1930, **9**, 421.
338. W. G. Burgers, " Broadening of the Diffraction Lines in Different Layers with W ", *Z. Physik*, 1929, **58**, 11.
339. W. A. Wood, " Broadening of the Diffraction Lines not Determined by Reduction in the Grain Size ", *Nature*, 1932, **129**, 760.
340. J. Hengstenberg and H. Marks, " Reduction in the Intensity of Interferences of Higher Order ", *Naturwiss.*, 1929, **17**, 443.

Cold Deformation and Physical and Chemical Properties

341. H. J. Gough, D. Hanson and S. J. Wright, " Density, Al ", *Aeronaut. Research Ctte. R. and M.*, 1926 (M 40), No. 1024.
342. G. Sachs and H. Shoji, " Density, Al, a-Brass ", *Z. Physik*, 1927, **45**, 776.
343. D. Hanson and M. A. Wheeler, " Density, Al ", *J. Inst. Metals*, 1931, **45**, 229.
344. P. Goerens, " Density, Modulus of Torsion, Rate of Diffusion, Electrical Resistance, Magnetic Properties, Fe ", *Ferrum*, 1912, **10**, 65.
345. T. Ishigaki, " Density, Fe, Steel ", *Sci. Rep. Tôhoku Imp. Univ.*, 1926, **15**, 777.
346. K. Tamaru, " Density, Fe, Steel ", *Sci. Rep. Tôhoku Imp. Univ.*, 1930, **19**, 437.
347. T. Ueda, " Density, Electrical Resistance, Fe, Cu, Brass ", *Sci. Rep. Tôhoku Imp. Univ.*, 1930, **19**, 473.
348. Ch. O'Neill, " Density, Cu ", *Mem. Proc. Manchester Lit. Phil. Soc.*, 1861, 243.
349. G. Kahlbaum and E. Sturm, " Density, Cu ", *Z. anorg. Chem.*, 1905, **46**, 217.
350. W. E. Alkins, " Density, Cu ", *J. Inst. Metals*, 1920, **23**, 381.
351. H. Wolff, " Density, W ", Dissertation. Danzig : 1923.
352. W. Geiss and J. A. M. van Liempt, " Density, W ", *Ann. Physik*, [4], 1925, **77**, 105.
353. J. Johnston and S. H. Adams, " Density, Bi ", *Z. anorg. Chem.*, 1912, **76**, 274.
354. M. Masima and G. Sachs, " Density, a-Brass Crystal ", *Z. Physik*, 1929, **54**, 666.
355. K. Honda and R. Yamada, " Modulus of Elasticity, Fe Crystal ", *Sci. Rep. Tôhoku Imp. Univ.*, 1928, **17**, 723.
356. T. Kawai, " Shear Modulus, Cu, Brass, Al, Ni ", *Sci. Rep. Tôhoku Imp. Univ.*, 1931, **20**, 681.
357. W. Müller, " Modulus of Elasticity ", Research VDI-Heft 211, 1918.

358. W. Jubitz, "Thermal Expansion, Fe, Bronze ", Z. techn. Physik, 1926, 7, 522.
359. C. J. Smithels, "Thermal Expansion, W," "Tungsten ". London : 1926.
360. W. U. Behrens and C. Drucker, "Specific Heat, Zn ", Z. physikal. Chem., 1924, 113, 79.
361. C. Chappel and M. Levin, "Specific Heat, Steel, Bronze ", Ferrum, 1912, 10, 271.
362. K. Honda, "Specific Heat, Fe, Steel ", Sci. Rep. Tôhoku Imp. Univ., 1924, 12, 347.
363. W. Geiss and J. A. M. van Liempt, "Specific Heat, Ni, W ", Z. anorg. Chem., 1928, 171, 317.
364. H. Hort, "Internal Energy, Fe ", Z.V.d.I., 1906, 45, 1831.
365. W. S. Farren and G. I. Taylor, "Energy of Deformation, Al ", Proc. Roy. Soc., 1925, 107, 422.
366. W. Rosenhain and V. H. Stott, "Internal Energy, Al, Cu ", Proc. Roy. Soc., 1933, 140, 9.
367. A. Ono, "Internal Energy, Alternating Bending Tests, Steel ", Trans. 3 internat. Congr. techn. Mech., Stockholm, 1930, 2, 305.
368. G. I. Taylor and H. Quinney, "Internal Energy, Cu, Steel ", Proc. Roy. Soc., 1934, 143, 307.
369. F. Koref and H. Wolff, "Heat of Solution, W ", Z. Elektrochem., 1922, 28, 477.
370. J. A. M. van Liempt, "Heat of Combustion, W ", Z. anorg. Chem., 1923, 129, 263.
371. G. Tammann and C. Wilson, "Galvanic Potential, Change of Colour ", Z. anorg. Chem., 1928, 173, 156.
372. E. Heyn and O. Bauer, "Rate of Diffusion ". Mitt. Kgl. Mat.-Prüf.-Amt. Gross-Lichterfelde, 1909, p. 57.
373. T. A. Eastick, "Rate of Diffusion, Cu ", Metal Ind. (Lond.), 1924, 6, 22.
374. J. Czochralski and E. Schmid, "Rate of Diffusion, Al, Cu ", Z. Metall-kunde, 1928, 20, 1.
375. J. Czochralski, "Disappearance of the Capacity for Etching after Torsion, Al Crystals ", Z.V.d.I., 1923, 67, 533.
376. G. Borelius, "Thermo-electric Force, Al, Ni, Cu, Ag, Au, Fe ", Ann. Physik, [4], 1919, 60, 381.
377. W. F. Brandsma, "Thermo-electric Force, Cu ", Z. Physik, 1928, 48, 703.
378. W. Geiss and J. A. M. van Liempt, "Electrical Resistance and Temperature Coefficient, Mo, Ni, Pt, W ", Z. Physik, 1927, 41, 867.
379. L. Addicks, "Electrical Resistance, Cu ", Amer. Inst. Electr. Engr.: New York, Nov. 1903.
380. K. Takahasi, "Electrical Resistance, Cu, Ag ", Sci. Rep. Tôhoku Imp. Univ., 1930, 19, 265.
381. E. Grüneisen and E. Goens, "Thermal Conductivity, Cu ", Z. Physik, 1927, 44, 615.
382. V. Caglioti and G. Sachs, "Proportion of Elastic Energy in the Increase in the Internal Energy ", Z. Physik, 1932, 74, 647.
383. M. Masima and G. Sachs, "Heat Development during the Gliding of Brass Crystals ", Z. Physik, 1929, 56, 394.
384. M. Masima and G. Sachs, "Change in Conductivity during the Gliding of Brass Crystals ", Z. Physik, 1928, 51, 321.
385. W. Geiss and J. A. M. van Liempt, "Matthiessen Law ", Z. anorg. Chem., 1925, 143, 259.
386. G. Tammann, "Colour, Au–Ag–Cu Alloys ", Z. anorg. Chem., 1919, 107, 1 ; see especially p. 115.
387. G. Tammann, "Limits of Resistance ". "Lehrbuch der Metallkunde ". 3rd ed., Leipzig : 1923.
388. W. Steinhaus, "Magnetic Properties ". "Handbuch der Physik ". Berlin : 1927, 15, 202.
389. H. J. Seemann and E. Vogt, "Susceptibility ", Ann. Physik, [5], 1929, 2, 976.

390. K. Honda and Y. Shimizu, " Susceptibility ", *Sci. Rep. Tôhoku Imp. Univ.*, 1931, **20**, 460.
391. A. Kussmann and H. J. Seemann, " Susceptibility ", *Z. Physik*, 1932, **77**, 567.
392. G. Tammann and K. L. Dreyer, " Rate of Phase Transformation, Sn ", *Z. anorg. Chem.*, 1931, **199**, 97.
393. G. Wassermann, " Rate of Phase Transformation, Co ", *Metallwirtschaft*, 1932, **11**, 61.
394. C. v. Hevesy and A. Obrutschewa, " Rate of Self-diffusion in Pb ", *Nature*, 1925, **115**, 674.
395. R. Wilm, " Acceleration of the Rate of Nucleation of Supersaturated Alloys ", *Metallurgie*, 1911, **8**, 223.
396. E. Schmid and G. Wassermann, " Acceleration of the Rate of Nucleation of Supersaturated Alloys ", *Metallwirtschaft*, 1930, **9**, 421.
397. U. Dehlinger and L. Graf, " Axial Ratio of the Tetragonal Au–Cu Crystals ", *Z. Physik*, 1930, **64**, 359.
398. K. Schäfer, " Atomic Distribution, Lattice Constants, Fe–Al Alloys ", *Naturwiss.*, 1933, **21**, 207.
399. A. Smekal, " Structurally Sensitive and Insensitive Properties ", *Como Congress Documents*, 1927, **1**, 181.

G. RECRYSTALLIZATION

400. J. Czochralski, " Historical Particulars ", *Z. Metallkunde*, 1927, **19**, 316.
401. G. Tammann, " Number of the Nuclei and the Linear Speed of their Growth are Decisive Factors in the Speed of Recrystallization ", " Lehrbuch der Metallkunde ".
402. H. Hanemann, " Recrystallization on the Glide Plane ", *Z. Metallkunde*, 1926, **18**, 16.
403. J. Czochralski, " Recrystallization after Torsion and Reverse Torsion of Al [Al twin (?)] ", *Z. Metallkunde*, 1925, **17**, 1.
404. M. Polanyi and E. Schmid, " Recovery and Recrystallization of Sn Crystals. Different Types of Deformation.   Orientation Relationship of the Newly Formed Crystals ", *Z. Physik*, 1925, **32**, 684.
405. A. E. van Arkel and M. G. van Bruggen, " Hardening and Recrystallization, Al ", *Z. Physik*, 1927, **42**, 795.
406. A. E. van Arkel and J. J. A. Ploos van Amstel, " Hardening and Grain Size after Recrystallization, Sn ", *Z. Physik*, 1928, **51**, 534.
407. A. E. van Arkel, " Identical Grain Size Accompanies the Same Hardening Despite Initially Different Grain Structure, Al ", *Z. Metallkunde*, 1930, **22**, 217.
408. W. G. Burgers (and J. J. A. Ploos van Amstel), " Capacity of Al Crystals for Recrystallization Dependent on the Number of Glide Systems Involved ", *Z. Physik*, 1933, **81**, 43.
409. A. E. van Arkel and J. J. A. Ploos van Amstel, " No Recrystallization in the Transition Zone between Stretched and Unstretched in the Case of Sn Crystals without Glide Bands ", *Z. Physik*, 1930, **62**, 46.
410. G. Sachs, " Recrystallization after Torsion and Reverse Torsion, Al ", *Z. Metallkunde*, 1926, **18**, 209.
411. P. Beck and M. Polanyi, " Recrystallization after Bending and Back Bending, Al ", *Naturwiss.*, 1931, **19**, 505; *Z. Elektrochem.*, 1931, **37**, 521.
412. H. J. Gough, D. Hanson and S. J. Wright, " Reverse Deformation Follows along Different Glide Planes from the Original Deformation ", *Aeronaut. Research Ctte., R. and M.*, 1924, No. 995.
413. R. Karnop and G. Sachs, " Recrystallization of Al after Stretching and Reverse Straining (Compression) ", *Z. Physik*, 1928, **52**, 301.
414. H. Bohner and R. Vogel, " Grain Size Dependent on Initial Grain Size. Significance of Recovery for Recrystallization. Polycrystal ", *Z. Metallkunde*, 1932, **24**, 169.

415. J. Czochralski, " Increase in the Number of Nuclei with Increasing Temperature ". " Moderne Metallkunde." Berlin : 1924.
416. A. E. van Arkel and M. G. van Bruggen, " Recrystallization. Pure Al. Surface Crystallization ", Z. Physik, 1928, **51**, 520.
417. E. Schmid, " Recrystallization in Deformation Twins. Rate of Growth, Al ". According to unpublished work carried out in collaboration with M. Altenberger, 1930.
418. G. Tammann and W. Crone, " Rate of Growth, Pb, Ag, Zn ", Z. anorg. Chem., 1930, **187**, 289.
419. J. Czochralski, " Rate of Growth, Sn Recrystallization Diagram ", Metall. u. Erz, 1916, **13**, 381.
420. R. Karnop and G. Sachs, " Constants of Rate of Growth ; Al Polycrystal ", Z. Physik, 1930, **60**, 464.
421. A. E. van Arkel and J. J. A. Ploos van Amstel, " Suspension of the Capacity for Growth as a Result of Slight Deformation ", Z. Physik, 1930, **62**, 43.
422. M. Polanyi and E. Schmid, " Orientation Relationship of Sn Crystals before and after Recrystallization ", Verhandl. deut. physikal. Ges., 1923, **4**, 27.
423. H. C. H. Carpenter and C. F. Elam, " Recrystallization of Stretched Al Crystals ", Proc. Roy. Soc., 1925, **107**, 171.
424. K. Tanaka, " Recrystallization of Stretched Al Crystals ", Mem. Coll. Sci. Kyōtō Imp. Univ., 1928, **11**, 229.
425. W. G. Burgers and J. C. M. Basart, " Oriented Recrystallization of Slightly Deformed Al Crystals ", Z. Physik, 1928, **51**, 545.
426. W. G. Burgers and J. C. M. Basart, " Recrystallization of Al Crystals after Heavy Deformation. Grain Growth ", Z. Physik, 1929, **54**, 74.
427. W. G. Burgers, " Recrystallization of Compressed and Rolled Al Crystals ", Z. Physik, 1930, **59**, 651.
428. W. G. Burgers and P. C. Louwerse, " Recrystallization of Compressed and Rolled Al Crystals ", Z. Physik, 1931, **67**, 605.
429. W. Feitknecht, " Grain Growth ", J. Inst. Metals, 1926, **35**, 131.
430. R. Karnop and G. Sachs, " Effect of Impurities on the Speed of Grain Growth, Al ", Metallwirtschaft, 1929, **8**, 1115.
431. C. Agte and K. Becker, " Temperature of Recovery Determined by X-rays and Mechanically. Matthiessen Law of Recovery ", Z. techn. Physik, 1930, **11**, 140.
432. R. Karnop and G. Sachs, " Temperature of Recovery Determined by X-rays and Mechanically ", Z. Physik, 1927, **42**, 283.
433. A. E. van Arkel and W. G. Burgers, " Resolution of the Kα Doublet ", Z. Physik, 1928, **48**, 690.
434. G. Sachs, " Changes in the Physical Properties Produced by Annealing ". " Grundbegriffe der mechanischen Technologie der Metalle." Leipzig : 1925.
435. P. Goerens, " Effect of Duration and Temperature of Annealing on the Tensile Strength and Elongation of Drawn Mild Steel Wire ", Ferrum, 1913, **10**, 226.
436. G. Tammann and F. Neubert, " Changes in Properties Produced by Annealing ", Z. anorg. Chem., 1932, **207**, 87.
437. G. Tammann, " Recovery from the Effects of Cold Working ", Z. Metallkunde, 1932, **24**, 220.
438. G. Tammann and K. L. Dreyer, " Various Behaviour of the Cu and Pt Group During Recovery ", Ann. Physik., [5], 1933, **16**, 111.
439. G. Tammann, " Recovery from the Effects of Cold Working ", Z. Metallkunde, 1934, **26**, 97.
440. G. Tammann, " Release of Gas During the Heat Treatment of Worked Metals ", Z. anorg. Chem., 1920, **114**, 278.
441. O. Werner, " Study of Recovery and Recrystallization by Radioactive Methods (Release of Gas) ", Z. Elektrochem., 1933, **39**, 611 ; Z. Metallkunde, 1934, **26**, 265.

Bibliography

442. F. Credner, " Formation of Cavities During Grain Growth ", *Z. physikal. Chem.*, 1913, **82**, 457.

VII. PLASTICITY AND STRENGTH OF IONIC CRYSTALS

443. H. Seifert, " Mechanism of Deformation. In Landolt-Börnstein-Roth-Scheel ", *Physikalisch-chemische Tabellen, Berlin*, 1927, Suppl. vol. **1**, 35.
444. J. Stark, " Selection of the Glide Elements and Cleavage Planes ", *Jb. Radioakt. u. Elektron*, 1915, **12**, 279.
445. A. Johnsen, " Selection of the Glide Elements and Cleavage Planes. Simple Shear ", *Ergeb. exakt. Naturwiss.*, 1922, **1**, 270.
446. G. Tammann and W. Salge, " Octahedral Glide of NaCl at High Temperatures. Decrease in the Yield Point with Rising Temperature ", *Neues Jahrb. Mineral. Geol. Paläon.* 1927, Suppl. **57**, 117.
447. F. Heide, " New Twinning and Glide Elements with Rising Temperature and Increased Compression ", *Z. Krist.*, 1931, **78**, 257.
448. M. J. Buerger, " Glide on Cube Planes, NaCl ", *Amer. Mineralogist*, 1930, **15**, No. 2–5–6.
449. E. Rexer, " Glide on Cube Planes, NaCl ", *Z. Physik*, 1932, **76**, 735.
450. O. Mügge, " Glide on Cube Planes, NaCl ", *Zentr. Mineral.*, 1931, A, 253.
451. S. Dommerich, " Glide on Cube Planes, NaCl ", *Z. Physik*, 1934, **90**, 189.
452. E. Schmid and O. Vaupel, " Lattice Rotation during Dodecahedral Glide. Stress–Strain Curves NaCl (Annealed). Normal Stress Law; Ultimate Tensile Stress Surface ", *Z. Physik*, 1929, **56**, 308.
453. W. Voigt, " Elastic Shear Strain in the Twinning of Calcite ", *Wiedemanns Ann.*, 1899, **67**, 201.
454. H. Tertsch, " Types of Cleavage ", *Z. Krist.*, 1930, **74**, 476.
455. W. D. Kusnetzow (in collaboration with W. M. Kudrjawzewa), " Impact Cleavage. (110) Cleavage Plane of NaCl is only an Apparent Plane ", *Z. Physik*, 1927, **42**, 302.
456. H. Tertsch, " Various Types of Cleavage, NaCl ", *Z. Krist.*, 1931, **78**, 53.
457. H. Tertsch, " Various Types of Cleavage, Lead Sulphide ", *Z. Krist.*, 1933, **85**, 17.
458. H. Tertsch, " Various Types of Cleavage, Anhydrite ", *Z. Krist.*, 1934, **87**, 326.
459. L. Tokody, " (110) Cleavage, Stepwise, NaCl ", *Z. Krist.*, 1930, **73**, 116.
460. H. Tertsch, " Energy Requirements for Impact Cleavage of NaCl along (100) and (110) Plane ", *Z. Krist.*, 1932, **81**, 264, 275.
461. A. Sella and W. Voigt, " Importance of Non-eccentric Application of Load in Tensile Testing. Normal Stress Law ", *Wiedemanns Ann.*, 1893, **48**, 636.
462. A. Joffé, M. W. Kirpitschewa and M. A. Lewitsky, " X-ray Determination of the Onset of Plasticity and of its Dependence on Temperature. Dependence of the Tensile Strength on Temperature ", *Z. Physik*, 1924, **22**, 286.
463. I. W. Obreimow and L. W. Schubnikoff, " Determination of the Yield Point with Double Refraction Due to the Stress. Internal Stresses ", *Z. Physik*, 1927, **41**, 907.
464. F. Blank, " Yield Point and Tensile Strength of NaCl of Various Origin and Different Previous History (Annealing) ", *Z. Physik*, 1930, **61**, 727.
465. F. Rinne, " Double Refraction Due to Stress, NaCl ", *Z. Krist.*, 1925, **61**, 389.
466. W. Schütze, " Stages of Deformation, Yield Points of Potassium Halides. Normal Stress Law, Progress of Fracture ", *Z. Physik*, 1932, **76**, 135.
467. M. A. Lewitsky, " Determination of the Yield Point. Bending of NaCl ", *Z. Physik*, 1926, **35**, 850.
468. A. Smekal, " Determination of the Yield Point by Photochemical Colouring ". 8. *internat. Kongr. Photographie, Dresden*, 1931, p. 34.
469. G. F. Sperling, " Normal Stress Law, NaCl ", *Z. Physik*, 1932, **74**, 476.
470. W. Schütze, " Shear Stress Law, Potassium Halides; Normal Stress Law ", *Z. Physik*, 1932, **76**, 151.

471. W. Voigt, " Elastic Shear Strain at the Yield Point ", *Ann. Physik.*, [4], 1919, **60**, 638.

472. E. Jenckel, " Dependence of the Yield Point, Stress–Strain Curve and Tensile Strength on the Cross-section, NaCl ", *Z. Elektrochem.*, 1932, **38**, 569.

473. W. Metag, " Effect of Admixtures on the Limits of Cohesion of NaCl ", *Z. Physik*, 1932, **78**, 363.

474. F. Blank and A. Smekal, " Influence of $PbCl_2$ in NaCl ", *Naturwiss.*, 1930, **18**, 306.

475. A. Edner, " Effect of Admixtures on the Limits of Cohesion of NaCl ", *Z. Physik*, 1932, **73**, 623.

476. H. Schönfeld, " Effect of Ternary Admixtures on the Limits of Cohesion of NaCl ", *Z. Physik*, 1932, **75**, 442.

476a. E. Poser, " Photochemically Determined Elastic Limit is Apparently not Affected by Structure ", *Z. Physik*, 1934, **91**, 593.

477. W. Theile, " Dependence on Temperature of the Yield Point, Stress–Strain Curve, and Tensile Strength of NaCl ", *Z. Physik*, 1932, **75**, 763.

478. H. Ekstein, " Torsion of NaCl ". Unpublished work.

479. W. Ewald and M. Polanyi, " Bending of NaCl; No Change in the Modulus of Elasticity by Cold Working ", *Z. Physik*, 1924, **28**, 29.

480. A. Joffé and A. Lewitsky, " Bending of NaCl ", *Z. Physik*, 1925, **31**, 576.

481. M. Polanyi and G. Sachs, " Bending Yield Point; Bauschinger Effect, NaCl ", *Z. Physik*, 1925, **33**, 692.

482. L. Milch, " Torsion of NaCl at Elevated Temperatures ", *Neues Jahrb. Mineral. Geol. Paläont.*, 1909, **I**, 60.

483. R. Gross, " Torsion of NaCl at Elevated Temperatures ", *Z. Metallkunde*, 1924, **16**, 18.

484. K. Przibram, " Compression of NaCl ", *Wien. Ber.*, IIa, 1932, **141**, 63.

485. K. Przibram, " Compression, Brinell Hardness, NaCl, KCl, KBr ", *Wien. Ber.*, IIa, 1932, **141**, 645.

486. K. Przibram, " Compression, Brinell Hardness NaBr, NaI, KI ", *Wien. Ber.*, IIa, 1933, **142**, 259.

487. W. Voigt, " After-effect and Bauschinger Effect, NaCl ", Dissertation. Königsberg : 1874.

488. W. Ewald and M. Polanyi, " Bauschinger Effect in the Bending of NaCl ", *Z. Physik*, 1924, **31**, 139.

489. L. Sohncke, " Normal Stress Law ", *Poggendorfs Ann.*, 1869, **137**, 177.

490. E. Schmid, " Normal Stress Law ", *Proc. Internat. Congr. Appl. Mech.*, Delft, 1924, p. 342.

491. W. Boas and E. Schmid, " Constant Normal Dilatation Excluded as a Condition of Fracture ", *Z. Physik*, 1929, **56**, 516.

492. A. Smekal, " Constant Elastic Energy Excluded as a Condition of Fracture ", " Handbuch der physik. u. techn. Mechanik ", 1931, **2/4**.

493. H. Müller, " Dependence of the Tensile Strength on the Cross-section, NaCl ", *Physikal. Z.*, 1924, **25**, 223.

494. E. Rexer, " Tensile Strength of CaF_2 ", *Z. Krist.*, 1931, **78**, 251.

495. W. Voigt, " Tensile Strength of CaF_2 ", *Wiedemanns Ann.*, 1893, **48**, 663.

496. G. Heyse, " Tensile Strength of $SrCl_2$ ", *Z. Physik*, 1930, **63**, 138.

497. W. Burgsmüller, " Tensile Strength of NaCl between $-190°$ C. and $+90°$ C. ", *Z. Physik*, 1933, **80**, 299.

498. W. Burgsmüller, " Tensile Strength of NaCl with Additions of $SrCl_2$ between $-252°$ C. and $+20°$ C.", *Z. Physik*, 1933, **83**, 317.

499. K. Steiner and W. Burgsmüller, " Tensile Strength of NaCl at $4·2°$ Abs.", *Z. Physik*, 1933, **83**, 321.

500. W. Ende and E. Rexer, " Cinematographic Determination of the Duration of Fracture. According to A. Smekal ". " Handbuch der physik. u. techn. Mechanik ", 1931, **4/2**.

501. A. Sella and W. Voigt, " Bending of NaCl ", *Wiedemanns Ann.*, 1893, **48**, 636, especially 652.

502. W. Ewald and M. Polanyi, " Plastic Deformations in the Bend Test ", *Z. Physik*, 1925, **31**, 746.

503. W. Voigt, "Torsional Strength of NaCl", *Wiedemanns Ann.*, 1893, **48**, 657.
504. A. Smekal, "Normal Stress Law as a Law of Strain Strengthening", *Metallwirtschaft*, 1931, **10**, 831.
505. P. Niggli, "Hardness". "Lehrbuch der Mineralogie". Berlin : 1924, **1**.
506. A. Reis and L. Zimmermann, "Hardness of Crystals", *Z. physikal. Chem.*, 1922, **102**, 298.
507. A. Martens, "Scratch Hardness". "Handbuch der Materialienkunde für den Maschinenbau". 1898, p. 235.
508. F. Exner, "Investigations on the Hardness of Crystal Planes". Vienna : 1873.
509. H. Tertsch, "Anisotropy of Grind Hardness", *Anz. Akad. Wiss. Wien*, 1934, No. 17; *Z. Krist.*, 1934, **89**, 541.
510. F. Rinne and W. Riezler, "Cone Indentation Hardness of NaCl, AgBr, AgI at Different Temperatures", *Z. Physik*, 1930, **63**, 752.
511. F. Rinne and W. Hofmann, "Cone Indentation Hardness of NaCl, KCl", *Z. Krist.*, 1932, **83**, 56.
512. E. G. Herbert, "Pendulum Hardness Test", *Engineer*, 1923, **85**, 390.
513. W. D. Kusnetzow and E. W. Lawrentjewa, "Pendulum Hardness Test", *Z. Krist.*, 1931, **80**, 54.
514. W. Schmidt, "Pendulum Hardness Test". Paper read to the Deutsche Mineralogische Gesellschaft. Berlin : 1931.
515. P. Rehbinder, "Reduction in Hardness Due to the Adsorption of Polar Molecules", *Z. Physik*, 1931, **72**, 191.
516. W. Voigt, "Coefficients of Elasticity of Ionic Crystals". "Lehrbuch der Kristallphysik." Leipzig : 1928.
517. P. W. Bridgman, "Coefficients of Elasticity of the Alkali Halide Crystals", *Proc. Amer. Acad.*, 1929, **64**, 19.
518. J. Hengstenberg and H. Mark, "Lattice Constants and Lattice Disturbances, KCl", *Z. Physik*, 1930, **61**, 435.
519. R. Gross, "Change in the Density of NaCl after Deformation and Heat Treatment. Increase in the Rate of Solution as a Result of Deformation. Recovery", *Z. Metallkunde*, 1924, **16**, 344.
520. S. Konobejewski and I. Mirer, "Internal Stresses Shown by Asterism of the Laue Photographs", *Z. Krist.*, 1932, **81**, 69.
521. N. J. Seljakow, "Lattice Distortion Due to Internal Stresses NaCl", *Z. Krist.*, 1932, **83**, 426.
522. A. Smekal, "Internal Stresses Determined by Photo-electric Measurements", *Physikal. Z.*, 1933, **34**, 633. Internat. Confer. Phys. London, 1934.
523. R. Brill, "Lattice Disturbances in Compressed KCl Crystals", *Z. Physik*, 1930, **61**, 454.
524. U. Dehlinger, "Lattice Disturbances and Broadening of the Lines", *Z. Krist.*, 1927, **65**, 615; *Z. Metallkunde*, 1931, **23**, 147.
525. E. Rexer, "Identity of the Absorption Spectra of NaCl Coloured by Sodium Vapour and by Irradiation", *Physikal. Z.*, 1932, **33**, 202.
526. K. Przibram, "Coloration Theory, NaCl", *Z. Physik*, 1923, **20**, 196.
527. A. Smekal, "Coloration of Bent NaCl Crystals", *Z. VDI*, 1928, **72**, 667.
528. H. J. Schröder, "Coloration of NaCl", *Z. Physik*, 1932, **76**, 608.
529. K. Przibram, "Coloration of NaCl Crystals after Compression", *Z. Physik*, 1927, **41**, 833.
530. E. Jahoda, "More Intensive Coloration of Melt-grown Crystals, NaCl", *Wien. Ber.*, IIa, 1926, **135**, 675.
531. K. Przibram, "More Intensive Coloration of Melt-grown Crystals, NaCl", *Wien. Ber.*, IIa, 1928, **137**, 409.
532. A. Smekal, "More Intensive Coloration of Annealed NaCl and KCl Crystals. Electrolytic Conductivity of Crystals Grown from the Melt is Greater than that of Solution Crystals", *Z. Physik*, 1929, **55**, 289.
533. K. Przibram, "Coloration of Calcite Twins", *Z. Physik*, 1931, **68**, 403.

534. A. Joffé and E. Zechnowitzer, " Electrolytic Conductivity of NaCl after Deformation ", Z. Physik, 1926, **35**, 446.
535. Z. Gyulai and D. Hartly, " Electrolytic Conductivity of NaCl after Deformation ", Z. Physik, 1928, **51**, 378.
536. F. Quittner, " Effect of Electrolytic Conductivity on Plastic Deformation, NaCl ", Z. Physik, 1931, **68**, 796.
537. A. Joffé, " Effect of Impurities on Electrolytic Conductivity, NaCl ", Ann. Physik., [4], 1923, **72**, 461.
538. F. Rinne, " Recrystallization of NaCl ", Z. Krist., 1925, **62**, 150.
539. K. Przibram, " Recrystallization of NaCl ", Wien. Ber., IIa, 1929, **138**, 353.
540. K. Przibram, " Recrystallization of NaCl ", Wien. Ber., IIa, 1930, **139**, 255.
541. K. Przibram, " Recrystallization Film, NaCl ", Z. Elektrochem., 1931, **37**, 535.
542. H. G. Müller, " Recrystallization of NaCl ", Physikal. Z., 1934, **35**, 646.
543. K. Przibram, " Recrystallization Diagram—Irradiation Retards Formation of Nuclei ", Wien. Ber., IIa, 1933, **142**, 251.
544. K. Przibram, " Irradiation Retards Recrystallization ", Wien. Ber., IIa, 1932, **141**, 639.

Joffé Effect

545. E. Hentze, " Bending of NaCl under Water ", Z. Kali, 1921, No. 4, 6, 8, 9.
546. A. Joffé, M. W. Kirpitschewa and M. A. Lewitsky, " Joffé Effect. Interpretation : Dissolution of Strength-reducing Cracks ", Z. Physik, 1924, **22**, 286.
547. W. Ewald and M. Polanyi, " Increase in the Rate of Flow on the Compressed Side of Wetted Crystals when Subjected to Bending. No After-effect. Interpretation : Elimination of Impediments " (" Verriegelungen "), Z. Physik, 1924, **28**, 29.
548. A. Joffé and M. A. Lewitsky, " After-effect of Wetting under Tension ", Z. Physik, 1925, **31**, 576.
549. W. Ewald and M. Polanyi, " Increased Rate of Flow of Wetted Crystals under Bending ", Z. Physik, 1925, **31**, 746.
550. M. A. Lewitsky, " Identical Yield Point Values when Bending Wetted Crystals. Reduction of the Shear Hardening ", Z. Physik, 1926, **35**, 850.
551. E. Schmid and O. Vaupel, " Joffé Effect on Tempered Crystals and on Specimens whose Axis is Parallel to One of the Body Diagonals : Interpretation : Penetration by Water Brings about Internal Changes in the Crystal ", Z. Physik, 1929, **56**, 308.
552. G. F. Sperling, " Identical Yield Point under Tension. Normal Stress Law Valid for Wetted Specimens. Effect on Tempered Crystals ", Z. Physik, 1932, **74**, 476.
553. K. H. Dommerich, " Identical Critical Shear Stress of Crystals Subjected to Tension under Solvents. Interpretation : Strain Strengthening ", Z. Physik, 1933, **80**, 242.
554. S. Dommerich, " Critical Shear Stress of the Cubic Glide System in NaCl ", Z. Physik, 1934, **90**, 189.
555. L. Piatti, " Large Extensions when Uniform Solution Occurs ", Nuovo Cimento, 1932, **9**, 102; Z. Physik, 1932, **77**, 401.
556. A. Smekal, " Slight Discoloration of Crystals Fractured under Water ", Naturwiss., 1928, **16**, 743.
557. E. Rexer, " Increased Strength of NaCl Crystals in a Solution of H_2SO_4, and of KI Crystals in H_2O and Methyl Alcohol ", Z. Physik, 1931, **72**, 613.
558. U. Heine, " Effect of a Solution of H_2O on KCl Crystals ", Z. Physik, 1931, **68**, 591.
559. K. Wendenburg, " Influence of the After-effect on Strength is Dependent on the Solvents ", Z. Physik, 1934, **88**, 727.

560. A. Joffé and M. A. Lewitsky, " ' Sphere ' Test ", Z. Physik, 1926, **35**, 442.

561. M. Polanyi, " Interpretation : Strain Strengthening Plastic Deformation in the ' Sphere ' Test ", Naturwiss., 1928, **16**, 1043.

562. E. Orowan, " Interpretation : Depth of the Crack Depends on the Dimensions of the Crystal ", Z. Physik, 1933, **86**, 195.

563. E. Jenckel, " High Strength with Thin Crystals ", Z. Elektrochem., 1932, **38**, 569.

564. A. Smekal, " Interpretation : Removal of the Cracks by Solution Facilitates Plasticity and Strain Strengthening ", Phys. Rev., 1933, **43**, 366.

565. A. Smekal, " Increase in Ionic Conductivity as a Result of Wetting. Interpretation : Penetration of the Water ", Physikal. Z., 1931, **32**, 187.

566. E. Schmid and O. Vaupel, " Scratch Hardness of Wetted NaCl Crystals. Constancy of the Lattice Constants and Density ", Z. Physik, 1930, **62**, 311.

567. N. Dawidenkow and M. Classen-Nekludowa, " Scratch Hardness of Wetted NaCl Crystals ", Physikal. Z. Sowjetunion, 1933, **4**, 25.

568. R. B. Barnes, " Penetration of Water Proved by Ultra-red Absorption ", Phys. Rev., 1933, **44**, 898; Naturwiss., 1933, **21**, 193.

569. A. Joffé, " Normal Strength, if Narrow Strips of Crystals are Protected against Solution ", Internat. Confer. Phys., London, 1934.

VIII. THEORIES OF CRYSTAL PLASTICITY AND CRYSTAL STRENGTH

570. M. Born, " Calculation of Strength ". Atomtheorie des festen Zustandes, Berlin, 1923.

571. F. Zwicky, " Theoretical Tensile Strength of NaCl Based on Lattice Calculations ", Physikal. Z., 1923, **24**, 131.

572. M. Born and J. E. Mayer, " New Calculation of Strength ", Z. Physik, 1932, **75**, 1.

573. M. Polanyi, " Estimating the Tensile Strength ", Z. Physik, 1921, **7**, 323.

574. W. Ewald and M. Polanyi, " Constancy of the Modulus of Elasticity to the Point of Fracture ", Z. Physik, 1924, **28**, 29.

575. J. Frenkel, " Calculation of the Shear Strength ", Z. Physik, 1926, **37**, 572.

576. M. Polanyi and E. Schmid, " Estimating the Shear Strength. Dependence of the Rate of Flow on Temperature Regarded as Recovery ", Naturwiss., 1929, **17**, 301.

577. W. Voigt, " Significance of Structural and Thermal Inhomogeneities ", Ann. Physik, 4, 1919, **60**, 638.

578. A. Smekal, " Theory of Vacant Lattice Sites. Hardening Due to Increasingly Defective Orientation ", Handbuch der physik. u. techn. Mechanik, 1931, **4/2**, 116.

579. A. Smekal, " Theory of Vacant Lattice Sites ", Handbuch der Physik, 2nd ed., 1933, **24**, 795.

580. C. E. Inglis, " Stress Distribution at a Crack ", Trans. Inst. Naval Architects, 1913, **55**, 219.

581. A. A. Griffith, " Crack Formation Determined by Balance of Energy ", Phil. Trans. Roy. Soc., 1921, **221**, 163; Proc. Internat. Congr. Appl. Mech., 1924, 55.

582. G. Masing and M. Polanyi, " Calculation of Crack Length, Zn. Tensile Strength Increases with the Fineness of the Grain ", Z. Physik, 1924, **28**, 169.

583. E. Orowan, " Crack Length Interpreted with the Aid of Plastic Glide ", Internat. Confer. Phys., London, 1934.

584. E. Orowan, " Crack Length a Function of the Section. Hardening Caused by Subdivision of Cracks ", Z. Physik, 1933, **86**, 195.

585. E. Orowan, " Tensile Tests on Mica ", Z. Physik, 1933, **82**, 235.

586. G. I. Taylor, " Low Shear Strength Accounted for by Presence of Cracks; Hardening by Subdivision of Cracks ", *Trans. Faraday Soc.*, 1928, **24**, 121.
587. A. T. Starr, " Shear Stress Distribution at a Crack ", *Proc. Phil. Soc. Cambridge*, 1928, **24**, 489.
588. E. Schmid, " Strengthening Effect of Holes and Notches : Mg Crystals ", *Z. Elektrochem.*, 1931, **37**, 447.
589. G. I. Taylor, " Glide Caused by Migration of ' Dislocations ' through the Crystal. Estimating the Shear Strength of the Crystal Mosaic. Theory of Work-hardening Curve ", *Proc. Roy. Soc.*, 1934, **145**, 362, 388.
590. H. G. J. Moseley and C. G. Darwin, " Mosaic Structure ", *Phil. Mag.*, 1913, **26**, 210. For further literature see (579).
591. M. Polanyi, " ' Dislocations ' in the Lattice ", *Z. Physik*, 1934, **89**, 660.
592. E. Orowan, " Glide Originates at Nuclear Points. Thermal and Structural Inhomogeneities Combined to Account for Crystal Glide ", *Z. Physik*, 1934, **89**, 605, 614, 634.
593. F. Zwicky, " Secondary Structure ", *Proc. Nat. Acad. Sci.*, 1929, **15**, 816; *Helv. Phys. Acta*, 1930, **3**, 269.
594. R. H. Canfield, " Objections to a Secondary Structure ", *Phys. Rev.*, 1930, **35**, 114.
595. E. Orowan, " Objections to a Secondary Structure ", *Z. Physik*, 1932, **79**, 573.
596. M. Renninger, " The Mosaic Structure of Rock Salt ", *Z. Krist.*, 1934, **89**, 344; P. P. Ewald, *Internat. Confer. Phys.*, London, 1934.
597. A. Smekal, " Adsorption and the Formation of Solid Solutions ", *Physikal. Z.*, 1934, **35**, 643.
598. A. Goetz, " Adsorption of Small Additions in Bi Crystals ", *Internat. Confer. Phys.*, London, 1934. Also gives particulars of literature on block structure.
599. R. Becker, " Thermal Inhomogeneities ", *Physikal. Z.*, 1925, **26**, 919.
600. R. Becker, " Thermal Inhomogeneities. Explanation of the Shape of the Recrystallization Diagram. Plasticity Due to Atomic Rearrangement ", *Z. techn. Physik*, 1926, **7**, 547.
601. R. Becker and W. Boas, " Dependence of Speed of Flow on Temperature : Cu ", *Metallwirtschaft*, 1929, **8**, 317.
602. G. T. Beilby, " The Amorphous Layer Hypothesis ", *J. Inst. Metals*, 1911, **6**, 5.
603. W. Rosenhain, " The Amorphous Layer Hypothesis ", *Internat. Z. Metallogr.*, 1914, **5**, 228.
604. G. Tammann, " Thermodynamic Grounds for Rejecting the Amorphous Layer Hypothesis. Hardening as a Result of Atomic Changes ", *Lehrbuch der Metallographie*.
605. J. Czochralski, " Displacement Hypothesis ", *Moderne Metallkunde*, Berlin, 1924.
606. P. Ludwik, " Interference Hypothesis ", *Z. V.d.I.*, 1919, **63**, 142.
607. O. Mügge, " Glide Accompanied by Bending ", *Neues Jahrb. Mineral. Geol. Paläont.*, 1898, **1**, 155.
608. M. Polanyi, " Glide Accompanied by Bending ", *Z. Krist.*, 1925, **61**, 49.
609. U. Dehlinger, " Disoriented Atoms. Derivation of the Recrystallization Temperature ", *Ann. Physik*, 5, 1929, **2**, 749; *Z. Metallkunde*, 1930, **22**, 221.
610. L. Prandtl, " Model Illustrating the Stability of Disorientations. Hysteresis ", *Z. angew. Math. u. Mech.*, 1928, **8**, 85.
611. H. J. Gough, D. Hanson and S. J. Wright, " Atomic Rearrangement the Cause of Hysteresis ", *Aeronaut. Research Cttee. R. and M.*, 1924, No. 995.
612. W. Geiss and J. A. M. van Liempt, " Work Hardening Caused by Atomic Deformation ", *Z. Physik*, 1927, **45**, 631.
613. G. Tammann, " Work Hardening Caused by Atomic Deformation ", *Z. Metallkunde*, 1934, **26**, 97.

614. A. E. van Arkel and W. G. Burgers, " Atomic Deformation Caused by Lattice Distortion ", *Z. Physik*, 1928, **48**, 690.
614a. J. A. M. van Liempt, " Recrystallization Temperature Explained ", *Z. anorg. Chem.*, 1931, **195**, 366; *Rec. Trav. Chim.*, 1934, **53**, 941.
615. G. Tammann and W. Crone, " Distribution of Grain Size ", *Z. anorg. Chem.*, 1930, **187**, 289; G. Tammann, *Z. Metallkunde*, 1930, **22**, 224.
616. v. Göler and G. Sachs, " Distribution of Grain Size ", *Z. Physik*, 1932, **77**, 281; A. Huber, " Distribution of Grain Size in Metallographic Specimens ", *Z. Physik*, 1935, **93**, 227.
617. F. Koref, " Amorphous Plasticity of Tungsten Coils ", *Z. techn. Physik*, 1926, **7**, 544.
618. E. Schmid and G. Wassermann, " Amorphous Plasticity of Copper and Aluminium Wire ", *Z. Metallkunde*, 1931, **23**, 242.
619. W. Rohn, " Creep Strength in Relation to Temperature and Previous Treatment ", *Z. Metallkunde*, 1932, **24**, 127.
620. G. Wassermann, " Amorphous Plasticity in the Case of the γ–α Phase Transformation of Nickel Steel ", *Arch. Eisenhüttenwesen*, 1932/33, **6**, 347.

IX. THE PROPERTIES OF POLYCRYSTALLINE TECHNICAL MATERIALS IN RELATION TO THE BEHAVIOUR OF THE SINGLE CRYSTAL

621. B. Sander, *Gefügekunde der Gesteine*, Vienna, Julius Springer, 1930.
622. W. Schmidt, *Tektonik und Verformungslehre*, Berlin, Gebr. Bornträger, 1932.
623. E. Schmid and G. Wassermann, " Determination and Description of the Textures. Bibliography ", *Handbuch der physik. u. techn. Mechanik*, 1931, **4/2**, 319.
624. E. Schröder and G. Tammann, " Study of Rolling Textures by Means of the Chladni Resonance Figures ", *Z. Metallkunde*, 1924, **16**, 201; G. Tammann and W. Riedelsberger, *Z. Metallkunde*, 1926, **18**, 105.
625. K. Becker, R. O. Herzog, W. Jancke and M. Polanyi, " Fibre Texture of Hard Drawn Wires ", *Z. Physik*, 1921, **5**, 61; M. Ettisch, M. Polanyi and K. Weissenberg, *Z. Physik*, 1921, **7**, 181; *Z. Physikal. Chem.*, 1921, **99**, 332.
626. R. O. Herzog and W. Jancke, " Texture of Cellulose Fibres ", *Z. Physik*, 1920, **3**, 196; R. O. Herzog, W. Jancke and M. Polanyi, *Z. Physik*, 1920, **3**, 343; P. Scherrer in *Kolloidechemie*, by Zsigmondy, 3rd ed., 1920.
627. F. C. Nix and E. Schmid, " Casting Textures ", *Z. Metallkunde*, 1929, **21**, 286.
628. E. Schmid and G. Wassermann, " Recrystallization of Very Pure Aluminium Wire ", *Z. techn. Physik*, 1928, **9**, 106.
629. E. Schmid and G. Wassermann, " Texture of Hard Drawn Wires (Inhomogeneity of the Texture) ", *Z. Physik*, 1927, **42**, 779.
630. W. A. Wood, " Inhomogeneity of the Drawing Texture of Cu ", *Phil. Mag.*, 1931, **11**, 610.
631. F. Wever, " Representation of Textures by Means of Pole Figures ", *Z. Physik*, 1924, **28**, 69.
632. v. Göler and G. Sachs, " Rolling and Recrystallization Textures of Face-centred Cubic Metals ", *Z. Physik*, 1927, **41**, 873, 889; **56**, 1929, 477, 485.
633. M. A. Valouch, " Rolling Texture of Zn ", *Metallwirtschaft*, 1932, **11**, 165.
634. K. Weissenberg, " ' Circular ' and ' Spiral ' Fibre Textures ", *Z. Physik*, 1921, **8**, 20.
635. E. Schmid and G. Wassermann, " Texture of Drawn Mg and Zn Wires ", *Naturwiss.*, 1929, **17**, 312.
636. G. v. Vargha and G. Wassermann, " Inhomogeneity of the Rolling Texture of Al ", *Metallwirtschaft*, 1933, **12**, 511.

637. E. Schmid and G. Wassermann, " Inhomogeneity of the Rolling Texture of Zn. Anisotropy of Zn Sheets ", Z. Metallkunde, 1931, **23**, 87.
638. W. A. Sisson, " Identity of Structure of Drawn and Rolled Sheets (Steel) ", Metals and Alloys, 1933, **4**, 192.
639. G. v. Vargha and G. Wassermann, " Identity of Structure of Drawn and Rolled Wires (Al, Cu) ", Z. Metallkunde, 1933, **25**, 310.
640. R. Glocker and H. Widmann, " Three Examples of Crystal Arrangement in Recrystallization Textures. No Recrystallization Texture with Al ", Z. Metallkunde, 1927, **18**, 41.
641. E. Schmid and G. Wassermann, " Recrystallization Texture of Al Sheet; Anisotropy of the Mechanical Properties ", Metallwirtschaft, 1931, **10**, 409.
642. R. Glocker and E. Kaupp, " Recrystallization Texture of Ag ", Z. Metallkunde, 1924, **16**, 377.
643. W. Köster, " Recrystallization Texture of Cu Sheet ", Z. Metallkunde, 1926, **18**, 112.
644. J. A. Ewing and W. Rosenhain, " Glide in a Deformed Fe-polycrystal ", Phil. Trans. Roy. Soc., 1900, **193**, 353.
645. R. Straumann, " Identity of the Rolling and Recrystallization Texture of Zn Sheet. Anisotropy of Thermal Properties of Sheets ", Deut. Uhrmacherztg., 1931, No. 2 et seq. Cf. also Helv. Phys. Acta, 1930, **3**, 463.
646. E. Schiebold and G. Siebel, " Identity of the Rolling and Recrystallization Texture of Mg Sheet ", Z. Physik, 1931, **69**, 458.
647. G. Kurdjumow and G. Sachs, " Rolling and Recrystallization Texture of Fe ", Z. Physik, 1930, **62**, 592.
648. J. Czochralski, " Deformation of the Crystals in the Polycrystal ", Moderne Metallkunde, Berlin, 1924.
649. E. Seidl and E. Schiebold, " Deformation of the Crystals in the Polycrystal ", Z. Metallkunde, 1925, **17**, 221; 1926, **18**, 241.
650. M. Polanyi and E. Schmid, " Stretching of Bi-crystals Sn and Bi. Increased Capacity for Recrystallization in the Elongated Bi-crystal ", Z. techn. Physik, 1924, **5**, 580.
651. Unpublished work (1930) by E. Schmid, assisted by O. Vaupel and especially by W. Fahrenhorst. Continuation of twins in the neighbouring grains.
652. R. v. Mises, " Achievement of any Desired Shape by Glide on Five Different Systems ", Z. angew. Math. u. Mech., 1928, **8**, 161.
653. E. Schmid, " Difference in the Work Hardening of the Single and Polycrystal of Cubic and Hexagonal Metals ", Physikal. Z., 1930, **31**, 892.
654. R. Karnop and G. Sachs, " Stress–Strain Curves of Single and Polycrystals of Al ", Z. Physik, 1927, **41**, 116.
655. E. Schmid and G. Siebel, " Stress–Strain Curves of Mg Single Crystals ", Z. Elektrochem., 1931, **37**, 447.
656. E. Schmid and G. Siebel, " Alternating Torsion of Mg Bi- and Tricrystals ", Metallwirtschaft, 1934, **13**, 353.
657. H. J. Gough, S. J. Wright and D. Hanson, " Alternating Torsion of an Al Tri-crystal ", Aeronaut. Research Cttee., R. and M., (M41), 1926, No. 1025.
658. H. J. Gough, H. L. Cox and D. G. Sopwith, " Alternating Torsion of Al Bi-crystals ", J. Inst. Metals, 1934, **54**, 193.
659. Z. Jeffries and R. S. Archer, " Increase in Work Hardening with Decreasing Grain Size ", Chem. Met. Eng., 1922, **27**, 747.
660. W. A. Wood, " Increase in Hardness with Decreasing Grain Size ", Phil. Mag., 1930, **10**, 1073.
661. G. Masing and M. Polanyi, " Increase in Tensile Strength with Decreasing Grain Size ", Z. Physik, 1924, **28**, 169.
662. v. Göler and G. Sachs, " Cast Structure where Mould Is Intricate ", Z. V.d.I., 1929, **71**, 1353.
663. E. Schmid and G. Wassermann, " Interpretation of the Deformation Textures of Hexagonal Metals ", Metallwirtschaft, 1930, **9**, 698.

664. M. L. Fuller and G. Edmunds, " Rolling Texture of a Zn Alloy ", *Amer. Inst. Min. Met. Eng. Techn. Publ.*, 1934, No. 524.
665. F. Körber, " Interpretation of Deformation Textures ", *Mitt. K. W. Inst. Eisenforsch.*, 1922, **3**, 1; *Z. Elektrochem.*, 1923, **29**, 295; *Stahl u. Eisen*, 1928, **48**, 1433.
666. F. Wever and W. E. Schmid, " Interpretation of Deformation Textures ", *Mitt. K. W. Inst. Eisenforsch.*, 1929, **11**, 109; *Z. Metallkunde*, 1930, **22**, 133. W. E. Schmid, *Z. techn. Physik*, 1931, **12**, 552.
667. M. Polanyi, " Interpretation of Deformation Textures ", *Z. Physik*, 1923, **17**, 42.
668. G. Sachs and E. Schiebold, " Interpretation of the Deformation Textures ", *Naturwiss.*, 1925, **13**, 964.
669. W. Boas and E. Schmid, " Interpretation of Deformation Textures ", *Z. techn. Physik*, 1931, **12**, 71.
670. W. G. Burgers, " Interpretation of Recrystallization Textures ", *Metallwirtschaft*, 1932, **11**, 251.
671. U. Dehlinger, " Interpretation of Recrystallization Textures ", *Metallwirtschaft*, 1933, **12**, 48.
672. W. Voigt, " Averaging of Elastic Properties : Electrical Resistance ", *Lehrbuch der Kristallphysik* (Supplement).
673. D. A. G. Bruggeman, " Averaging of Elastic Properties ", Diss. Utrecht, 1930 (J. B. Wolters).
674. A. Reuss, " Averaging of Flow Limit and Elastic Properties ", *Z. angew. Math. u. Mech.*, 1929, **9**, 49.
675. A. Huber and E. Schmid, " Averaging of Elastic Properties ", *Helv. Phys. Acta*, 1934, **7**, 620.
676. W. Boas, " Averaging of Elastic Properties, Sn ", *Helv. Phys. Acta*, 1934, **7**, 878.
677. N. S. Akulow, " Averaging of Magnetic Properties ", *Z. Physik*, 1930, **66**, 533.
678. W. Boas and E. Schmid, " Comparison between the Calculated Elastic Properties and the Experimental Averaging of the Coefficient of Expansion and the Specific Resistance ", *Helv. Phys. Acta*, 1934, **7**, 628.
679. E. N. da C. Andrade and B. Chalmers, " Averaging of the Specific Resistance ; Change in Resistance Due to Re-orientation ", *Proc. Roy. Soc.*, 1932, **138**, 348.
680. G. Sachs, " Condition of Flow for the Polycrystal Derived from the Shear Stress Law ", *Z. V.d.I.*, 1928, **72**, 734.
681. P. Ludwik and R. Scheu, " Effect of the Mean Principal Stress on the Probability of Flow ", *Stahl u. Eisen*, 1925, **45**, 373.
682. W. Lode, " Effect of the Mean Principal Stress on the Probability of Flow ", *Z. Physik*, 1926, **36**, 913.
683. A. T. Huber, " Energy of Deformation as a Measure of the Probability of Flow ", *Czasopismo technizne*, Lemberg, 1904; R. v. Mises, *Götting. Nachr*, 1913, 582.
684. G. Sachs, " Interpretation of Elastic After-effect, Hysteresis, Bauschinger Effect ", *Grundbegriffe der mechanischen Technologie der Metalle*, Leipzig, 1925.
685. E. Heyn, " Theory of the ' Concealed Elastic Stresses ' ", *Met. u. Erz*, 1918, **15**, 411; *KWG-Festschrift*, 1921.
686. G. Masing, " Structural Interpretation of the ' Concealed Elastic Stresses ' ", *Z. techn. Physik*, 1922, **3**, 167.
687. H. v. Wartenberg, " Theory of the After-effect ", *Verhandl. deut. physikal. Ges.*, 1918, **20**, 113.
688. R. Becker, " Theory of the After-effect ", *Z. Physik*, 1925, **33**, 185.
689. H. J. Gough and D. Hanson, " Appearance of Glide Bands at Stresses Far Below the Limit of Proportionality ", *Proc. Roy. Soc.*, 1923, **104**, 538.
689a. D. A. G. Bruggeman, " Calculation of the Elastic Moduli of Cubic Metals of Different Structure ", *Z. Physik*, 1934, **92**, 561.

690. J. Weerts, " Elastic Anisotropy of Rolled and Annealed Cu Sheet ", Z. Metallkunde, 1933, **25**, 101.
691. E. Goens and E. Schmid, " Elastic Anisotropy of Rolled Fe Sheet ", Naturwiss., 1931, **19**, 520.
692. W. Fahrenhorst, K. Matthaes and E. Schmid, " Anisotropy of the Mechanical Properties of Recrystallized Cu Sheet ", Z. V.d.I., 1932, **76**, 797.
693. v. Göler and G. Sachs, " Anisotropy of the Mechanical Properties of Recrystallized Sheets of Cubic Metals ", Z. Physik, 1929, **56**, 495.
694. F. Saeftel and G. Sachs, " Analysis of the Cupping Test ", Z. Metallkunde, 1925, **17**, 155.
695. K. Kaiser, " Cupping Test, Cu Sheet ", Z. Metallkunde, 1927, **19**, 435.
696. O. Dahl and J. Pfaffenberger, " Cupping Test, Fe Sheet. Anisotropy of Magnetization ", Z. Physik, 1931, **71**, 93.
697. L. Weiss, " Layer-type Structure of Drawn Al Bars ", Z. Metallkunde, 1927, **19**, 61.
698. R. Glauner and R. Glocker, " Texture and Corrosion Resistance of Cu Sheet ", Z. Metallkunde, 1928, **20**, 244.
699. W. Jubitz, " Change in the Thermal Expansion Due to Deformation of Mg, Zn, Cd ", Z. techn. Physik, 1926, **7**, 522.
700. G. Masing, " Change in the Thermal Expansion Due to Deformation, Zn, Cd ", Z. Metallkunde, 1928, **20**, 425.
701. H. Sieglerschmidt, " Anisotropy of Poisson's Ratio and Young's Modulus of Zn Sheet ", Z. Metallkunde, 1932, **24**, 55.
702. G. Edmunds and M. L. Fuller, " Anisotropy of the Bending Capacity of Zn Sheet ", Trans. Amer. Inst. Min. Met. Eng., 1932, **99**, 75.
703. W. Schmidt, " Significance of the Crystalline Structure in the Behaviour of Mg Alloys ", Z. Metallkunde, 1933, **25**, 229. See also Z. Metallkunde, 1931, **23**, 54.

INDEX